# Forest Policy and Governance in the United States

This new textbook provides an up-to-date and comprehensive introduction to both the policy background and contemporary governance of forests in the United States.

Starting with a history of the development of forest policies and conservation agencies, the book then explores the diversity of forest owners, users, and uses and examines emerging approaches to forest governance that cross traditional jurisdictional and property boundaries. It tackles key contemporary issues such as the forest–water nexus, the conservation of threatened and endangered species, and the challenges of managing fire, insect, and disease dynamics under a changing climate. Key focal areas include the emergence of collaborative approaches to forest governance, community–forest relationships, changes to corporate timberland ownership, and contemporary governance mechanisms such as certification and payments for ecosystem services. This text raises the "big questions" about the distribution of rights and responsibilities in forest management, the tensions between equity and efficiency, and how to sustain a diversity of forest values under the pressures of ecological and social complexity. Written in an accessible and engaging manner, this textbook provides a timely synthesis of both the foundations and current trends and issues in forest policy and governance in the United States with a strong emphasis on illustrative real-world cases.

*Forest Policy and Governance in the United States* is essential reading for students in forest and natural resource policy courses and will be of great use to students in environmental governance courses. It will also be of interest to policymakers and professionals working in forest conservation and in the forest industry.

**Jesse Abrams** is an Assistant Professor of Natural Resource Policy and Sustainability in the Warnell School of Forestry and Natural Resources and Savannah River Ecology Laboratory, University of Georgia, USA.

"*Forest Policy and Governance in the United States: An Introduction* incorporates deep historical research, integration of theoretical frameworks, and an inclusive approach to understanding the complexity of forest policy. The details of some of today's most important trends and debates are made accessible through numerous real-world examples. This text fills the enormous need for an updated work on US forest policy and governance."

**Dr. Courtney Schultz**, *Associate Professor of Forest and Natural Resource Policy and Director of the Public Lands Policy Group, Colorado State University*

"In *Forest Policy and Governance in the United States: An Introduction*, Jesse Abrams takes us beyond a simplistic mountaintop view of America's forested landscapes to the diverse array of ownerships, governance institutions, management approaches, uses and users, social values, environmental dynamics, and policies operating beneath the canopy. This comprehensive and outstanding volume is essential reading for anyone engaged with forest issues in the United States today."

**Dr. Susan Charnley**, *Research Social Scientist, Pacific Northwest Research Station, USDA Forest Service*

"*Forest Policy and Governance in the United States: An Introduction* is an important new survey of key topics in forest policy and governance. It includes a variety of perspectives, examples from every region, and broad consideration of issues impacting both public and private lands in an engaging, easy-to-read style."

**Mindy S. Crandall**, *Assistant Professor of Forest Policy Oregon State University*

"We need more students trained in the science and art of forest and natural resource policy. And we need policy textbooks and case studies students want to read. *Forest Policy and Governance in the United States: An Introduction* provides a clear and complete treatment of a complex topic and is written in an engaging style that will hold students' interest."

**Dennis Becker**, *College of Natural Resources, University of Idaho*

# Forest Policy and Governance in the United States

## An Introduction

Jesse Abrams

Routledge
Taylor & Francis Group
LONDON AND NEW YORK
earthscan
from Routledge

Designed cover image: Getty Images

First published 2023
by Routledge
4 Park Square, Milton Park, Abingdon, Oxon OX14 4RN

and by Routledge
605 Third Avenue, New York, NY 10158

*Routledge is an imprint of the Taylor & Francis Group, an informa business*



*British Library Cataloguing-in-Publication Data*
A catalogue record for this book is available from the British Library

*Library of Congress Cataloging-in-Publication Data*
Names: Abrams, Jesse, author. Title: Forest policy and governance in the United States: an introduction/Jesse Abrams.
Description: New York, NY: Routledge, 2023. | Includes bibliographical references and index.
Identifiers: LCCN 2022028185 (print) | LCCN 2022028186 (ebook) | ISBN 9780367489571 (hardback) | ISBN 9780367489557 (paperback) | ISBN 9781003043669 (ebook)
Subjects: LCSH: Forest policy–United States. | Forests and forestry–United States. | Environmental management–United States.
Classification: LCC SD565 .A27 2023 (print) | LCC SD565 (ebook) | DDC 634.90973–dc23/eng/20220822
LC record available at https://lccn.loc.gov/2022028185
LC ebook record available at https://lccn.loc.gov/2022028186

ISBN: 978-0-367-48957-1 (hbk)
ISBN: 978-0-367-48955-7 (pbk)
ISBN: 978-1-003-04366-9 (ebk)

DOI: 10.4324/9781003043669

Typeset in Goudy
by Deanta Global Publishing Services, Chennai, India

To Vero

# Contents

# Acknowledgments

I express my sincere gratitude to all of the colleagues, friends, and students who contributed in ways large and small to the preparation of this text. Special thanks to Courtney Schultz, Erin Kelly, Mindy Crandall, Gwen Trice, John Daigle, Joe Parsons, Chase Cook, Susan Charnley, Michelle Steen-Adams, and Heidi Huber-Stearns.

# Introduction

The panorama from atop the east moraine of Wallowa Lake in northeastern Oregon is an impressive sight (Figure I.1). The jagged, snowcapped Wallowa mountains, covered in a deep blue-green forest of pine, fir, spruce, larch, Douglas-fir, and aspen, loom over the state's second-deepest lake. Nearly all of the mountain range is owned and managed by the federal government as part of the Wallowa-Whitman National Forest; the higher terrain is included in the Eagle Cap Wilderness, a legal designation that prohibits roads, machines, and most commercial activities. Federal ownership ends where the moraines begin; much of the west moraine is dotted with summer homes and their surrounding private parcels. A belt of coniferous forest covers the southern half of the east moraine; although there had been detailed plans to install homes with million-dollar views here on what was once industrial timberland, a campaign led by local organizations succeeded in raising money to purchase the land and convert it to a community-owned forest. On the terminal moraine at the north end of the lake sits another parcel saved from residential development that has since been designated a state preserve.

The land conservation projects on both the east and terminal moraines were financed in part by the Nez Perce Tribe, whose people have inhabited this region for thousands of years and who once counted the Wallowa region among its most important territories. The final resting place of Old Chief Joseph of the Wallowa Band of the Nez Perce is on the terminal moraine, close to a traditional sockeye salmon fishing site where the lake empties into the Wallowa River. Although the tribe does not hold any reservation lands in the county, it maintains ownership of a 16,000-acre wildlife area in Joseph Canyon called Héte'wits Wétes, or "Precious Lands"; the tribe took ownership as the result of a 1995 settlement with the Bonneville Power Administration.

Just north of the lake sits Joseph, the rural town named for the younger Nez Perce Chief. The old Boise-Cascade and R-Y Timber sawmills can still be seen on the town's western side; they were both shuttered in 1994 when the Wallowa-Whitman National Forest sharply reduced timber harvests to conserve salmon species listed under the Endangered Species Act. On the horizon to the north lie forests that once belonged to the Boise-Cascade corporation when it was structured as a company that owned both timberland and mills. Those lands have

*Figure I.1* Looking toward the Wallowa Mountains from atop the east moraine of Wallowa Lake. Photograph by author.

since been bought, sold, and resold by various timber investment management organizations.

Places to hunt wild animals, places to gather huckleberries and edible mushrooms, places that provide water for people and salmon, places that provide habitat for endangered species, places that provide wood for harvest, places that soak up carbon and remove it from the atmosphere, places for recreation, places for inhabitation, places to graze livestock, places to build memories and strengthen social ties, places for solitude and reflection, places for scientific study, places that carry the stories and the bones of prior generations, places for worship—the forests of Oregon's Wallowa County are all of these and more. Their fate, and the fate of those who use, enjoy, and depend on these and all forests in the United States, is ultimately tied to the decisions we as people make through our roles as individuals, through our communities, and through our political institutions. We decide what uses of the forest are most important; we decide who has access to the benefits of forests (and upon what moral or philosophical grounds this access is justified); we decide where and whether we will respond to fires, insect outbreaks, and other events; we decide under what conditions forests can be altered or converted to other uses. We even decide on the processes that we will use to make decisions on such matters. Understanding forest governance in the United States means understanding the complex and varied ways in which our systems for making and enacting decisions affect forests, and how the dynamics of those forests affect the ways we make decisions.

## Forests in the United States

The United States is endowed with a richness and diversity of forests that almost no other country can claim: towering temperate rainforests of the Pacific Northwest and Southeast Alaska, montane forests of the Rocky and Sierra Nevada ranges, deep North Woods forests stretching from Minnesota to Maine, diverse temperate hardwood and mixed pine-hardwood forests in the Appalachian region, the productive and biodiverse southern pine belt, the ponderosa and piñon-juniper forests of the arid Southwest, tropical forests of the Caribbean and South Pacific, and myriad other forest types in settings that range from urban backyards to remote landscapes. These forests play host to an almost unimaginable variety of plant, animal, fungal, and other species. They experience climates ranging from humid tropical to warm semi-arid to near-arctic and are shaped by disturbance agents that include hurricanes, tornadoes, fires, insects, ice storms, windstorms, floods, and avalanches, just to name a few.

The richness and diversity of forest ecosystems in the United States is matched by a diversity of human relationships with forests. Human inhabitation and use of forests in the United States carries a history stretching back well into the last ice age. Forests as we know them today have been shaped by thousands of years of human influence, with fire as the principal (but not the only) management tool until the very recent past. In the last few centuries, humans have introduced new technologies, new institutions of land ownership and management, and new species from around the globe. Along with these changes have come new pressures on forests: fragmentation, conversion to non-forest uses, global climate alterations, the introduction of novel insect and disease species, and other stressors. In this text, we will consider how human decisions regarding ownership, management, and the distribution of rights and responsibilities have shaped forests in the United States over time and how our social and political institutions have struggled to reconcile the varied pressures and demands currently being brought to bear on them. We will begin each chapter by revisiting the people and forests of Wallowa County as a starting point, a means of illustrating the variety of forests, policies, governance approaches, and their real-world effects.

## Forest Governance

This book explores the means by which people claim ownership of, make decisions about, and collectively set management direction for forests in the United States. To do so, we engage with the concept of governance, a term that includes formal legal and policy mechanisms as well as informal processes of negotiation, contestation, and deal-making that affect publicly valued resources. In other words, governance includes, but is not limited to, the institutions and processes that normally fall under the category of "government."[1] According to Darla Munroe and colleagues, governance consists of "collective decision-making, involving a constellation of multiple actors, with the goal of negotiating some acceptable outcome among competing interests."[2] When looking at contemporary forest governance, this constellation of actors often includes non-governmental organizations (NGOs), communities defined in various ways, businesses, individuals, governments at multiple scales (such as local, state, tribal, federal), and other entities. Talking about governance implies, as geographers Gavin Bridge and Tom Perreault explain, engaging with "the process of imagining, challenging and producing collective environmental futures."[3] By focusing on governance, we will take a broad and integrative perspective on the ways that practices and ideas regarding land, people, and property have influenced past forest management—and how they continue to influence the production of our environmental futures.

One place to begin thinking in terms of governance is by considering the different categories of property in forests and the legal and cultural expectations associated with each. Property systems set rights and responsibilities, and therefore are an important starting point for understanding who makes decisions, by what processes, and under what justifications. Those accustomed to property regimes in the United States will immediately recognize the two broad categories of private and public ownership. We often assume that on private property the landowner's discretion is unquestioned and that on public land the general public's values and interests hold sway. The truth is more complicated.[4] Any private forest owner knows that they are subject to various legal obligations that can limit their decision-making discretion. For example, the presence of a federally listed threatened or endangered species will limit certain management activities; the presence of water connected to a "navigable waterway" (as defined by administrative regulations and interpreted by court decisions) will require the owner to comply with provisions of the federal Clean Water Act; and state- or local-level policies pertaining to permitting, forestry practices, use of chemical pesticides or fertilizers, road-building, residential development, taxation, and other considerations might shape or constrain their options as well.

On public lands, by contrast, decisions are made by forest management agencies (such as the US Forest Service for national forests or parallel organizations for state forestlands) under the direction set by various formal policies. Additionally, individuals or businesses often maintain private claims to these public lands, including rights of access to or across forestlands, rights of exclusivity to harvest timber, minerals, livestock forage, or other resources, or rights of long-term inhabitation. Some recreational uses may be allowed while others are not. Firewood harvesting may be allowed, but only at certain times and in designated places. In addition to the legal mechanisms that dictate the aforementioned layers of public and private claims on forestlands are various informal mechanisms that help maintain some forms of access (and block others): some forest users may take advantage of the remoteness of forests to engage in activities that are formally prohibited. Others may attempt to secure their access by claiming legitimacy, knowledge, tradition, propriety, or other categories that reflect cultural norms rather than legal requirements.[5]

Beyond the familiar categories of public and private forestland, we should also consider two other categories that are frequently, yet incorrectly, lumped together. The first is what is called "open access"—in essence, a property regime defined by the absence of enforced rights of exclusion, in which anyone is free to access any resource in any quantity at any time. This is the scenario erroneously referred to as the "commons" in Garrett Hardin's now-famous thought piece and population-control manifesto "The Tragedy of the Commons."[6] A true commons, by contrast, is a property owned collectively among members of a defined community and governed by rules that are both crafted and enforced by that community. Scholarship by recipient of the Nobel Prize in Economics[7] Elinor Ostrom and her colleagues has sparked a vibrant body of research in recent decades demonstrating that these common-property systems may be well-managed over very long time periods under the right conditions, in many cases through rules that are socially rather than legally enforced.[8] Scholarship such as Ostrom's compels us to consider forest governance in terms of both the formal (legal) rules as well as the less formal but still important social rules that pertain to private, public, commons, and open-access forestlands. It also directs us to consider how these rules influence the long-term productivity and sustainability of forests and their constituent parts. Each of these categories will be explored further as this text proceeds.

## Forest Governance in the United States: Starting with Locke

"Thus in the beginning all the world was America." These famous words from John Locke's *Second Treatise of Government*, published in 1689, capture an important insight into how British colonists of that era viewed American land and their place within it. Locke was personally involved in the administration of the Carolina Colony and his *Second Treatise* served, in part, to justify British colonialism in North America.[9] In his treatise on property, Locke drew a distinction between two different worlds: in one, human labor invested in agricultural cultivation secured the blessings of individual private property; in the other, the land remained in its natural, minimally productive state, owned in common and contributing little to societal well-being.[10] The second model represented, to Locke, most of "America" at that time—a vast landscape owned (in his mistaken view) by no one in particular, waiting to be transformed by the power of European settlement, privatization, and improvement.

Although Locke recognized that the real-world America was inhabited by Indigenous populations, he saw them as residents rather than as owners of the land, "tenant[s] in common" with "no knowledge of enclosure" who harvested nature's bounty but (again, in his mistaken view) did little to mix their labor with the soil. According to Locke, the Indigenous peoples of the Americas might have had property in the animals they hunted or the native fruits they collected, but the land itself was a great uncultivated commons and the people living on it therefore had no legitimate claim to it. Under Locke's view, the enclosure of land and its transformation to productive agriculture were the actions that created—and ideologically justified—private property. Locke grounded these ideas in natural law, creating an image of private land ownership that was isolated from both society and government (to say nothing of its isolation from ecological processes). The landowner and the land they owned were the only components of this model. Locke's ideas were highly influential on the landed Americans that revolted against British rule in 1775-1776 to create a new republic independent of the British crown. Locke's ideas about property left their mark on the United States Constitution, on policies such as the various Homestead Acts, and on American conceptions of land and liberty.

Looking back on the history of America's forests from our current vantage point more than three centuries after the publication of the *Treatises*, the world appears much more complicated than Locke could have imagined. Locke had agriculture in mind when he spoke of mixing labor with soil to create property, raising questions about how his model applies to forestland—particularly forests that appeared to European eyes to be in their "primeval" state. It would be nearly two centuries following Locke's death before the first acts of future-oriented forest harvesting and replanting would take place in the nation formed by the British colonies. Forest management for most of the colonial and early Republican eras consisted merely in harvesting nature's bounty. How could Locke's ideas justify ownership of land when no labor was apparently being mixed with the soil?

More importantly, it is now clear that even Locke's basic premise was flawed: the original inhabitants of North America were in fact quite active in applying their labor to the land; many Indigenous peoples (including those inhabiting Locke's Carolina Colony) practiced agriculture, and the use of fire to manage landscapes was widespread. We now recognize that in the absence of regular fires, many forests in the US have become less resilient; some tree species depend on fire to release and germinate seeds, and some formerly abundant tree species have failed to reproduce in landscapes where fire has been excluded. The Indigenous inhabitants derided by Locke were not just mixing their labor with the soil—they were sustaining the very ecosystems that were so desired by the European colonists.

In addition, the recent history of forests in the United States shows that the establishment of individual private property has not always generated the kinds of beneficial outcomes that Locke predicted. The US Forest Service—and the hundreds of millions of national forest acres the agency is responsible for managing—were established in large part to serve as a check on the rampant deforestation and associated soil erosion, watershed degradation, and related problems that were commonplace on private forestlands nationwide in the nineteenth and early twentieth centuries. State and county governments came to own and manage substantial forestland estates following the abandonment of private forestlands after the first round of industrial harvesting, and many states went on to regulate the practice of forestry on private lands. The problem of poor management on private forestlands was so pervasive that many of the country's leading foresters and conservation advocates were calling for national legislation to regulate private forest practices as recently as the middle of the twentieth century.

Although the state of practice on private forestlands has improved greatly since that time, numerous policies designed to prevent private actions from degrading the air, water, and soil continue in effect. Landowners and their land uses are, after all, always part of larger watersheds and landscapes; land use decisions have repercussions beyond the boundaries of an individual parcel.

Even as Locke's ideas about land live on in the country's cultural and legal legacies, our institutions of forest ownership and management have continued to evolve. Tribal nations have advanced in their struggle to achieve recognition of the rights outlined in treaty language, to increase their stewardship over a variety of landscapes, and to reintroduce practices that had been suppressed by colonial powers. These practices, and the knowledge on which they are based, have increasingly come to be recognized for their potential to help resolve some of the complex dilemmas resulting from past forest management and the pressures of contemporary environmental change. A movement to secure both environmental and local economic benefits through community-owned forests has gained steam since the 1990s, building on a long tradition of common land tenure in the United States. Land owned by public entities serves as the nucleus for landscape-scale restoration and sustainable management initiatives in many landscapes. Contemporary forest management on private lands increasingly incorporates guidance, resources, and rules from external governmental and non-governmental entities, further complicating the ideal of the independent and autonomous landowner. Taken together, these varied examples of departure from the ideas of Locke and Hardin reflect a growing recognition of the complexity of both forests and people's interests and values related to forests. In part, they can be seen as responses to the past degradation of forests, the ongoing transformations associated with climate change, the growing scope and scale of forest disturbance events, and other elements of global environmental change that have altered forest patterns and processes and created new risks. And in part they are the outcome of changes in political power that have allowed a more diverse array of people to have a voice in governing forests.

## Organization of This Text

As a means of better understanding forest governance in the United States, this text begins with a brief review of the nation's forest history, starting with the practices common among Indigenous populations prior to Euro-American colonization, changes introduced by colonization, and the institutions established by American political leaders in the young nation (Chapter 1). Chapter 2 takes us from the beginnings of the American conservation movement in the late nineteenth century through the establishment of the national forest system and US Forest Service, exploring other changes to law and practice along the way. Chapter 3 completes this brief historical review by covering the period from the "environmental decade" of the 1970s through the turn of the twenty-first century, addressing some of the divisive political battles regarding both public and private forest management and the governance innovations that resulted. The policy and governance contexts for wildlife and water are introduced in Chapter 4, and the intersecting challenges of climate change, fire management, and management of insects and diseases are the focus of Chapter 5.

The chapters that follow explore the histories and contemporary roles and responsibilities of varied organizations, populations, and categories: state and county government entities (Chapter 6); Native American tribal nations, agencies, and managers (Chapter 7); individual and family owners of private forestlands (Chapter 8); corporate forestland owners (Chapter 9); and communities of place with their varied formal and informal relationships to forests (Chapter 10). We also explore a category of forest users that often remain "unseen" and have relatively little power to influence governance, but that influences forest dynamics through activities including harvesting of non-timber forest products, manual labor in the woods, and illegal and clandestine cultivation (Chapter 11). Chapter 12 examines some of the persistent tensions and policy controversies that remain relevant as of the date of this writing. Throughout these chapters, the focus is on forests for which management direction is, at least potentially, up for debate and therefore subject to varied policy and governance influences. In other words, the emphasis in this text is on private forests, community forests, tribal forests, and multiple-use public forests for which management direction is likely to be contested, with less attention dedicated to parks and strictly protected areas on which management options are more constrained.

## Focusing Concepts for Forest Governance

As we continue forward in our exploration, we can approach forest governance in the United States by making use of four focusing concepts: formal policies, informal practices, participation, and paradigms. Formal policies include forest-relevant laws passed by legislatures at national, state, or local levels as well as agency rules and regulations, case law associated with the judicial system, and the requirements and opportunities associated with various incentive and partnership programs. Formal policies are what likely come to mind for most of us when we think about forest policy: they are the most visible and are in some sense "tangible." We can read the text of legislation, follow it through committees and floor votes, and track its interpretation by the courts. Government agencies make these policies even more tangible by implementing them: for example, many governments own forests outright and hire their own teams of experts to manage them, providing direct control over forest resources and providing a demonstration of practices to other owners. Governments also attempt to sway behavior on lands that are not under their direct control, levying fines or jail time for those who act in violation of the law, spending money on incentive programs, or announcing and endorsing policies through public information campaigns. Scholars sometimes refer to punitive policies (enforced through fines, payments, and the possibility of jail time) as "sticks" and incentive-based policies as "carrots." Still other policies are referred to as "sermons" if their intent is to drive changes in behavior through advice, information, or publicity.

Informal practices intersect in complex ways with formal policies: in cases where communities devise their own means to regulate the use of forests, these may act to replace (or in some cases complement) formal laws and regulations. Informal practices may not be as obvious as formal policies; in some cases, people go to great lengths to ensure they stay hidden. In other cases, informal practices seem invisible because they have become the accepted and expected ways of doing things. For example, Wallowa County residents had long used the terminal moraine for hiking and other low-impact recreational activities even when it was privately owned; although there was no formal system of giving or receiving permission, the landowners did not object so long as visitors behaved responsibly. Not until those lands were slated to become residential homesites did residents and interested communities feel the need to take action to conserve the land and secure their continued access to it.

This brings us to the next concept, participation. Questions of governance ultimately depend on who has the authority to participate. Whereas private landowners and government foresters were long considered the only entities justified in making decisions about forests, recent decades have seen an expansion in the kinds of people and interests considered relevant to decision-making. In some cases, communities of place and interest have devised their own forms of participation that complement official protocols even as they deviate from the assumptions guiding those protocols. Participation is inseparable from the concept of power, a complicated term that refers to the ways in which people act within social structures to achieve their aims. The expansion of participation in governance is one outcome of changes in the balance of power since the middle of the twentieth century.

Finally, all of the previous concepts are informed by paradigms: mental models and narratives that form the moral and cognitive foundations upon which policies, practices, and forms of participation come into being. Paradigms inform our collective understandings of what uses of forests, what models of property, and what approaches to governance are worthy of social and legal endorsement. They help determine what obligations those who benefit from forests have to the land and to other people. For example, John Locke's conceptions of property and Garrett Hardin's (mis)use of the concept of "commons" have served as paradigms, offering stories about the land and our relationship to it that were later used to set policies, inform practices, and justify the participation or exclusion of various populations in making decisions. Like informal practices, paradigms often escape notice because of their "taken-for-grantedness." It is typically not until paradigms change or stand in conflict with one another that we begin to see their contents and are able to examine their influences on patterns and processes of governance.

## Beginning and Ending with Complexity

A theme running throughout this book is the tension between the past and future of forest governance. Through the first decades of the twenty-first century we are coming to see ever more clearly the complexities of conserving, utilizing, and sustaining forests—complexities that were mostly invisible when many of our policies and paradigms regarding forest ownership and management were first put into place. A case in point is the national forest system and the US Forest Service created to manage it. In the early years of the twentieth century, the Forest Service's focus on treating trees as renewable rather than as resources to be harvested and abandoned was a radically forward-looking idea, but one that attempted to achieve its outcomes through simplifying the forest. The notion of integrating or balancing the sustained production of timber with non-timber uses came along decades later, and not until closer to the end of the century did the Forest Service begin to confront the problems created by decades of fire exclusion, conversion of mature forests to younger forests, and watershed alteration. More recently, the Forest Service has had to incorporate resilience to climate change into its planning and management and has engaged tribal, state, county, private, and non-profit entities in forest governance in new and innovative ways. Likewise, private forest policies traditionally had a focus on individual owners as isolated entities. Only more recently have we seen an increase in programs that reflect thinking and managing at larger scales encompassing multiple owners. Whereas earlier policies addressed individual issues (like water quality or the avoidance of forest conversion) in isolation, newer programs and practices often integrate across issues. Across forest types and property systems, those who own and manage forests have had to confront the complexities of a diverse and growing society with myriad, often conflicting, values and priorities for forests. In short, our forest policies, practices, and institutions are constantly being challenged to adapt to a socially and ecologically complex world. Understanding how we got to that level of complexity, and how our policies have or have not adapted in response, is the focus of the pages that follow.

## Notes

1 Lemos & Agrawal, 2006.
2 Munroe et al., 2019, p. 56.
3 Bridge & Perreault, 2009, p. 485.
4 Geisler, 2000.
5 Ribot & Peluso, 2003.
6 Hardin, 1968.
7 Technically the Sveriges Riksbank Prize in Economic Sciences in Memory of Alfred Nobel.
8 Ostrom, 1990.
9 Armitage, 2004; Hsueh, 2006.
10 Ince, 2018.

## References

Armitage, D. (2004). John Locke, Carolina, and the *Two Treatises of Government*. *Political Theory*, *32*(5), 602–627.

Bridge, G., & Perreault, T. (2009). Environmental governance. In N. Castree, D. Demeritt, D. M. Liverman, & B. Rhoads (Eds.), *A companion to environmental geography* (pp. 475–497). Wiley-Blackwell.

Geisler, C. (2000). Property pluralism. In C. Geisler & G. Daneker (Eds.), *Property and values: Alternatives to public and private ownership* (pp. 65–86). Island Press.

Hardin, G. (1968). The tragedy of the commons. *Science, 162*(3859), 1243–1248.

Hsueh, V. (2006). Cultivating and challenging the common: Lockean property, Indigenous traditionalisms, and the problem of exclusion. *Contemporary Political Theory*, 5(2), 193–214.

Ince, O. U. (2018). *Colonial capitalism and the dilemmas of liberalism*. Oxford University Press.

Lemos, M. C., & Agrawal, A. (2006). Environmental governance. *Annual Review of Environment and Resources, 31*(1), 297–325.

Munroe, D. K., Batistella, M., Friis, C., Gasparri, N. I., Lambin, E. F., Liu, J., Meyfroidt, P., Moran, E., & Nielsen, J. Ø. (2019). Governing flows in telecoupled land systems. *Current Opinion in Environmental Sustainability*, 38, 53–59.

Ostrom, E. (1990). *Governing the commons: The evolution of institutions for collective action*. Cambridge University Press.

Ribot, J. C., & Peluso, N. L. (2003). A theory of access. *Rural Sociology*, 68(2), 153–181.

# 1 Early Forest Policy and Governance

Two observations stand out in the accounts of nineteenth century White settlers moving across the Wallowa region: they typically remarked on the ease with which they were able to travel through the open, grassy forests with their horse-drawn carriages and they often complained about the smoky conditions resulting from fires set by the Native peoples inhabiting those landscapes.[1] That those open forest conditions might be the direct result of Native burning was, apparently, a connection lost on many observers. Indigenous communities had been using and managing the Wallowa region for at least 10,000 years prior to the arrival of the first non-Native fur trappers and missionaries, and their extensive application of fire had helped to shape area forests in ways that were poorly understood by the new populations.[2] The fact that the Plateau cultures inhabiting these landscapes had developed intricate and detailed understandings of rights and obligations regarding land and resources also went largely unrecognized. These systems of rights and obligations helped peoples such as the Nimiipu (Nez Perce) survive in a sometimes harsh and unforgiving landscape in which many food resources were only seasonally available.

By the beginning of the nineteenth century, Plateau peoples had already seen their worlds turned upside-down by the arrival of diseases like smallpox that spread in advance of colonial populations, as well as by the introduction of European horses obtained through continentwide trading networks. The decades following Euro-American colonization brought not only changes in land use but also in land governance: the combinations of rights and obligations, formal procedures and informal conflict resolution mechanisms, understandings of boundaries and of ownership, all grounded in paradigms that contrasted with those of the Plateau cultures. Parallel changes had already been instituted across much of the rest of the US by this time. When the US Army forced the Wallowa Band of the Nez Perce out of their ancestral homeland in 1877, it represented one of the last major acts of Native dispossession in a project initiated centuries earlier. The transformations that came to America's forests as a result of this process of colonization were complex in both their causes and consequences. In this chapter, we broadly outline the major trends in forest governance during the precolonial and colonial eras and in the first century following the American Revolution.

DOI: 10.4324/9781003043669-1

## Precolonial Forest Management and Use

Prior to the European colonization of North America, fire was the primary tool used by varied Indigenous peoples for managing the landscape according to their own interests: encouraging habitat for wildlife species important to their diets; corralling those same species to facilitate hunting; keeping travel corridors open; selecting for favored plant species (and against less favored species); consuming fuels that could otherwise contribute to deadly high-intensity conflagrations; for use in warfare; for communication; and in myriad other applications.[3] This is not to say that all forests were uniformly open or that the effects of Native burning were felt equally on every acre continentwide.[4] There remains an ongoing debate regarding where, and to what extent, Native burning shaped forests in precolonial North America;[5] ample evidence exists of closed-canopy conditions in some precolonial forests, indicating a relative absence of fire.[6] It is also the case that lightning has long been an ignition source in many landscapes, meaning that human activity cannot always be assumed to be the ultimate cause of fire in forests.[7] It is well established, however, that Indigenous fire use was frequent, widespread, and that—in concert with distinctly nonhuman forces of change—it shaped forest conditions and ecologies over millennia throughout much of the present-day United States.[8] According to fire historian Stephen Pyne, Indigenous fire use was ubiquitous throughout most of the present-day United States; although wetter areas may have escaped the regular influence of fires set by Native populations, "naturally drained landscape was nearly everywhere burned."[9]

The North American continent was populated by a dizzying variety of Indigenous cultures during thousands of years of inhabitation prior to European colonization. Forests were centrally important to many of these cultures, and they remain important to many tribal communities in the present (see Chapter 7). In addition to providing habitat for plant, animal, and fungal species of vital importance to Native diets, forests traditionally provided wood for shelters, canoes, tools, fishing platforms, baskets, cooking utensils, baby boards, ceremonial masks, medicines, ropes, hunting instruments, burial boxes, totems, fuelwood, and many other uses. Trees served as food sources by producing acorns, walnuts, chestnuts, pine seeds, and other edible products. In many cases, these species were selectively cultivated and maintained through the use of fire. One review of Indigenous land uses in eastern North America concluded,

> as a result of forest management, a variety of economically beneficial plants and animals were often found in greater abundance near Native American settlements than in the surrounding landscape…these stands of managed forests were created, managed and maintained by pruning, planting and protecting desired trees, and by burning the forest understorey.[10]

The relationships between precolonial Indigenous populations and their forests were characterized neither by passive use nor by benign neglect; rather, forests were actively modified through human agency and in return strongly shaped patterns of inhabitation and livelihood (Box 1.1).

Describing the patterns of forest governance among Indigenous populations prior to European colonization is no simple task. Patterns of resource access and use varied across the range of cultures that inhabited and managed what is now the United States in the years prior to colonization, and these patterns were irrevocably altered as a result of the diseases, warfare, demographic changes, and new technologies introduced by colonial populations. Fundamental differences in paradigms between colonial and Native populations add even more complexity to the story: whereas the settler European culture stressed individualism and acquisitiveness, precolonial Native cultures were characterized by the primacy of kinship networks, expectations of reciprocity, and obligations toward the living and nonliving worlds.[11] Despite the challenges in defining precolonial Indigenous forest governance using contemporary Western terms, we can say with confidence that John Locke was not at all accurate in his characterization of Native Americans as passively living off of nature's bounty without management and without property relations. Had he bothered to look closely, Locke might have recognized at least some familiar principles at play in the systems used by Indigenous peoples to define claims to property.

For example, in reviewing resource patterns among the Nimiipu of the Wallowa Region and the broader Interior Northwest, anthropologist Deward Walker concluded that while there is no evidence that individual private property in land existed prior to Euro-American colonization, families and villages did possess stewardship over certain sites.[12] The concept of "stewardship" in this case includes both rights of access and obligations to care and use without waste. Small groups and individuals might have held something like private property in the things they personally constructed via their own labor, such as the wooden fish traps installed in strategic instream locations, but they did not possess lands or resources in the abstract. Villages and extended kin groups could be said to own specific rights to particular resources at particular times, and inseparably attached to these rights were associated obligations. For the most part, other members of the Nimiipu, as well as members of allied tribes, could make use of plant, animal, and other resources within the territory of a given band or village so long as they did not engage in wasteful practices.[13] These resources were not passively provided by the bountiful hand of nature, however. The precolonial Nimiipu used fire for purposes ranging from hunting to improving huckleberry production to encouraging the growth of plants like camas and biscuitroot that were central to their diet. Anthropologist Alan Marshall referred to these stewarded foodscapes as "unusual gardens"—unusual, in part, because the role of the Nimiipu in actively cultivating them went unrecognized for generations by colonial populations, including by many anthropologists.[14]

Fire was also used by the Indigenous peoples in what is now southern New England to increase the production of desirable foods—in this case, wildlife species such as elk, deer, and turkey in addition to berries and other edible plants.[15] As environmental historian William Cronon describes in his classic history

*Changes in the Land*, burning in this region was used extensively to convert forest to land suitable for agriculture. Paralleling patterns of territory and property among the Nimiipu and many other North American Indigenous groups, the Native peoples of New England recognized territorial boundaries in which certain villages retained privileged access, and individuals could be said to possess those things (tools, for example) in which their own labor was directly invested. Cronon explains:

> Any village member, for instance, had the right to collect edible wild plants, cut birchbark or chestnut for canoes, or gather sedges for mats…fish and shellfish could generally be taken anywhere, although the nets, harpoons, weirs, and tackle used to catch them—and hence sometimes the right to use the sites where these things were installed—might be owned by an individual or a kin group.[16]

Families held exclusive rights to tend and harvest agricultural plots, but these plots shifted in location over time, in contrast to the model of fixed individual property recognized and instituted by European colonists. Here, as elsewhere in precolonial America, rights of access and use depended upon ecological context, and there was ultimately an understanding that the benefits that people derived from that access would be shared in accordance with expectations of reciprocity and obligations to both humans and nonhumans alike. It is for this reason that Deward Walker argues that "stewardship," rather than "ownership," is the concept that best represents Indigenous property relations.

Patterns of forest use and governance changed dramatically under the multiple influences of European colonization: the introduction of diseases that took heavy tolls on Native populations; the warfare against Indigenous peoples that typically preceded the spread of colonial populations in the Americas; the enslavement of Indigenous individuals to serve the interests of colonial powers;[17] and the introduction of new technologies and markets that altered traditional activities and forms of social organization. Native population numbers declined sharply in the decades following Euro-American colonization, and all but a few scattered territories were ultimately appropriated by the settler colonists. Yet it is important to recognize that Native peoples, cultures, and land uses did not disappear; they survived nationwide, sometimes in places that were forsaken and forgotten by colonial populations and in the face of sustained efforts to erase them. These points will be covered in more detail in Chapter 7.

**BOX 1.1 PEOPLE, FIRE, AND FORESTS**

In 2019, archaeologists working in Idaho located the oldest evidence of human inhabitation of the Americas to date, a site whose origins were traced back to a time over 16,000 years before the present day.[18] Just a year later, cave discoveries in Mexico raised the possibility of an even earlier date: as much as 33,000 years ago[19]—a figure given credence by still another discovery in New Mexico in 2021 dated to 23,000 years ago.[20] The North American landscape would have appeared much different 16,000 to 33,000 years ago than it does now, with continental glaciers still covering most of present-day Canada and parts of the northern United States. The kinds of boreal forest systems that we now associate with northern latitudes blanketed the Midwest and Mid-Atlantic region as recently as 17,000 years ago.[21] Cold-tolerant spruce and pine grew in what is now coastal North Carolina. Tree species that are currently common and widespread may have been much less so tens of thousands of years ago: oaks were still spreading northward from what had been a small range in the South during the glacial maximum. Ponderosa pine—the pine species with the widest current distribution in North America—was likely restricted to a much smaller range in present-day California, southern Arizona, and southern New Mexico.[22] If, as seems likely, the original inhabitants of North America regularly used fire as a tool of environmental management from the time of their arrival on the continent, it means that North American forest systems as we know them today are, at least partially, the products of human inhabitation and use.[23] For this reason, many scholars argue that humans and human-ignited fires were not "introduced" into America's forests; rather, culture, fire, and forests developed and were co-produced together.[24]

## Colonial Forest Governance on the Eastern Seaboard

The precolonial eastern seaboard is sometimes imagined as an unbroken expanse of dense and impenetrable forest, but the historical evidence paints a very different picture. Early European surveyors and colonists in eastern North America frequently remarked on the open condition of the landscape in their journals and reports: the abundant open, parklike forests made for easy travel on foot or on horseback.[25] Early British colonists in southern New England complained of their challenges in obtaining firewood, noting that they often had to seek it out in wetlands where the soil was inhospitable to the ubiquitous fires set by the Native populations. In what is now the southeastern US, the precolonial mosaic of forests of different densities, treeless prairies, and fields actively cultivated to crops supported vast herds of bison that may have ranged as far south as the Tampa Bay region of Florida.[26]

In many places, the dense, dark forests that are often imagined to be the "natural" condition of eastern North America appeared only after Native populations were decimated by warfare and introduced diseases. Colonists arriving decades after the first wave of encounters between European and Indigenous populations found a landscape radically different than the one that had dominated for centuries prior to their arrival. They often had to expend substantial effort to clear dense forests in places that were earlier reported as being open and lightly forested.[27] Researchers have concluded that, prior to the ecological transformations instituted by European settlers, "open forests rather than closed forests once were the most abundant ecosystem in eastern North America."[28] In addition to radically changing fire regimes, the newcomers from Europe introduced new plants, animals, and technologies as well as new paradigms that implied changes in the governance of forests.

Members of early European colonies along the eastern seaboard of the North American continent experienced land and landed resources in abundance even though they often struggled to establish permanent settlements. Simply surviving was no easy feat: Spanish expeditions often ended in disaster; famines killed scores of early British colonists in places like Jamestown and Plymouth as well as in other colonies that were subsequently abandoned. British colonies often began with communal models of land and resource ownership but soon transitioned to models with stronger private property dimensions.[29] Early British and Spanish colonies depended heavily upon the agricultural and forest knowledge of Indigenous peoples,[30] and the latter were subject to being captured and enslaved to feed, clothe, and enrich colonial populations.

Early British colonists in America had complex and sometimes contradictory experiences with forests: on the one hand, trees represented impediments to establishing their imported model of agriculture; on the other, they furnished colonists with a luxuriant abundance of heating, cooking, and building materials compared to the inhabitants of the wood-scarce British Isles.[31] Beyond their immediate uses to settlers, the continent's forest resources were especially prized for their contributions to maritime trade: eastern white pine provided straight poles for masts, oak and cedar were used for hulls, and seemingly endless forests of longleaf, slash, pitch, and other pines supplied naval stores used to seal wooden vessels and preserve ropes and sails. In the great pinelands of southern colonies such as Virginia and the Carolinas, the production of such "naval stores" was a centrally important economic activity, and one that relied heavily upon the labor of enslaved people brought to the American continent from Africa.

The allure of America's forest resources was central to British interests in colonization: the British Isles had been heavily deforested in the sixteenth century, leaving the kingdom reliant on imported wood to supply its naval fleet. Logs were among the first products shipped back to England from the Jamestown settlement after its founding in 1607, and exports of wood and wood products were central to colonial economies from New England to Georgia.[32] Enormous quantities of wood were consumed in the production of colonial export products such as iron and potash as well as shingles, staves, and a variety of shipping containers and related products.[33] Far from simply eking out a subsistence living, European colonists were engaged in transforming natural resources into commodities for international markets from their earliest days on the North American continent.

Wood products were so important to the Plymouth Colony that colonists began to regulate use and trade in forests within the first years following settlement. In 1626, less than six years after the arrival of the *Mayflower*, colonial leaders decreed what may have been the first formal forest policy in colonial America, that "no man of what condition soever sell or transport any manner of workes… [that] may tend to the destruction of timber…without the consent approbation and liking of the Governour and council."[34] Later regulations required well-wooded settlements in the Plymouth Colony to help supply timber to settlements where wood was scarce. Colonists throughout seventeenth century New England devised village laws to prevent the waste of valuable timber resources, especially on common lands that were vital sources of fuelwood and timber for construction, and to protect forest owners' property in trees.[35] A 1681 Pennsylvania policy famously required settlers to leave one acre of trees standing for every five acres cleared.[36] The existence of laws such as these shows that, even from their earliest days on the continent, Euro-American colonists recognized the need for some degree of regulation of the management and disposition of forests.

While many of these colonial policies sought to maintain forests and forest resources for the benefit of settlers, the English powers had their own interests in regulating forest use due to the importance of trees to British naval interests. The large, straight eastern white pines of New England were especially coveted. From 1691 through the American Revolution, the British Crown claimed possession of eastern white pine trees outside of settled townships in New England. The first decree applied only to trees 24 or more inches in diameter, and later decrees applied to all white pines. This "broad arrow" policy (so called because large white pines would be marked with a hatchet in the form of an arrow or crow's foot) was initially ignored by the colonial population but eventually served as a major source of tension between colonists and England. Royal claims to white pine trees created enough resentment that on multiple occasions New England colonists beat and otherwise harassed the king's surveyors and other officials enforcing the policy; one such event in 1772 became known as the Pine Tree Riot. Anger over the broad arrow policy may even have contributed to motivations for the American Revolution itself.[37]

Wood harvest, consumption, and export were not the only targets of colonial forest governance. When they were not being employed to clear land for agriculture, fires threatened colonial farms and structures as well as valuable standing timber; fire regulations began to appear in the early decades after establishment of the first British colonies.[38] Stephen Pyne notes that "By the time of the Revolution, only Maryland, Virginia, South Carolina, and Georgia lacked protective [fire] legislation."[39] Before the establishment of professionalized wildland firefighting organizations, the task of fire control fell to settler populations. In wooded and rural settlements nationwide, local residents typically formed ad-hoc fire brigades as needed to protect their own and their neighbors' properties. The colonies also established laws regarding wildlife, protecting favored species such as deer while incentivizing the elimination of unwanted species like wolves and bobcats.[40]

**BOX 1.2 THE FIRST FEDERAL FOREST RESERVES**

Civilian and military leaders of the newly independent United States recognized, as did their British forebears, the value of forest resources to military readiness. The sturdy live oak native to coastal areas of the southeastern US was a crucial component of the first warships constructed under the presidency of George Washington. This species, along with eastern red cedar, continued to be used in naval ships through the Civil War and both were considered resources of vital national importance. The first federal appropriations to purchase and hold forests for military supply purposes were made in 1799; President Adams used these funds to purchase Grover and Blackbeard Islands on the Georgia coast, where live oaks grew in abundance. An 1817 law allowed for the inventory and reservation of federal lands that contained live oak and cedar. The first federal naval forest reserves were created with the withdrawal from disposal of eight Louisiana islands in 1820. Two years later, a law was passed prohibiting the cutting or export of trees important for naval uses from the newly established Florida territory, and subsequent laws expanded the reservation and purchase authorities of the president. Over 240,000 acres of federal naval reserves were eventually established in Alabama, Florida, Louisiana, and Mississippi, including a tree farm in Deer Point, Florida, on which live oaks were to be cultivated from acorns. These reservations were heavily degraded by timber trespassers despite the existence of federal timber agents assigned to patrol and monitor them. The program of naval forest reserves was dismantled after the Civil War following the transition to ironclad warships.[41]

## Forest Governance in a Young Nation

The American Revolution ended direct British control and administration of the eastern seaboard south of Canada, but it did little to diminish the influence of British ideas of law and property in the young nation. The United States continued to apply the common law tradition from England, under which the courts in effect create legal principles through their resolution of questions about the dimensions of rights, including rights to landed property. In this system, concepts like nuisance and trespass are legally enforceable because they have long been recognized as inherent dimensions of property ownership even if they are not codified in any formal legislation. Yet the young nation introduced new formal institutions as well, such as a written constitution positioned as the "supreme law of the land." The Constitution of 1787 also included language authorizing the federal government to own and dispose of land (Article IV), and this authority came to be essential not only in administering territorial expansion but also in securing forest resources considered to be of strategic national importance (Box 1.2).

Freed from the control of the British crown, the states proceeded to dispense with the remnants of British feudal property relations and institute a system in which land could be bought and sold freely as a market commodity.[42] At the same time, prevailing law concerning landed property reflected an understanding of both rights and responsibilities—as well as substantial limitations on the extent to which owners of unenclosed and "unimproved" lands could exclude others. Forests, in contrast to farms, were often seen as part of the common resource endowments of a community or region. For example, laws in several New England states expressly recognized public rights to hunt, fish, trap, cut firewood, and collect non-timber resources such as berries and wild fruits on unenclosed forestlands and to cross private lands to fish and hunt on "great ponds" of at least ten acres in size.[43] Laws such as these made sense in light of a Lockean view of property: because the landowner had done little to mix their labor with the soil, their claim to exclusive property in native forestlands was weak at best. The understanding of "unimproved" forests as a kind of commons would persist to various degrees through both formal and informal institutions throughout the country and across time (Box 1.3).

Another implication of the Americans' victory in the Revolutionary War was an official nullification of the long-ignored Proclamation of 1763, under which Britain forbade colonization west of the crest of the Appalachian Mountains. Under laws such as the Land Ordinance of 1785, the federal government took a central role in administering the settlement and disposition of lands ceded by the original colonies and by Britain. This entailed surveying the territory and imposing a grid system for inventory and disposal: surveyed townships were divided into 36 sections measuring one square mile each. As time went on, land purchased from other colonial powers, established through treaty, or appropriated following wars with Indigenous or colonial nations also fell into the hands of the federal government.[44] The task of overseeing the disposal of these lands to private individuals and businesses was eventually carried out by the General Land Office, a federal agency created in 1812 within the Interior Department.

Following the Revolution, Euro-American populations continued to spread their colonial model ever further inland. The US Government applied its military resources to defeating and displacing Native tribes and transferring conquered lands to settler colonists. The pattern would be repeated countless times in subsequent decades: treaties were made only to be broken by the US Government when settlers ran out of available land to colonize or when resources of value were identified on tribal lands. For example, the Cherokee people of southern Appalachia had already seen their territory progressively reduced via ten separate treaties between 1721 and 1777. In the years following the Revolution, the American Government continued to renegotiate its commitments to recognize Native land and sovereignty as settlers pushed to claim Cherokee lands and resources, especially following the discovery of gold near Dahlonega, Georgia, in 1828. The State of Georgia attempted to regulate mining and other activities on Cherokee lands, leading eventually to legal challenges over the scope of state authority on Native lands. Both the State of Georgia and President Andrew Jackson refused to abide by an 1832 Supreme Court decision that found state law regulating Cherokee land to be unconstitutional. Georgia opened all remaining Cherokee properties for distribution by lottery in 1832; only White males and widows were eligible to receive lands in the lottery.[45] In 1838, following the signing of a treaty of questionable legitimacy, federal troops began rounding up Cherokee families to force them to walk the Trail of Tears to Indian Territory (present-day Oklahoma), during which over four thousand Cherokee perished. This was just one of many federal acts in which treaties were broken and Indigenous people forcibly removed to western territories to establish private property claims for White settlers.[46]

A central justification for the removal of the Cherokee and other Native peoples was that they would not make productive use of the land the way that White property owners would. This argument was leveled in spite of the fact that the Cherokee had adopted numerous aspects and technologies of Euro-American culture and property relations—including private property, written constitutions, a written newspaper, and, in some instances, the enslavement of African Americans. It is also the case that many of the White settlers who came to occupy lands wrested from tribes like the Cherokee did not fully adopt Locke's ideal of individual private ownership. For example, in the southern Appalachian Mountains that were once the homeland of the Cherokee, a distinctive culture of land use developed among White populations that in some ways resembled the New Mexican ejido model (Box 1.3) more than it did Locke's freehold ideal. Although mountain families held their own individual farmsteads as private property, much of the higher forested terrain was managed as a vast commons—accessible to local people under expectations of reasonable use and the consideration of others' needs. On these lands, local families grazed hogs and cattle, collected chestnuts, acorns, and wild medicines, hunted game, harvested firewood, and managed extensively with fire. These land use patterns continued until they were disrupted by the collapse of the chestnut economy due to the introduced chestnut blight and, later, by the arrival of industrial timbering operations.[47]

Southern Appalachia was just one of many regions that developed a culture of fire use over many generations. The South, overall, gained a reputation for frequent use of fire; its generous application by both White and Black forest users improved forage for wildlife and domestic livestock, stimulated the production of important mast resources such as acorns and chestnuts, and helped prevent uncontrollable wildfires. Another locus of "fire culture" was found in northern California where hot, dry summers were capable of incinerating local forests of mixed conifer and hardwood species. Frequent fire use by White settler populations was part of a strategy of self-preservation, protection of property, and securing the kinds of wildlife and forage benefits noted above. Fires were also used to cover up evidence of poaching or other illegal activities, to generate demand for firefighting work among unemployed laborers, and as a tool of resistance against private and public landowners.[48]

## BOX 1.3 COLONIAL FOREST GOVERNANCE IN THE SPANISH SOUTHWEST

Spanish colonial populations established permanent settlements in Florida before the first permanent British settlements appeared in Virginia and New England. Spanish settler populations began to colonize what is now the US Southwest before the end of the sixteenth century. The Upper Rio Grande Valley in present-day southern Colorado and northern New Mexico was one of the early centers of Spanish influence in what is now the United States. This region was first colonized by Juan de Oñate and hundreds of accompanying settlers in 1598. As they did elsewhere in the Americas, the settlers subjugated Indigenous populations and placed them into forced labor.[49] Spanish settler populations continued to live in the region except for the twelve-year period following the 1680 Pueblo Revolt, a coordinated uprising of the Native Puebloans who had been badly mistreated under colonial rule.[50] Although agriculture and livestock raising were the primary livelihood activities of Spanish settlers and their descendants, forests—including riparian cottonwood forests, arid woodlands of piñon and juniper, and higher-elevation forests dominated by ponderosa pine and other coniferous species—represented important resources for heating, cooking, construction, and crafting of agricultural and other tools. Piñon pine was especially valuable as a source of nutritionally and commercially valuable piñon nuts.[51]

Under the Spanish system, continued by the Mexican Government following its independence from Spain, communities of settlers were typically granted specified territories that would be owned and managed in common. Individual households retained private property in their homesteads and irrigated agricultural fields, but woodlands, forests, and dry pasturelands formed part of common properties called ejidos.[52] These came with a substantially different set of rights and responsibilities from the individual private property model that came to prevail throughout much of the British colonies.[53] As historian Robert MacCameron notes, "Spanish law strictly outlawed such hallmarks of the Anglo-American frontier as land speculation and absentee ownership."[54] Each ejido community developed its own rules regarding both access and obligations for ejidal lands and waters. These were not the anarchic commons of Garrett Hardin's imagination, but rather community assets that were managed according to enforced rules of conduct.

Mexico lost 55% of its territory to the United States as a result of the Mexican–American War. The 1848 Treaty of Guadalupe Hidalgo that ended the war included a provision committing the US to honor existing land grants, but this remained largely an empty promise. The Spanish and Mexican grants represented some of the most valuable properties in the Southwest in terms of agricultural, water, and timber resources, and it was easy for Anglo officials and businessmen to appropriate them through means both legal and nefarious.[55] Substantial expanses of what had once been land grants eventually came to be incorporated in the Carson, Santa Fe, and Cibola National Forests, provoking tensions over the legitimacy of federal ownership that remain relevant to this day.[56]

## Scaling up Timber Harvesting

From the earliest days of Euro-American settlement, forest clearing had been essential for the establishment of agriculture in the humid eastern states. The wood removed was used for fuel, fencing, potash production, and myriad other applications. Through the middle of the nineteenth century, forests continued to play a central role in the economy of the United States, supplying not only raw materials for building, heating, and cooking but also furnishing energy for industry and transportation: raw wood and wood-derived charcoal were used in metallurgy and to power steamships and other steam-powered equipment. Shipbuilders from Maine to the Carolinas continued to churn out ever-larger vessels employed in global commerce and trade as the nineteenth century advanced. The vast majority of the nation's energy consumption in the mid-nineteenth century was based on fuelwood—this at a time when England had already switched to the cheaper and more energy-dense coal for its heating and industrial needs.[57] At this point, timbering was the second-largest industry in the nation, outstripping even cotton.[58] The scale of timber cutting and processing is perhaps hard to fathom, but a statistic from 1840 Census provides a glimpse: it counted 31,650 water-powered sawmills in the US at that time.[59]

Because of its superior value for construction of all kinds and because of its buoyancy (allowing it to be transported by water), eastern white pine was the most sought-after commercial species—this led New England, and later New York and Pennsylvania, to be the nation's dominant wood suppliers in the early to middle decades of the nineteenth century. As the large, straight white pines from these states were logged out, the industry shifted to new regions and worked to develop valuable products from less-favored species such as spruce and hemlock. Lumbering operations progressively moved from the original colonies to territories west of the Appalachians, increasingly taking place on landscapes of mixed private and federal ownership; timber trespass was widespread and often difficult to control.[60]

Timber harvesting during the nineteenth century was a primitive and often dangerous activity, focused on extracting the straightest, most commercially valuable trees without consideration for what was left behind or what might come next. The typical model of nineteenth century logging consisted of camps that existed

*Figure 1.1* A logjam on the Big Blackfoot River, Montana, 1899. Image: National Archives.

just long enough to pluck the choicest trees from a particular landscape, sending them on to mills via massive log "rafts." Timber workers were frequently sent to unclog logjams that developed as the wood was sent downstream (Figure 1.1), a task that was even more dangerous than felling trees in the woods.[61] Once the biggest and best trees had been harvested, logging operations moved on to uncut forests to repeat the cycle anew.

Harvesting accelerated as a result of new technologies and new demands created by expanding cities as Euro-American colonization spread westward. The pine forests of Michigan, Wisconsin, and Minnesota fueled the construction needs of prairie cities like Chicago as well as newer settlements along points westward. It was here in the Upper Midwest that the forest industry transitioned into something less chaotic and more organized. The German immigrant Frederick Weyerhaeuser spearheaded a series of agreements and contractual arrangements among scattered mill owners that allowed for billions of board feet of white pine from Wisconsin's Chippewa River drainage to be processed by downriver mills (including Weyerhaeuser's own in Rock Island, Illinois) over a few short decades.[62] In addition to setting Weyerhaeuser on the way to be one of the wealthiest people in the country, the arrangement brought order to what had been a patchwork of forest owners, mill owners, and transportation intermediaries. The model piloted in the Chippewa drainage showed the potential advantages of controlling the entire supply chain from forestland through harvesting, processing, and distribution—a model known as "vertical integration." The industrialization of timbering operations was helped along by advances in logging, hauling, and milling technology that allowed for economies of scale in liquidation of the Lake States' vast pineries.[63]

As a result, the Upper Midwest white pine forests that were still considered to be inexhaustible as late as 1870 were completely cut over by the end of the century, replaced by an abandoned wasteland ripe for runaway wildfire.[64] Historian William Cronon explains,

> As the loggers finished their work in the forests they had consumed, they left behind a literal wasteland. Great piles of slash—small timber, branches, and other debris that had little economic value—remained on the ground where they fell, sometimes in piles ten to fifteen feet high. They accumulated over a vast area, turned brown in the summer heat, and waited for the dry season, when a spark might set them alight.[65]

Fires under such conditions were a constant threat, and they burned hotter and spread faster than the fires that were typical of these landscapes prior to Euro-American colonization. The 1871 Peshtigo Fire, attributed in large part to this combination of factors, remains the deadliest wildfire in US history to date, killing over 1,000 people and incinerating entire villages in Wisconsin's Oconto, Kewanee, and Door Counties. Equally ferocious wildfires would periodically consume similarly abused landscapes in the upper Midwest, Northeast, and Northwest through the first half of the twentieth century.

Logging practices of the time left not only degraded forests in their wake, but often devastated nearby communities that found their watersheds destroyed, their previously navigable waterways clogged with sediment, and their topsoil heavily eroded even as they faced new threats of runaway conflagrations in the cutover landscapes. When a particular piece of land no longer contained commercially valuable timber, it was typically abandoned by its owners. The reasons for this pattern were simple: it made little economic sense to continue to pay taxes for abused, cutover lands that would fail to produce another timber crop for decades (and then only if they were spared from the ubiquitous post-logging fires). From the perspective of the nineteenth-century timber industry, there were always additional unexploited forests just over the horizon.

Forests in the Pacific Northwest and South were next in line once timber barons had cut their way through the Northeast and Upper Midwest, and it was largely northern timber companies that came to own the South's valuable pinelands and the Northwest's lucrative pine and Douglas-fir forests.[66] Rail lines were extended throughout vast forests of southern longleaf, slash, and loblolly pine that had been spared agricultural clearing because of their poor soils. In many places, the longleaf pines that had already been worked for their naval stores were harvested as timber (Figure 1.2). Swampy forests of centuries-old cypress trees were also targeted, with many of these stands converted to rice cultivation in

*Figure 1.2* Hauling pine by mule team, Georgia, 1903. Note the "cat faces" on some of the trees from naval stores production. Image: US Forest Service.

the aftermath of logging.[67] Throughout the South, shortsighted logging practices contributed to the runaway soil erosion that was exacerbated by cotton cultivation and other agricultural practices.[68] The technological innovations and corporate organizational forms that first appeared in the Lake States were transferred to the South, converting it into "the epitome of industrial capitalism in the lumber industry."[69] In just a few short decades, millions of acres of southern forestlands were liquidated—as usual with little thought given to what was left behind.

## Early Impulses toward Conservation

As the colonial project continued from east to west, most American political and business leaders saw forests and other elements of the nonhuman world in largely utilitarian terms and as practically inexhaustible. However, by the middle of the nineteenth century, writers, poets, and painters in the Romantic tradition were beginning to depict forests and other unpeopled landscapes as places of beauty and sublimity. Themes of the American "wilderness"—and of the struggle to subdue it—became central in the creation of an American identity among the mostly White beneficiaries of the era's land and resource policies.[70] As forest loss accelerated, so did the growth of a sentimental, romantic, and spiritual regard for forests as places of refuge from the corrupting influences of an industrializing society. The New England transcendentalist movement is especially notable in this regard; it placed nature at the center of a spiritual vision in which humans could directly experience transcendental truths. Henry David Thoreau, protégé of Ralph Waldo Emerson, is a pivotal figure in American conservation history for his writings such as *Walden* (1854). This work provided an intimate and detailed account of the author's experiences living a simple, low-technology life in the woods outside of Concord, Massachusetts, over a period of two years. In Thoreau's vision, a modern society obsessed with commerce and material acquisitiveness failed to understand what it was losing by subjugating nature to the advance of industry.

Writing in less spiritual and more practical terms, the scholar and diplomat George Perkins Marsh called attention to the dire consequences of contemporary land uses in his 1864 book *Man and Nature*. Its warnings of the contributions of forest and soil loss to the decline of civilizations drew on Marsh's extensive travels worldwide and his deep historical research. This text was the first to consider environmental degradation as a phenomenon across space and time and connect it to the fate of social development. Marsh's narrative of people as blindly destructive agents whose instincts could only be corralled by scientifically informed government intervention would come to have a lasting influence on the conservation movement that arose in the coming decades.[71] What Marsh and Thoreau had in common was their belief that the land needed protection; both called for the creation of preserves that would protect forests in their wild and "primitive" states.[72]

A complementary social movement of the mid-nineteenth century attempted to rectify the ills of America's burgeoning industrial cities by introducing the benefits of a cultivated version of nature to urban environments. Frederick Law Olmsted's creation of Central Park as a verdant, thoroughly engineered natural oasis in New York City in the late 1850s, was a defining moment in the nationwide parks movement; Olmsted went on to design parks from Quebec to California. One of his projects, now known as Yosemite National Park, began as a grant of federal land to the state of California in 1864 in the midst of the Civil War. The Yosemite Valley Grant Act stated that the uses of the deeded land were "for public use, resort, and recreation," echoing the animating concepts behind New York City's Central Park. Far from being constructed on unoccupied terrain, parks such as these went hand in hand with dispossession. Native American communities were displaced, their lands appropriated, and their resource access cut off to establish national parks nationwide.[73] The construction of Central Park began with the displacement of entire communities such as Pigtown and Seneca Village. These were some of the only places that African American families owned homes in New York; property ownership brought not only financial independence but also offered voting rights to men.[74]

## Multiple Experiences, Multiple Histories

The relationship between parks and displacement highlights the fact that conservation history is inseparable from the broader context of American history.

The story of America's changing relationships to its forests is often told by focusing on those with access to economic, political, and cultural influence: political leaders, government scientists, diplomats, founders of professional societies. Yet Americans' forest experiences and values have always been diverse. For example, Dorceta Taylor details the deeply transcendentalist spirituality of Harriet Tubman, who had to navigate forests, swamps, and marshes in her numerous travels to rescue enslaved people and lead them to freedom on the Underground Railroad. Tubman's connections to forests were personal and spiritual, and her life—and the lives of her "passengers" on the Underground Railroad—depended upon her ability to survive and traverse the landscape without being detected.[75]

In experiencing the forest as a conduit to freedom, Tubman may have been continuing one thread in the varied experiences of African Americans in the "New World." For as long as there was slavery in the Americas, there existed communities of escaped and freed former enslaved people who took to isolated wooded landscapes to establish the kind of autonomy that was impossible to attain under a legal and political system that viewed them as property. These so-called "maroon" communities were common in some of the most inhospitable settings in the South, places such as the Great Dismal Swamp of Virginia and North Carolina, the Santee River watershed of South Carolina, the cypress swamps of Louisiana, and the forests and marshlands of Florida.[76] While many maroon communities focused on subsistence, others were known for launching raids on plantations, in some cases helping to free people who were enslaved there. Members of maroon communities also fought in the Revolutionary and Civil Wars—first in support of the British, who promised to free them upon a British victory, then alongside Union forces and Confederate deserters.[77] Writers such as Carolyn Finney caution, however, that we should not idealize the difficult lives of maroon community members who lived in some of the continent's most unforgiving locations and who were always subject to the possibility of being pursued and killed by White landowners and law enforcement agents.[78] Yet Finney also maintains that maroon landscapes represented sites of resistance and resilience in the face of legal systems of oppression.

Following the end of legal slavery, different segments of society continued to experience vastly different kinds of access to the benefits of forests. For example, systems of debt peonage and convict leasing allowed forest owners to continue to profit from the involuntary labor of African Americans, including many who had previously toiled under conditions of enslavement.[79] Debt peonage, which also ensnared poor unskilled Whites, forced workers into labor to pay a debt owed to the employer; the terms of the employment ensured that the laborer could rarely if ever pay off the debt. While formally illegal, this practice was widespread in the turpentine industry and was actively supported by state policies throughout the South that created a presumption of guilt for indebted laborers.[80] Through convict leasing, states and counties leased incarcerated individuals to labor in slave-like conditions for no pay. The proliferation of laws against vagrancy, loitering, and other minor offenses disproportionately applied to African Americans ensured a steady supply of labor in the pinelands.[81] It was common for individual sheriffs to have contracts to deliver a specified number of convicts to supply individual operators.[82]

The forest experiences of diverse populations of Americans go well beyond conditions of slavery, dispossession, and forced labor, but highlighting these experiences is nevertheless important. These histories serve as a contrast with stories like Thoreau's two-year experiment in self-reliance or the aesthetic appreciation of natural landscapes in popular art and literature. They demonstrate that concepts like "property" and "conservation" can look very different according to the experiences and perspectives of diverse social groups. The following chapters provide a deeper exploration of some of these varied and diverse experiences and understandings.

## Conclusion

After centuries of treating both the land and people as thoroughly expendable resources, the leaders of the United States were faced with twin crises by the middle of the nineteenth century. One crisis would rip the young nation in two in a civil war that lasted for four years, claiming hundreds of thousands of lives and ultimately leading to the abolition of legal slavery. The other crisis created a movement that developed more slowly, not coming to full fruition until the following century. Marsh's publication of *Man and Nature* in 1864 at the height of the Civil War and President Lincoln's signing that same year of the Yosemite Valley Grant Act signaled a watershed change in the nation's orientation towards its forests. The American conservation movement that was just coming into being drew attention to the linkages between the uses of land and the fate of the larger society. As this movement progressed, it would confront both the potential and the limitations of federal power.

## Questions for Discussion

1. How do you usually imagine the state of America's forests prior to European colonization? Does this image include the frequent and widespread use of fire as described in this chapter?
2. Did America's Indigenous nations have conceptions of property prior to European colonization? How did these conceptions differ from the ideas imported from Europe?
3. How did the concept of property in land change throughout the first centuries of American history?
4. What evidence is there that commons-type property systems persisted in the United States even centuries after the first permanent European settlements?
5. How did timber harvest practices change between the colonial period and the mid-nineteenth century when the first industrial timber firms were founded? What were the environmental consequences of these changes?

## Notes

1 Hessburg & Agee, 2003; Langston, 1995; Robbins, 1997.
2 Langston, 1995.
3 M. K. Anderson & Moratto, 1996; Condie & Raish, 1981; Fowler & Konopik, 2007; Pyne, 1982.

4 H. R. Delcourt & Delcourt, 1997.
5 Matlack, 2013; Stambaugh et al., 2015; Vale, 2002.
6 Cronon, 2011; Hanberry et al., 2020.
7 Allen, 2002.
8 Abrams, 1992; H. R. Delcourt & Delcourt, 1997; Denevan, 1992; Lafon, 2010; Munoz et al., 2014; Nowacki & Abrams, 2008; Pyne, 1982; Stewart, 2002; M. Williams, 1989, pp. 43–49.
9 Pyne, 1982, pp. 79–80.
10 Munoz et al., 2014, p. 2200.
11 Brooks, 2008; LaDuke, 1999; Norgaard, 2019; Trosper, 2009.
12 Walker Jr, 1967.
13 Marshall, 1977.
14 Marshall, 1977; see also M. K. Anderson, 2005.
15 Cronon, 2011, pp. 50–51.
16 Cronon, 2011, p. 63.
17 Reséndez, 2016.
18 Davis et al., 2019.
19 Ardelean et al., 2020.
20 These are not definitive dates of earliest inhabitation; we may yet learn that humans were present in the Americas even earlier.
21 P. A. Delcourt & Delcourt, 1987, 2004.
22 R. S. Anderson, 1989; Shinneman et al., 2016; J. W. Williams et al., 2004.
23 Pyne, 2021, p. 77.
24 M. K. Anderson, 2005; P. A. Delcourt & Delcourt, 2004; Denevan, 1992.
25 Pyne, 1982; Rostlund, 1957; M. Williams, 1989, pp. 40–48.
26 Rostlund, 1960.
27 Nowacki & Abrams, 2008; Rostlund, 1957.
28 Hanberry et al., 2020, p. 2.
29 Merchant, 2002.
30 Hsueh, 2006.
31 Cox et al., 1985.
32 Rutkow, 2012.
33 Cox et al., 1985, pp. 15–16.
34 Rutkow, 2012, pp. 21–22.
35 Cox et al., 1985, p. 44; McCullough, 1995.
36 Clepper, 1945.
37 Andrews, 1947; Rutkow, 2012.
38 Cox et al., 1985, p. 46.
39 Pyne, 1982, p. 164.
40 Leopold, 2018; Newfont, 2012.
41 Snell, 1983.
42 Freyfogle, 2003, pp. 51–52.
43 Acheson & Acheson, 2010; Freyfogle, 2003, p. 23.
44 Merchant, 2002.
45 H. D. Williams, 1989.
46 Saunt, 2020.
47 Newfont, 2012.
48 Jacoby, 2001.
49 Dunbar-Ortiz, 2007.
50 Knaut, 1997.
51 MacCameron, 1994.
52 Gonzales, 2003.
53 However, note that a distinction between private and common lands was likewise typical of many early New England villages (McCullough, 1995).

54 MacCameron, 1994, p. 29.
55 deBuys, 1985; Gonzales, 2003.
56 Raish, 1996.
57 Pyne, 1982, p. 189; Rutkow, 2012, p. 70.
58 Rutkow, 2012, p. 108.
59 Cox et al., 1985, p. 64.
60 Cox et al., 1985, pp. 139–142.
61 M. Williams, 1989, pp. 171–173.
62 Rutkow, 2012, pp. 110–115.
63 M. Williams, 1989, pp. 193–230.
64 Cronon, 1991, pp. 200–206.
65 Cronon, 1991, p. 202.
66 Rutkow, 2012, p. 181.
67 Cox et al., 1985, p. 96.
68 Cox et al., 1985, p. 101.
69 M. Williams, 1989, p. 263.
70 Cronon, 1995; M. Williams, 1989, pp. 18–21.
71 Jacoby, 2001, p. 13.
72 M. Williams, 1989, p. 18.
73 Spence, 1999.
74 Taylor, 2011.
75 Taylor, 2011.
76 Aptheker, 1939; Bledsoe, 2017.
77 Aptheker, 1939.
78 Finney, 2014.
79 Barnett, 2019.
80 Shofner, 1981.
81 Blackmon, 2008.
82 Shofner, 1981.

## References


Abrams, M. D. (1992). Fire and the development of oak forests. *BioScience*, *42*(5), 346–353.

Acheson, J., & Acheson, J. (2010). Maine land: Private property and hunting commons. *International Journal of the Commons*, *4*(1), 552–570.

Allen, C. D. (2002). Lots of lightning and plenty of people: An ecological history of fire in the upland Southwest. In T. Vale (Ed.), *Fire, native peoples, and the natural landscape* (pp. 173–194). Island Press.

Anderson, M. K. (2005). *Tending the wild: Native American knowledge and the management of California's natural resources*. University of California Press.

Anderson, M. K., & Moratto, M. J. (1996). Native American land-use practices and ecological impacts. In *Sierra Nevada ecosystem project: Final report to Congress* (Vol. 2, pp. 187–206). University of California, Centers for Water and Wildland Resources.

Anderson, R. S. (1989). Development of southwestern ponderosa pine forests: What do we really know? In A. Tecle, W. W. Covington, & R. H. Hamre (Eds.), *Multiresource management of ponderosa pine forests* (pp. 15–22). US Department of Agriculture, Forest Service, Rocky Mountain Research Station.

Andrews, H. N. Jr. (1947). The King's pines. *Historical New Hampshire*, March, 1–14.

Aptheker, H. (1939). Maroons within the present limits of the United States. *The Journal of Negro History*, *24*(2), 167–184.

Ardelean, C. F., Becerra-Valdivia, L., Pedersen, M. W., Schwenninger, J.-L., Oviatt, C. G., Macías-Quintero, J. I., Arroyo-Cabrales, J., Sikora, M., Ocampo-Díaz, Y. Z. E., Rubio-Cisneros, I. I., Watling, J. G., de Medeiros, V. B., De Oliveira, P. E., Barba-Pingarón, L., Ortiz-Butrón, A., Blancas-Vázquez, J., Rivera-González, I., Solís-Rosales, C., Rodríguez-Ceja, M., … Willerslev, E. (2020). Evidence of human occupation in Mexico around the last glacial maximum. *Nature*, *584*(7819), 87–92.

Barnett, J. P. (2019). *Naval stores: A history of an early industry created from the South's forests* (General Technical Report SRS-240). USDA Forest Service Southern Research Station.

Blackmon, D. A. (2008). *Slavery by another name: The re-enslavement of Black Americans from the Civil War to World War II*. Doubleday.

Bledsoe, A. (2017). Marronage as a past and present geography in the Americas. *Southeastern Geographer*, *57*(1), 30–50.

Brooks, L. (2008). *The common pot: The recovery of Native space in the Northeast*. University of Minnesota.

Clepper, H. (1945). Rise of the forest conservation movement in Pennsylvania. *Pennsylvania History: A Journal of Mid-Atlantic Studies*, *12*(3), 200–216.

Condie, C. J., & Raish, C. (1981). Indigenous and traditional use of fire in southwestern grassland, woodland, and forest ecosystems. In Proceedings of the Ninth International Symposium on Society and Resource Management, Bloomington, IN (pp. 26–35).

Cox, T. R., Maxwell, R. S., Thomas, P. D., & Malone, J. J. (1985). *This well-wooded land: Americans and their forests from colonial times to the present*. University of Nebraska Press.

Cronon, W. (1991). *Nature's metropolis: Chicago and the Great West*. W. W. Norton & Company.

Cronon, W. (1995). The trouble with wilderness: Or, getting back to the wrong nature. In W. Cronon (Ed.), *Uncommon ground: Toward reinventing nature* (pp. 69–90). W. W. Norton & Company.

Cronon, W. (2011). *Changes in the land: Indians, colonists, and the ecology of New England*. Hill and Wang.

Davis, L. G., Madsen, D. B., Becerra-Valdivia, L., Higham, T., Sisson, D. A., Skinner, S. M., Stueber, D., Nyers, A. J., Keen-Zebert, A., Neudorf, C., Cheyney, M., Izuho, M., Iizuka, F., Burns, S. R., Epps, C. W., Willis, S. C., & Buvit, I. (2019). Late upper paleolithic occupation at Cooper's Ferry, Idaho, USA, ~16,000 years ago. *Science*, *365*(6456), 891–897.

deBuys, W. (1985). *Enchantment and exploitation: The life and hard times of a New Mexico mountain range*. University of New Mexico Press.

Delcourt, H. R., & Delcourt, P. A. (1997). Pre-columbian Native American use of fire on southern Appalachian landscapes. *Conservation Biology*, *11*(4), 1010–1014.

Delcourt, P. A., & Delcourt, H. R. (1987). *Long-term forest dynamics of the temperate zone: A case study of late-quaternary forests in eastern North America*. Springer-Verlag.

Delcourt, P. A., & Delcourt, H. R. (2004). *Prehistoric Native Americans and ecological change: Human ecosystems in eastern North America since the Pleistocene*. Cambridge University Press.

Denevan, W. M. (1992). The pristine myth: The landscape of the Americas in 1492. *Annals of the Association of American Geographers*, *82*(3), 369–385.

Dunbar-Ortiz, R. (2007). *Roots of resistance: A history of land tenure in New Mexico*. University of Oklahoma Press.

Finney, C. (2014). *Black faces, White spaces: Reimagining the relationship of African Americans to the great outdoors*. University of North Carolina Press.

Fowler, C., & Konopik, E. (2007). The history of fire in the southern United States. *Human Ecology Review, 14*(2), 165–176.

Freyfogle, E. T. (2003). *The land we share: Private property and the common good*. Island Press.

Gonzales, P. B. (2003). Struggle for survival: The Hispanic land grants of New Mexico, 1848–2001. *Agricultural History, 77*(2), 293–324.

Hanberry, B. B., Bragg, D. C., & Alexander, H. D. (2020). Open forest ecosystems: An excluded state. *Forest Ecology and Management, 472*, 118256.

Hessburg, P. F., & Agee, J. K. (2003). An environmental narrative of inland northwest United States forests, 1800–2000. *Forest Ecology and Management, 178*(1), 23–59.

Hsueh, V. (2006). Cultivating and challenging the common: Lockean property, Indigenous traditionalisms, and the problem of exclusion. *Contemporary Political Theory, 5*(2), 193–214.

Jacoby, K. (2001). *Crimes against nature: Squatters, poachers, thieves, and the hidden history of American conservation*. University of California Press.

Knaut, A. L. (1997). *The Pueblo revolt of 1680: Conquest and resistance in seventeenth-century New Mexico*. University of Oklahoma Press.

LaDuke, W. (1999). *All our relations: Native struggles for land and life*. South End Press.

Lafon, C. W. (2010). Fire in the American South: Vegetation impacts, history, and climatic relations. *Geography Compass, 4*(8), 919–944.

Langston, N. (1995). *Forest dreams, forest nightmares: The paradox of old growth in the Inland West*. University of Washington Press.

Leopold, B. D. (2018). History of wildlife policy and law through colonial times. In B. D. Leopold, W. B. Kessler, & J. L. Cummings (Eds.), *North American wildlife policy and law* (pp. 19–26). Boone and Crockett Club.

MacCameron, R. (1994). Environmental change in colonial New Mexico. *Environmental History Review, 18*(2), 17–39.

Marshall, A. G. (1977). *Nez Perce social groups: An ecological interpretation* [PhD Thesis]. Washington State University.

Matlack, G. R. (2013). Reassessment of the use of fire as a management tool in deciduous forests of eastern North America. *Conservation Biology, 27*(5), 916–926.

McCullough, R. (1995). *The landscape of community: A history of communal forests in New England*. University Press of New England.

Merchant, C. (2002). *The Columbia guide to American environmental history*. Columbia University Press.

Munoz, S. E., Mladenoff, D. J., Schroeder, S., & Williams, J. W. (2014). Defining the spatial patterns of historical land use associated with the Indigenous societies of eastern North America. *Journal of Biogeography, 41*(12), 2195–2210.

Newfont, K. (2012). *Blue Ridge commons: Environmental activism and forest history in western North Carolina*. University of Georgia Press.

Norgaard, K. M. (2019). *Salmon and acorns feed our people: Colonialism, nature, and social action*. Rutgers University Press.

Nowacki, G. J., & Abrams, M. D. (2008). The demise of fire and "mesophication" of forests in the eastern United States. *BioScience, 58*(2), 123–138.

Pyne, S. J. (1982). *Fire in America: A cultural history of wildland and rural fire*. Princeton University Press.

Pyne, S. J. (2021). *The Pyrocene: How we created an age of fire, and what happens next*. University of California Press.

Raish, C. (1996). Historic land use and grazing patterns in northern New Mexico. In D. W. Shaw & D. M. Finch (Eds.), *Desired future conditions for southwestern riparian ecosystems: Bringing interests and concerns together* (pp. 189–197). US Department of Agriculture, Forest Service, Rocky Mountain Research Station.

Reséndez, A. (2016). *The other slavery: The uncovered story of Indian enslavement in America.* Houghton Mifflin Harcourt.

Robbins, W. G. (1997). *Landscapes of promise: The Oregon story, 1800–1940*. University of Washington Press.

Rostlund, E. (1957). The myth of a natural prairie belt in Alabama: An interpretation of historical records. *Annals of the Association of American Geographers*, *47*(4), 392–411.

Rostlund, E. (1960). The geographic range of the historic bison in the southeast. *Annals of the Association of American Geographers*, *50*(4), 395–407.

Rutkow, E. (2012). *American canopy: Trees, forests, and the making of a nation* (1st Scribner hardcover ed.). Scribner.

Saunt, C. (2020). *Unworthy republic: The dispossession of Native Americans and the road to Indian Territory*. WW Norton & Company.

Shinneman, D. J., Means, R. E., Potter, K. M., & Hipkins, V. D. (2016). Exploring climate niches of ponderosa pine (Pinus ponderosa Douglas ex Lawson) haplotypes in the western United States: Implications for evolutionary history and conservation. *PLOS ONE*, *11*(3), e0151811.

Shofner, J. H. (1981). Forced labor in the Florida forests 1880–1950. *Journal of Forest History*, *25*(1), 14–25. JSTOR.

Snell, C. W. (1983). *A history of the Naval Live Oak Reservation Program, 1794–1880: A forgotten chapter in the history of American conservation.* US Department of the Interior, National Park Service, Technical Information Center.

Spence, M. D. (1999). *Dispossessing the wilderness: Indian removal and the making of the national parks*. Oxford University Press.

Stambaugh, M. C., Varner, J. M., Noss, R. F., Dey, D. C., Christensen, N. L., Baldwin, R. F., Guyette, R. P., Hanberry, B. B., Harper, C. A., Lindblom, S. G., & Waldrop, T. A. (2015). Clarifying the role of fire in the deciduous forests of eastern North America: Reply to Matlack. *Conservation Biology*, *29*(3), 942–946.

Stewart, O. C. (2002). *Forgotten fires: Native Americans and the transient wilderness* (H. T. Lewis & M. K. Anderson, Eds.). University of Oklahoma Press.

Taylor, D. E. (2011). Introduction: The evolution of environmental justice activism, research, and scholarship. *Environmental Practice*, *13*(4), 280–301.

Trosper, R. (2009). *Resilience, reciprocity and ecological economics: Northwest Coast sustainability*. Routledge.

Vale, T. (Ed.). (2002). *Fire, Native peoples, and the natural landscape*. Island Press.

Walker Jr, D. E. (1967). *Mutual cross-utilization of economic resources in the plateau: An example from aboriginal Nez Perce fishing practices.* Washington State University Laboratory of Anthropology.

Williams, H. D. (1989). Gambling away the inheritance: The Cherokee Nation and Georgia's gold and land lotteries of 1832–33. *The Georgia Historical Quarterly*, *73*(3), 519–539.

Williams, J. W., Shuman, B. N., Webb, T., Bartlein, P. J., & Leduc, P. L. (2004). Late-quarternary vegetation dynamics in North America: Scaling from taxa to biomes. *Ecological Monographs*, *74*(2), 309–334.

Williams, M. (1989). *Americans and their forests: A historical geography*. Cambridge University Press.

# 2 Forestry and the National Interest

Fifty-eight percent of present-day Wallowa County is under federal ownership. The US Forest Service is responsible for managing the Wallowa-Whitman National Forest, which accounts for the vast majority of the county's federal footprint. The first federal withdrawals of Wallowa County land from the public domain to create federal forest reserves occurred in the early 1900s.[1] Federal forest designations were prompted in part by the growing local presence of timber speculators ready to take advantage of the nation's land disposal laws to assemble a private timber empire from lands intended to be transferred to settler families.[2] As environmental historian Nancy Langston details, the arrival of federal foresters in this region came with a promise to make the forests productive, to control irrational land uses that were leading to overgrazing and watershed degradation, and to eliminate the fires that threatened standing timber and other valuable resources.[3]

The contemporary Wallowa-Whitman National Forest covers 2.4 million acres, a little more than one percent of a system of national forestlands and grasslands that encompasses 193 million acres nationwide. Nearly 40,000 permanent and seasonal federal employees work for the US Forest Service, carrying out duties related to timber management, recreation, fish and wildlife habitat management, range management, watershed management, fire and fuels management, planning, environmental analysis, research, administration of private landowner programs, and many other duties. How the federal government came to own and manage vast forested landscapes in places like Wallowa County and what this means for people and forests are the subject of this chapter. The story begins with the emergence of the conservation movement and of professional forestry in the US in the late nineteenth century.

## The Rise of American Conservation

> Any fool can destroy trees. They cannot run away; and if they could, they would still be destroyed—chased and hunted down as long as fun or a dollar could be got out of their bark hides, branching horns, or magnificent bole backbones...God has cared for these trees, saved them from drought, disease, avalanches, and a thousand straining, leveling tempests

DOI: 10.4324/9781003043669-2

> and floods; but he cannot save them from fools—only Uncle Sam can do that.

So wrote environmentalist, author, and Sierra Club co-founder John Muir in the closing years of the nineteenth century.[4] Elsewhere in his essay he detailed in his characteristic stirring prose the myriad ways that the nation's forests, especially the coniferous forests of the West, were being felled, burned, overgrazed, and otherwise laid to waste in the absence of sound government policy. The rush by individuals, timber barons, and railroad companies to profit from the nation's timber resources was quickly destroying the native forests that were, to Muir, the American equivalent of Europe's great cathedrals.

Even as federal powers were consumed by an effort to rebuild a nation torn apart by war in the final decades of the nineteenth century, the United States began a process of rethinking the proper role of government, higher education, and civil society in conserving the nation's forest heritage. This transformative period included establishment of the world's first national park at Yellowstone (1872), the first federal report on the condition of the nation's forests (1878), the first American Forestry Congress (1882), New York's establishment of the Adirondacks Reserve (1885, later to become Adirondack Park), the creation of the nation's first forestry school at Biltmore, North Carolina, and the first four-year forestry program at Cornell (both in 1898), and the founding of the American Forestry Association (1875), Audubon Society (1886), Sierra Club (1892), and Society of American Foresters (1900). The years between 1891 and 1898 saw passage of a law allowing for the federal reservation of forests in the public domain, a second law declaring what those forest reserves were to be used for, and the appointment of Gifford Pinchot as Chief of the Division of Forestry—the first US-born forester to hold that title. Pinchot and other vanguards of forest conservation agreed with Muir that more government action was needed to prevent short-sighted timber harvesting from destroying the nation's forests, even if they held divergent visions of the proper balance between timber production and nature preservation. As described in Chapter 1, the prevailing timbering practices in the US were more akin to timber mining than to what we now know as the practice of forestry, and they threatened to leave the growing country without a sustained source of wood for the future. The diverse ideas that animated forest policy change during those years—and the tensions between them—would continue to reverberate more than a century afterward.

When John Muir traversed the South on foot in 1867 on his famous thousand-mile walk from Kentucky to the botanical wonderland of Florida, he crossed a landscape in the grips of tumultuous change.[5] Behind the social, political, and economic reshuffling playing out in the wake of the Confederacy's defeat to Union forces was a no less significant ecological transformation. Prevailing cotton farming practices had sent centuries' worth of topsoil into southern waterways in just a few decades; the forces of erosion were carving canyons into landscapes where fertile rolling hills had predominated just a few years before. In Muir's still largely unroaded destination of Florida, hunters would soon arrive to ply the dark, malarial cypress swamps in pursuit of egrets, ibises, and other wading birds whose feathers featured prominently in hats fashionable with the urban elite of London, Paris, and New York. Practices like these would later spur the young Audubon Society to wage battle against the destruction of the nation's birdlife and against the millinery industry that benefited from it.[6] Forests throughout the Northeast, South, and Midwest were at this time being systematically liquidated with little thought given to what was left behind. Yet federal recognition of the "big tree" groves in California's Sierra Nevada in 1864 indicated that governments were ready to take a more active role in setting direction for the nation's forests. Muir would soon apply his skills as a writer to help advocate for an even more aggressive government response, one that included the designation and expansion of Sequoia, Yosemite, Mount Rainier, and other national parks.

The initial push to conserve forests in the United States came as part of a growing wave of concern in the latter part of the nineteenth century regarding the loss and degradation of wildlife, topsoil, water, and timber resources. Leading the forest conservation charge were scientists like Franklin Hough, Gifford Pinchot, Carl Schenck, Charles Sprague Sargent, and Bernhard Fernow, supported by writers like John Muir and civic groups that included garden clubs and other urban associations, many of them led and organized by women.[7] Despite important soil conservation work being done by the likes of George Washington Carver, much of the early conservation movement was motivated by interests of the White upper class in conserving huntable game species for sport.[8] Forest issues became increasingly central to the conservation movement under the influence of George Perkins Marsh's *Man and Nature* and mounting evidence of wasteful and destructive timber practices. The American Forestry Association was founded in 1875 with the goal of protecting the country's forests from waste and promoting tree planting; it merged shortly thereafter with the American Forestry Congress. This coalition of scientists, politicians, and lumber and railroad barons advocated for stronger governmental forestry agencies and the creation of government-owned forest reserves in the West.[9]

Another conservation milestone came in 1885 when the State of New York established a "forest preserve" on state lands in the Adirondack and Catskill Mountains. Ten years later the Adirondacks region would be preserved as a park via an amendment to the state constitution. However, the conservation ideals advanced by wealthy sportsmen and landowners often did not sit well with local people who had developed their own informal practices related to forests and the resources contained therein. As Karl Jacoby recounts, the establishment of state and private preserves in the Adirondacks engendered decades of conflict as local forest users defended their informal practices and traditional forms of access against the formal restrictions imposed by the state and wealthy landowners:

> Local people ripped down the 'no trespassing' signs that ringed the parks' boundaries. They cut the elaborate wire fences that enclosed many estates, hoping that the deer and other game within might escape. They set fire to private parklands, sometimes with the intention of damaging fences, other times with the apparent desire to make the preserve an

> undesirable refuge for game animals. And they shot at the guides hired to protect the private parks.[10]

These sometimes violent actions aimed at resisting the replacement of an informal commons-type forest tenure system with a more formalized system dominated by state and private ownership.

The early years of the new century ushered in a more active role for the federal government as well, following on the designation of Yellowstone as the first national park in 1872. Laws like the Lacey Act of 1900 and the Migratory Bird Treaty Act of 1918 indicated that federal authority would be used to solve problems of declining wildlife populations (Chapter 4) and demonstrated the political influence of organizations such as the Audubon Society. Private conservation groups even funded their own wardens to patrol nature reserves in pursuit of hunters and poachers.[11] At the same time, early wildlife conservation laws exacted heavy economic and livelihood tolls on recent immigrants, poor Whites, Native Americans, and African Americans who depended upon wild game to supplement their meager earnings.[12]

## Changes to Land and Ownership

In the late nineteenth century, the federal government was busy "disposing" of the vast public domain that had come under federal control as a result of land deals (such as the Louisiana and Gadsden Purchases), foreign wars, wars against and subsequent treaties with Native American nations, and through other means. The Homestead Act of 1862 is the best-known public land disposal law, but there were other versions of the Homestead Act as well as many lesser-known policies designed to transfer federal lands to private and state control. The varied purposes of these public land disposal laws included raising money for the federal government, encouraging industrial development, promoting agricultural settlement, making productive use of lands more suited to grazing or forestry, and recognizing claims made by settlers prior to the existence of a legal framework.[13] For example, the ill-fated 1873 Timber Culture Act provided 160 acres of public domain land in the Plains states in exchange for the planting and maintenance of trees on 40 acres; the Timber and Stone Act of 1878 allowed purchases of up to 160 acres of land unsuited for farming at the cost of $2.50 per acre.

The end of the Civil War also created a vanishingly brief opportunity for newly freed African American populations to participate in the benefits of land ownership. Following a series of aborted federal attempts to confiscate and redistribute lands belonging to southern landowners who had taken up arms against the government, General Sherman's Special Field Order Number 15 decreed that 400,000 acres of former plantation land along coastal South Carolina, Georgia, and Florida were to be "reserved and set apart" and the land distributed to formerly enslaved families in 40-acre parcels.[14] The plan was forged during Sherman's meeting with Black ministers in Savannah, Georgia, in January 1865 and its implementation via the Special Field Order gave rise to the now-famous promise of "forty acres and a mule." This radical vision was to be short-lived, however, as President Johnson rescinded the order in the fall of the same year, returning these lands to the hands of former slaveowners.[15] The Freedman's Bureau and the Southern Homestead Act of 1866 were also animated by a vision of land distribution to African Americans but were ultimately limited in their ability to change longstanding patterns of property ownership. Even the Port Royal Experiment on the South Carolina Sea Islands—designed to foster Black land ownership and autonomy—saw only limited success as wealthy White northerners ultimately acquired most of the land abandoned by slaveowners during the Civil War.[16]

In the western states and territories, the majority of former public domain lands ended up as the property of powerful (and almost exclusively White) owners, in spite of the yeoman imagery associated with the Homestead Act.[17] In part this was by design, as public domain lands were used as payment to incentivize the construction of transcontinental railroads and other infrastructure, and in part this was the result of the widespread land fraud that resulted from a weak and easily corruptible federal agency, Interior's General Land Office, being placed in charge of a vast estate filled with timber, mineral, and other valuable resources.[18] Western forestlands were especially susceptible to fraud and graft. In the 1907 tell-all *Looters of the Public Domain*, Stephen A. D. Puter, the self-described "King of the Oregon land fraud ring," described how he and others consolidated control over vast expanses of forestland:

> Thousands upon thousands of acres, which included the very cream of the timber claims in Oregon and Washington, were secured by Eastern lumbermen and capitalists, the majority of whom came from Wisconsin, Michigan and Minnesota, and nearly all of these claims, to my certain knowledge, were fraudulently obtained.[19]

Just as Southern and Midwestern forests in the late nineteenth century were transformed by changes in markets and technologies, the lush temperate forests of the Pacific Northwest began to meet a similar fate as well-connected lumbermen took advantage of corrupt officials and weak federal oversight to amass landed estates of almost incalculable value.

Seeing the forest destruction plaguing the South, East, and Midwest, and well aware that the same logging practices were starting to arrive in the western forests that were quickly falling out of federal ownership, conservation leaders pushed for more aggressive governmental action. The notion of an impending "timber famine" was central to their argument: if the nation were to allow the same fate to befall western forests as their eastern counterparts, it would be left with no timber for its continued domestic needs. Their efforts paid off in 1891 when an obscure legislative rider was added to an appropriations bill authorizing the president to reserve any areas of federally owned public domain forest from disposal to private hands. This seemingly modest legislative language, which came to be known as the Forest Reserve Act, established a role for extensive federal forest ownership and management and led ultimately to what we now know as the National Forest System. Presidents Harrison and Cleveland made active use of this authority, together establishing 39 million acres as "forest reserves" during their mandates and provoking sharply critical reactions from western timber, grazing, and mining interests as well as western elected officials.[20] Twenty-one million acres of this total were designated in a single act by President Cleveland just prior to the end of his second term, drawing a backlash from western legislators who threatened to rescind the forest reserve authority and undo reserves that had already been established.

Despite a growing consensus among conservation leaders about the imperative to reserve forests from the ravages of prevailing logging practices, there was no clear legislative mandate for the purposes and uses of the federally reserved lands for several years.[21] In both popular and official discourse, the category of forest reserves was often used interchangeably with national parks, and for a brief time the General Land Office interpreted its mandate as one of preventing any trespass or use of the reserves. To provide guidance to the president and Congress, the National Academy of Sciences convened a National Forest Commission charged with investigating the state of the forest reserves and recommending policy to guide their management. Although the commission members could not agree on whether federal reserves should be protected by the Army (mirroring the approach taken in Yellowstone National Park) or by a professional forestry agency, they did agree that more government forest ownership was needed.

With the encouragement of President McKinley, Congress passed the Forest Management Act (or "Organic Act") in 1897 for what would become the National Forest System, based in part on the recommendations of the Forest Commission. It specified that no forest reserve should be established except to serve the following purposes: "to improve and protect the forest within the reservation, or for the purpose of securing favorable conditions of water flows, and to furnish a continuous supply of timber for the use and necessities of citizens of the United States." The Act authorized the Secretary of Interior to regulate the "occupancy and use" of reserved lands and set out conditions for timber management, making it clear that scientific methods were to be instituted. Only mature timber was to be harvested, and each tree designated for harvest would be marked as such. In the span of a few years, the federal government created both a conservation estate and a mandate for its management. All that was missing was a competent federal agency to do the managing.

**BOX 2.1 OTHER FEDERAL FORESTLANDS**

The US Forest Service is not the only federal agency with authority to manage forests; multiple federal agencies have come to control diverse forestlands nationwide as the result of a patchwork of laws, transfers, and executive orders. President Theodore Roosevelt established Pelican Island in Florida as the nation's first federal bird reservation via a 1903 executive order and used executive power to establish national monuments under the authorities provided by the 1906 Antiquities Act. Congress established Yellowstone in 1872 as "a public park or pleasuring-ground for the benefit and enjoyment of the people." The National Park Service, created in 1916 in the wake of the Hetch Hetchy controversy, was established to provide for the preservation and, in some cases, restoration of some of the nation's most spectacular landscapes, including many dominated by forests: groves of giant sequoias in places like Yosemite and Sequoia/King's Canyon National Parks, centuries-old cedar, spruce, and maple forests in Olympic National Park, the ponderosa pine savannahs atop the rim of Grand Canyon National Park, and the diverse hardwood forests blanketing Great Smoky Mountains National Park, among others. Today, the National Park Service, along with the US Fish and Wildlife Service, Department of Defense, Department of Energy, and other non-Forest Service federal agencies collectively own and manage over 100 million acres of forestland nationwide.[22] One agency, the Bureau of Land Management (BLM), accounts for nearly half of that total. The BLM was created in 1946 by combining the Grazing Service with the General Land Office. At that time, the General Land Office was still in possession of millions of acres of public domain lands that were never privatized via the various disposal laws. Although the BLM is better known for managing vast areas of rangeland, its estate includes substantial expanses of piñon-juniper woodlands throughout the arid western US, tens of millions of acres of boreal forest in Alaska, and around 2.5 million acres of highly productive timberland in western Oregon. The BLM's holdings in Oregon are the legacy of grants of public domain lands made to developers of the Oregon and California Railroad, a freight line connecting Portland with southern Oregon and northern California, and the Coos Bay Wagon Road in southwestern Oregon.[23] The developers of both projects were found to have violated the terms of their land grants, resulting in the revestment of millions of acres in a characteristic checkerboard pattern to federal ownership in 1916 and 1919 (Figure 2.1). In the depths of the Great Depression, Congress charged the Department of the Interior with developing these "O&C Lands" under sustained yield principles to benefit local economies, protect watersheds, and provide for recreation; this charge passed to the BLM upon its creation.[24] Later in the twentieth century, the O&C lands were included in programs to conserve habitat for species such as the northern spotted owl and marbled murrelet that depend on old-growth forest conditions (see Chapter 3).

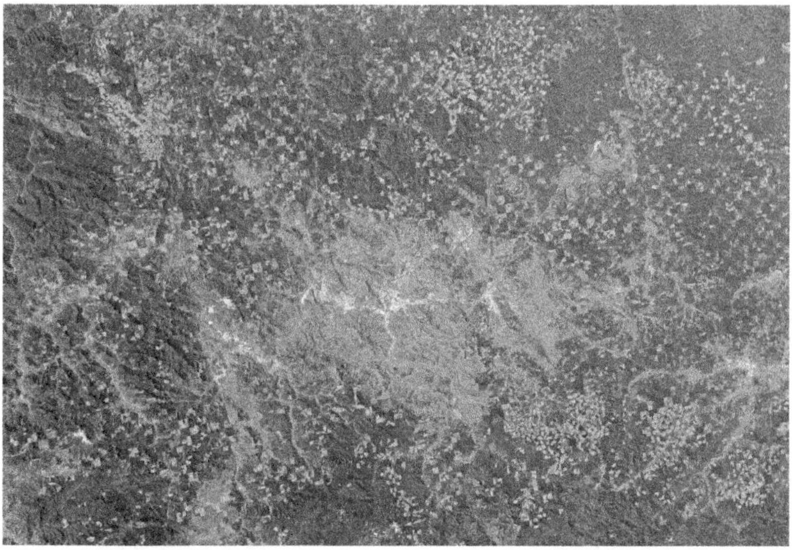

*Figure 2.1* A 2019 photograph from the International Space Station, centered on Roseburg, Oregon, shows the checkerboard pattern of Bureau of Land Management "O&C Lands" and nonfederal lands in western Oregon. Image: NASA.

## Pinchot and Roosevelt

Federal forest conservation in the early twentieth century was bolstered by the convergence of multiple trends and developments, including the advancement of professional forestry in the US, growing social concern over natural resource conservation and environmental preservation more broadly, and the unexpected ascendancy of Theodore Roosevelt to the presidency following the assassination of William McKinley. Central to this story is the enigmatic figure of Gifford Pinchot: the aristocrat who felt most at home walking or camping in deep forest; the champion of science known to carry on nearly daily conversations with his deceased wife; and the unparalleled Progressive Era conservation leader whose single-minded focus on scientific forestry came to define the agency he created (Figure 2.2).

Gifford Pinchot was born into a wealthy family and spent his childhood alternating between Europe, New England, New York City, and the family's Grey Towers estate in Pennsylvania. Gifford's father, James Pinchot, was inspired by George Perkins Marsh's *Man and Nature*, and it was James who encouraged his son to pursue a career in forestry.[25] Gifford attended school in France and was educated in the German forestry tradition, as no forestry schools existed in the

*Figure 2.2* Gifford Pinchot, the first US-born forester, in his office in Washington, D.C., in 1910. Image: US Forest Service.

US in the 1880s. Indeed, the practice of forestry barely existed in the US at that time. As described in Chapter 1, few forest operators gave a second's thought to what they left behind after harvesting the most valuable trees from a stand. In contrast, the German forestry tradition emphasized scientific management oriented toward "sustained yield"—the ideal of a perpetual supply of timber over time through the rational planning of forest growth and harvest.[26] After returning from Europe, Pinchot embarked on an experiment at the Vanderbilt family's Biltmore estate in North Carolina to demonstrate the value of scientific forestry. The Biltmore Experiment that Pinchot began in 1891 before ceding it to Carl Schenck is now considered to be the beginning of forestry in the US, the moment scientific management held out the promise of informing a continentwide process of resource management to secure timber supplies in perpetuity.

Pinchot leveraged his connections and his status as the nation's first US-born forester into positions with the US Department of Agriculture's Division of Forestry and was named Chief in 1898. Inspired by the German model of scientific forestry and by practitioners of this model like Sir Dietrich Brandis, Pinchot brought with him big ideas about the potential for federally owned forests and of an agency staffed by trained professionals to manage them. These ideas had already gained steam under Presidents Harrison, Cleveland, and McKinley, but it was the big-game hunter and Progressive Era champion Theodore Roosevelt who provided Pinchot the opportunity to fully realize his vision. Roosevelt and Pinchot were more than political partners; they were like brothers, known to box, wrestle, camp, swim, and hike together—sometimes in neglect or outright defiance of their official duties. In Pinchot, Roosevelt found the ideal figure to lead his forest conservation agenda.

High on this agenda was a plan to expand the footprint of the federal forest reserves and to transfer their ownership and management from the Department of Interior to the Department of Agriculture. Roosevelt already had the authority under the 1891 Forest Reserve Act language to retain more public domain land under permanent federal ownership; he used this authority aggressively in spite of fierce resistance from western congressmen opposed to the expansion of federal control. Western legislators successfully used a must-pass 1907 appropriations bill to revoke the president's power to establish forest reserves in Colorado, Idaho, Montana, Oregon, Washington, and Wyoming, but Pinchot and Roosevelt took advantage of the waning days before the bill was signed to designate another 16 million acres as national forests. Thirteen of these so-called "Midnight Reserve" proclamations included the transfer of tribal reservation land to the National Forest System without either the consent of tribal members or any compensation to the tribes whose land had been summarily taken.[27] By the time Roosevelt was forced to rescind his forest reserve authorities, over 150 million acres had been reserved in total by his predecessors and by Roosevelt himself.[28] Beyond his role enlarging the nation's forest reserves, Roosevelt's conservation legacy includes creating the nation's first federal bird reserves (precursors to the national wildlife refuge system), pushing for passage of the Antiquities Act of 1906 and using it aggressively to protect vulnerable landscapes, establishing new national parks, and expanding existing parks (Box 2.1).

Transfer of the forest reserves from the Department of Interior to the Department of Agriculture was especially important to Pinchot, who saw in the General Land Office a corrupt agency operating on the patronage system.[29] Just as troubling was the fact that this agency responsible for managing tens of millions of forested acres lacked any trained foresters. Pinchot's campaign to wrest control of the federal forest estate away from Interior was grounded in a network of business interests, legislators, and civil society organizations that came to support his Progressive Era vision of utilitarian scientific management. These entities and organizations helped generate what Daniel Carpenter calls "networked legitimacy" for the nascent US Forest Service.[30] The campaign came to fruition with the 1905 Transfer Act; a month following passage of the act, the Bureau of Forestry was renamed the US Forest Service, with Pinchot as its first Chief Forester. In 1907, the forest reserves were renamed "national forests" to make it clear that they would not be reserved from management (Box 2.2).

Upon gaining full control of the forest reserves, Pinchot could now go about crafting a federal forest management agency in his own image. He went so far as to pen instructions to himself in the form of a letter from Agriculture Secretary Wilson that would come to take on the status of a founding vision for national forest management. In it, he stated that "all the resources of the forest reserves are for use," that these lands were "for the benefit of the home-builder first of all," that "local questions will be decided upon local grounds," and that the federal forest estate was to be put to productive use for "the greatest good of the greatest number in the long run." Together, these ideas encapsulated Pinchot's understanding of "conservation," a term he credited himself with popularizing. Conservation stood in contrast not only to the wasteful practices of cut-and-run logging but also to the preservationist vision represented by John Muir and allied environmentalists who advocated for expansion of the national park model. Pinchot was emphatic that national forests and other public lands must be placed in the service of producing commodities useful to society and not preserved for recreation or aesthetic appreciation.[31]

**BOX 2.2 FEDERAL FORESTS AND NATIVE LANDS**

Three years before the Transfer Act, the Bureau of Forestry took management control over its first federal forests with passage of the Morris Act in 1902. This piece of legislation created the 225,000-acre Minnesota Forest Reserve (renamed the Minnesota National Forest in 1908 and then the Chippewa National Forest in 1928) in heavily forested northern Minnesota and granted management authority to the Bureau of Forestry. This was the first non-military forest reserve to be created through an act of Congress.[32] Importantly, it was composed of lands that had previously belonged to the Ojibwe people (also known as Chippewa or Anishinaabe).[33] The Leech Lake Reservation was established by the 1855 Treaty of Washington and later altered through subsequent treaties and laws. However, it was the Dawes Act of 1887 and Nelson Act of 1889 that wrested away tribal ownership of the bulk of these lands, as they did to tens of millions of acres of reservation lands nationwide (see Chapter 7 for more details on the Dawes Act). The Morris Act represented a compromise between members of Minnesota Women's Clubs (who pushed for creation of a park on former tribal lands), timber interests, wealthy sportsmen, and Gifford Pinchot himself. The Ojibwe people were largely excluded from the negotiations and reportedly felt betrayed by the ultimate outcome, with Chief Flatmouth publicly proclaiming, "The people that passed the Morris Bill are not honest."[34] Today, the Ojibwe and its members control only about four percent of the land within reservation boundaries; a much larger proportion is composed of state or federal conservation lands including the Chippewa National Forest.

## The Young Forest Service

Collectively, the Forest Reserve Act, the Organic Act, and the Transfer Act—along with reserve designations by four presidents—created a system of federal forestlands across the West; Pinchot created an agency of professional foresters to manage them. Federal foresters also took control of tropical forestlands on the island of Puerto Rico that had fallen under US ownership following the Spanish–American War along with lands in Minnesota carved out from territory previously committed by treaty to the Ojibwe people (Box 2.2). Gifford Pinchot and his agency of foresters went to work demonstrating scientific forestry practices on the national forests.

Other changes instituted by Pinchot included the regulation of grazing through a system of permits and fees, thereby ending the de facto system of open access on national forestlands.[35] He instituted a permitting system for the use of hydropower resources on federal forestlands, established the first experiment stations on the national forests, and—continuing a tradition started before the transfer of federal forestlands to the Department of Agriculture—hired trained foresters to advise private forest owners on proper management techniques.[36] Although Pinchot had hoped to ramp up a program of national forest timber sales, many timber companies pushed to keep federal timber off the market as a means of ensuring better prices for their own products. For this reason, some refer to the first decades of national forest management as the "custodial period," a time when systems were established but without any major expansion of active timber management.[37]

Indeed, the establishment of the National Forest System was not so much a rebuke of the prevailing timber industry as it was a means of resolving dilemmas widely identified by timber interests of the time: by removing timberlands from private disposal, the federal government helped to address the problems of overproduction, wasteful harvesting, and razor-thin profit margins that were taking their toll on both forests and corporate earnings.[38] Yet Pinchot's vision was always strongly utilitarian and he wanted to show that the national forests would contribute to prosperity through rational timber production. His belief that natural resources should be used for the material benefit of society famously led to a falling out with John Muir over the question of whether the Hetch Hetchy valley in Yosemite National Park should be dammed to provide drinking water to San Francisco. The conflict between natural resource conservationists, represented by Pinchot, and nature preservationists, represented by Muir, exposed a rift within the conservation movement—one that continues to the present day.

Pinchot imprinted the Forest Service with his vision of a rationalized forest and professionalized forestry agency and received the adoration of many young federal foresters in return. President Taft, Roosevelt's successor, was less smitten and set about reining in Pinchot's independence. Pinchot was ultimately ousted in 1910 after crossing a political line in his public criticism of Interior Secretary Ballinger, but his ideals lived on in the agency he created. Pinchot's vision for the Forest Service reflected both the best and the worst of the Progressive Era ideology. On the positive side of the ledger, the Forest Service represented a trained, professional bureaucracy. Many of the first wave of rangers came out of the Yale School of Forestry that Pinchot himself helped to establish. This spare, efficient agency managed forests according to a set of principles oriented toward broad social benefits—a far cry from the General Land Office with its patronage system and lack of trained foresters. At a time when forests nationwide were falling prey to poor logging practices driven by short-term economic motives, the national forests demonstrated that good management could yield wood products without threatening the viability or quality of future forests. In this sense, the success of Pinchot's vision was the triumph of science and effective public administration over corruption and short-sighted forest exploitation.

On the negative side, however, was a Forest Service agency culture that could seem opaque, condescending, or unresponsive to some. Pinchot, following his predecessor Bernhard Fernow, took inspiration from the Prussian approach to forest administration with its quasi-military bureaucratic style[39]; forest rangers were empowered to make decisions in line with their understanding of the needs of the forest and the "greatest good" philosophy but were expected to remain insulated from local and national politics. Some local forest users were understandably put off by the sudden arrival of Yale-educated foresters who justified sometimes unpopular decisions by claiming scientific expertise. The "greatest good" philosophy itself could also be problematic; its vision of a generalized public could erase the substantial differences among forest-using populations, including Hispanos (descendants of early Spanish colonizers) in the Southwest and Native Americans throughout the West, on whose traditional homelands national forests were established. The prevailing emphasis on eliminating waste through a paradigm centered on sustained yield principles had the effect of oversimplifying the very complex ecologies of diverse forest types and diverse human uses of forests.[40] The German forestry tradition in which Pinchot was steeped was itself reductionist and rationalizing to a fault. The ideal of a regulated forest managed for sustained yield timber production stood at odds with the more complex and multifaceted values that many Americans had regarding their forests. It made no room for species, processes, or activities lacking clear utilitarian values and had an especially large blind spot regarding the possible beneficial role of fire in forests.[41]

### The Fire Question

Chapter 1 discusses that fire has been a part of many of America's forests for millennia. As integral as fire was to many of the nation's forests, it was largely incompatible with the young Forest Service for three principal reasons. First, by damaging and consuming standing trees, it seemed to work against the concept of wise use and the project to make national forests maximally productive. Second, fire control was seen as a necessary ingredient in establishing sound forestry practices on nonfederal lands; few timber companies were willing to invest in replanting and long-term forestry in the face of the ever-present threat of destructive fire.[42] The timber industry saw much to like in maintenance of a federally subsidized firefighting brigade and successfully steered federal legislation toward fire control rather than forest practice regulation.[43] Third, the idea of protecting Americans and their forests from fire came to serve as a central justification for the continued existence of the US Forest Service once the threat of a timber famine appeared to subside. At this point in history, the young forestry profession in the United States was not only battling the kind of fire set by humans or ignited by lightning over thousands of years prior to European colonization; it was also confronting a new and especially destructive kind of fire enabled by the extensive expanses of slash and dead and damaged trees left behind by the wasteful timber harvesting practices of the time.

The US Forest Service began its stewardship of federal forestlands by waging both ideological and physical war against fire, casting it as the single greatest threat to forest conservation.[44] Broadcast fire, which did not respect property boundaries and could easily escape human control, was distinguished from the smaller fires used to consume slash left over after harvesting or to prepare a site for planting; the latter were still accepted as part of good forestry practice. The campaign for aggressive fire control was given a tragic boost in the form of the Big Blowup in 1910. That year saw numerous fires in the West, the largest of which was a fast-moving conflagration that swept across three million acres of forestland in northeastern Washington, northern Idaho, and northwestern Montana, consuming trees, towns, homes, people, wildlife, and everything else in its path. It left behind mile after mile of incinerated forests, clouded the skies over New England, and deposited soot on Greenland's ice fields. As Timothy Egan describes in his history of the fires, this was a turning point in the national government's treatment of forest fire.[45] The Big Blowup presented an opportunity for the young Forest Service to embrace a mission that might secure the agency's legitimacy and political support into the future after the ousting of the charismatic Pinchot.

At the same time, many forest owners and livestock producers continued to advocate for light burning even in the wake of the Big Blowup. Their reasons for continued fire use were pragmatic: regular fires kept brush down, reduced the threat of uncontrollable fire, increased forage, improved hunting, and helped control pests. Indeed, the New Jersey Fire Control Service had been applying controlled fire to the pine barrens region as early as 1928, adopting practices long used by cranberry growers.[46] At this time, livestock owners in the South were still treating forestland—even land formally owned by private timber firms—as a kind of commons, setting broadcast fires to improve forage and reduce insects and other pests.[47] Fire was also a centrally important activity for many Native populations, who continued to apply traditional land management practices to their much-diminished territories whenever possible.[48] Regardless of their benefits to local people, such practices were largely seen by forestry professionals as a reflection of "folk" wisdom with no basis in science or rational management. Even a young forester named Aldo Leopold derided the "light-burning propagandists" who would decimate future timber supplies with their insistence on the use of fire as a management tool;[49] his views on this issue would later undergo a revolution. Resolving the light-burning controversy remained at or near the top of the agenda for the federal foresters who came after Pinchot. Following a series of Forest Service-funded studies (some of which were suppressed when they suggested that fire could be beneficial), the agency came down on the side of the elimination of fire altogether.[50] Complete fire suppression—even in remote backcountry forests—was to be the official policy of the agency. Still, many forest owners, forest scientists, and even Forest Service employees continued throughout the middle decades of the twentieth century to push back against the prohibition on fire as a management tool.

## Establishment of Eastern National Forests

Only western forests could be federally reserved under the 1891 Forest Reserve language because the only remaining public domain lands were in the West. Forests east of the Rocky Mountains had been claimed by individuals, families, and businesses long before, and some of those lands were never part of the federal domain to begin with. Unlike many of the western national forests that had seen little if any prior timber harvesting, the bulk of the eastern forests had already been converted or cut over at least once, leaving little in the way of mature forest. Still, many conservation leaders of the early twentieth century called for the federal government to take ownership of eastern landscapes on which forests had been harvested and then abandoned. According to advocates, federal ownership would accomplish multiple objectives, including demonstrating scientific forestry in a part of the country where federal foresters were scarce and conserving the navigability of waterways important for transportation by protecting their headwaters. From a political perspective, the establishment of eastern national forests would also deliver the benefit of creating a nationwide political constituency in support of the National Forest System and the US Forest Service. Without the support of powerful eastern legislators, the agency was vulnerable to being underfunded or even abolished altogether, particularly as the West accounted for only a sliver of the voting population and, therefore, of congressional representation.

The constitutional justification for federal purchases of eastern forestland came through the Commerce Clause: protecting navigable waterways fell squarely within the federal government's power to regulate interstate commerce. The legislative vehicle for this program was the Weeks Act of 1911, which provided authority and funding for the federal government to acquire forestlands at the headwaters of navigable streams by purchase or exchange. The Act was built upon years of organizing by conservation advocates in southern Appalachia and New England.[51] A sense of urgency for the bill's passage came from catastrophic flooding on the Monongahela and Ohio Rivers as well as the Big Blowup in the northern Rocky Mountains, both of which were presented as evidence of the destruction caused by uncontrolled logging.[52] The first Weeks Act forests were established in North Carolina, including purchases from the Vanderbilt family's Biltmore estate where Pinchot and Schenck endeavored to demonstrate the viability of scientific forestry in the US. The Act was successful enough that it was expanded in 1924 with the Clarke-McNary Act, which dispensed with the requirement that purchased forests be located at the headwaters of navigable streams. These and other laws such as the Woodruff-McNary Act provided funding and authority to establish extensive national forestlands in states from Florida to Maine and from Texas to Minnesota. Over 20 million acres of land—including additions to western national forests—were ultimately purchased and held as national forestlands under these laws.

## Regulation or Cooperation with the Private Sector?

Pinchot's vision of a vast federal forest estate managed by professionals trained in scientific forestry became a reality with the establishment of the National Forest System, which was later expanded by the Weeks Act and subsequent laws. Yet many conservationists and forestry professionals in the early twentieth century (Pinchot perhaps most of all) felt that more should be done to stem the destructive logging practices still prevalent on private lands—both those owned by timber enterprises as well the smaller parcels owned mostly by farm families. National reports on American forestry conducted in 1920 (the Capper Report) and 1933 (the Copeland Report) detailed the extensive degradation and destruction taking place on forestlands nationwide—with nearly all the worst practices found on privately owned lands. As late as the 1940s, federal foresters could point to abandoned mill towns and cutover private forests in places like western Washington and Oregon as evidence of the cut-and-run practices that still predominated. According to federal investigators, private timber interests were not practicing sustained yield, and were therefore failing to sustain both forests and the communities that depended on them.[53]

Various proposed forest policies contained mechanisms to rein in private forestry practices, but none of these efforts were ultimately successful. Early versions of the Weeks Act included language requiring private forest management to conform to federal regulations; these provisions were removed just before the final House vote.[54] The short-lived National Industrial Recovery Act, a New Deal program that was eventually found to be unconstitutional, required the development of mandatory forest practice rules governing issues such as restocking and tree retention.[55] Although various alternative approaches to federal regulation of private timber practices were considered, these failed to gain enough political support to move forward in light of organized opposition from the timber industry.[56] Instead, federal policy embraced a cooperative model that included federal inducements to states to develop their forestry and firefighting programs and federal leadership in forest fire and insect control across ownerships. Later policies also provided technical assistance and economic support for family forest owners to improve the productivity and management of their forestlands.

In addition to opening the door for the establishment of national forests in eastern states, the Weeks Act also set in motion the model of cooperative forestry that would be expanded through later policies. The Weeks Act authorized $200,000 in matching funds for states that established fire protection policies and committed state funding to the cause.[57] The states were key partners in implementing a conservation agenda focused on eliminating fire, and the federal government stood ready to support them in their efforts—as well as to withhold funding from states that still permitted broadcast burning.[58] The office of State and Private Forestry Cooperation was established in the US Forest Service in 1908 and given a mission to promote sound forestry on nonfederal lands. A handful of states established forestry commissions in the waning years of the nineteenth century, and by the first decade of the twentieth century, some states were implementing systems of fire surveillance and forest protection. Private fire protective associations were also eligible for federal assistance through the cooperative model created by the Weeks Act.[59] The Clarke-McNary Act greatly expanded federal financial contributions to the cooperative fire control model established under the Weeks Act.

Many of the dominant forestry and conservation voices of this time, including leaders of the US Forest Service, American Forestry Association, and state forestry officials, identified fire control as the most pressing conservation need, even as some voices from the field continued to tout the benefits of light burning. The US Forest Service led the charge against fires large and small; in 1935 it adopted the "10 a.m. Policy" under which any new fire starts should be extinguished by 10:00 the following morning. Fire control efforts received a boost from the Civilian Conservation Corps (CCC), established by President Franklin Delano Roosevelt to provide work opportunities in the depths of the Great Depression. Over a period of nine years, the CCC helped to increase systems of fire surveillance and control through the construction of fire towers, trails, roads, firebreaks, and communications infrastructure. CCC crews also played a central role in direct suppression in support of the 10 a.m. policy. CCC fire efforts were not limited to federal lands; they also helped to increase state fire control capacity.[60] Federal agencies were at the same time engaged in an ideological war against fire even as they fought a physical war in the woods. The Smokey Bear campaign began in 1945 as one iteration of a decades-long anti-wildfire effort led by the US Forest Service and other federal entities. It soon grew into one of the most successful public relations efforts of all time with its simple message about the tragedy of wildfire. Generations of Americans came to see forest fire as destructive, menacing, a thing to be extinguished at all costs.

Cooperative efforts continued to expand through the twentieth century via policies such as the Cooperative Forest Management Act in 1950, which operated under the familiar model of federal contributions to states that invested in their own forestry agencies—in this case, focused on technical assistance to landowners and operators. The federal government also took a lead role in managing insect and disease outbreaks. While government entomologists had been investigating destructive forest pests since the close of the nineteenth century,[61] the Forest Pest Control Act of 1947 granted authority to the Forest Service over the detection and control of pests and diseases on federal, state, and private forestlands nationwide. An emerging ensemble of incentives, policies, technical advice, and technologies of detection and control increasingly operated in service to the common goal of creating efficient and productive—yet simplified—forests across the American landscape. Under this vision, fires, insects, and diseases would be eliminated, and wasteful harvest and management practices would be replaced by rational techniques, leading to increased timber production and greater societal well-being.

## World War II

World War II is often considered a turning point in US forest history for three principal reasons. First, the war effort itself required massive quantities of wood, resulting in the near exhaustion of supplies from much of the country's private lands and a subsequent increase in national forest harvests. Second, technologies developed during the war were used for domestic applications following the war's end. In the case of forest management, these included aircraft deployed for fire detection and suppression as well as an array of new chemical fertilizers, pesticides, and herbicides. Third, the postwar economic boom led to the spread of suburbs (including in places near national forestlands), widespread private automobile adoption, and a marked increase in demand for outdoor recreation.[62]

The postwar Forest Service transitioned out of its earlier "custodial" model of national forest management and toward a more aggressive model focused on increased timber production through intensive harvesting, site preparation, and reforestation practices. In recognition of the growing tensions between timber production and outdoor recreation, Congress passed the Multiple Use-Sustained Yield Act in 1960. The multiple uses identified for national forests in the Act included "outdoor recreation, range, timber, watershed, and wildlife and fish," and wilderness was explicitly stated to be consistent with the purposes of the Act. Sustained yield was defined by the Act as "the achievement and maintenance in perpetuity of a high-level annual or regular periodic output of the various renewable resources of the national forests without impairment of the productivity of the land." The Act recognized that timber was not the only resource that Americans valued on their national forestlands, yet it also signaled an expectation that the Forest Service would continue to sustain high levels of timber production. As Paul Hirt explains in his post-World War II history of the US Forest Service, the agency came to participate in a "conspiracy of optimism" in which ever-higher levels of current timber harvests were justified through increasingly optimistic projections of future growth and timber yield, thereby assuring compliance with sustained yield mandates—at least in theory.[63]

The midcentury timber industry continued to struggle with the sustainability of practices on both industrial and family forest lands, and as late as the 1950s, US Forest Service leadership was still advocating for federal regulation of private forestry practices. However, the industry was already undergoing a transition to its own version of sustained-yield practices in many parts of the country, and these trends continued apace following the war. Management practices were also improved as a result of New Deal-era changes to federal tax policy that encouraged selective harvesting and long-term timberland investments.[64] In addition, forestry associations were encouraging better forestry practices on a voluntary basis and coordinating with state forestry agencies.[65] In 1941, the Weyerhaeuser corporation established the nation's first certified "tree farm" on cutover land in Washington State, signaling a commitment to Pinchot's vision of growing trees as a renewable crop on private land.[66] The tree farm system quickly spread nationwide and eventually came to be extensively adopted by family forest owners.[67] As a result of fire control and broader adoption of forestry practices, in 1958 the US Forest Service was able to announce for the first time that annual timber growth on all private and public lands exceeded the annual harvest.[68]

## The Wilderness Question

As private industry grew less transient and more focused on long-term production and as the Forest Service embraced increasingly intensive timber production methods, the contrast between federal and industrial forestry practices all but disappeared in many places. One important difference, however, was that national forests still retained unroaded forests in some places—especially but not exclusively in steep, remote, and high-elevation landscapes. The postwar increase in recreation on national forests helped to cultivate a new generation of American outdoor enthusiasts who valued these lands primarily for the camping, fishing, hunting, hiking, horseback riding, paddling, and other recreational opportunities they offered, setting up tensions with the growing federal emphasis on timber production. The postwar acceleration in road construction on national forests, combined with an increase in automobile-centered recreation, also raised concerns among some forest managers and preservationists regarding the future of undeveloped areas. A growing movement to reserve some national forest landscapes from commercial development took shape through the organizing and activism of a new generation of foresters and scientists.

The first wilderness area was designated in New Mexico's Gila Mountains in 1924 as a result of the advocacy of Aldo Leopold. In a series of articles published in both popular and academic magazines, Leopold described his vision for protected wilderness areas across the national forest and national park systems. He defined wilderness as "a wild, roadless area where those who are so inclined may enjoy primitive modes of travel and subsistence, such as exploration trips by pack-train or canoe."[69] For Leopold and his contemporaries, wild places provided recreational, cultural, scientific, psychological, aesthetic, and conservation benefits that would only be diminished through further conversion to rationalized commodity-producing landscapes.[70] The Forest Service created its own program of "primitive areas" via the L-20 regulation in 1929; these were superseded ten years later by the U regulations that designated primitive areas as wilderness, wild, or recreation. The authority to establish, modify, or remove these designations was held by the Chief of the Forest Service and Secretary of Agriculture.[71] In practice, this meant that protections were often temporary and subject to the discretion of agency leaders. In pursuit of more lasting protections for wild places, Leopold went on to co-found The Wilderness Society in 1935 with Robert Marshall, Benton MacKaye, Harvey Broome, Robert Sterling Yard, and other advocates.

Howard Zahniser, who took over the helm of The Wilderness Society in 1945, wrote 66 different drafts of federal wilderness legislation starting in 1956. Although Zahniser died just months before seeing his vision realized, his years of tireless activism paid off in 1964 with Congressional passage of the Wilderness Act, which defined wilderness as "an area where the earth and its community of life are untrammeled by man, where man himself is a visitor who does not remain." In practical terms, roads, commercial activities, motorized vehicles and equipment, and permanent structures were prohibited in designated wilderness areas except in emergency situations. In creating the National Wilderness Preservation System, Congress took over from the Forest Service and other federal agencies[72] the power over wilderness designation and de-designation. Areas already designated by the Forest Service as wild, wilderness, or canoe areas became congressionally protected wilderness under the Act. These included iconic places such as the Bob Marshall Wilderness in Montana and the Boundary Waters Canoe Area in Minnesota.[73]

Upon signing the Wilderness Act, President Johnson remarked that wilderness preservation was needed in order to give future generations "a glimpse of the world as it was in the beginning, not just after we got through with it." His inspirational words belie the historically questionable assumptions upon which this concept of wilderness was built. Landscapes modified by people for thousands of years—and landscapes from which those people had only recently been expelled—became, under the wilderness paradigm, an area where "man himself is a visitor who does not remain" and "a glimpse of the world as it was in the beginning." Although the cultural assumptions animating the concept of wilderness remained mostly unquestioned for the first decades following passage of the Wilderness Act, prominent writers and thinkers have more recently raised criticisms of the wilderness concept as part of larger debates regarding conservation's potentially problematic legacy (Box 2.3).

### BOX 2.3 A RECKONING WITH CONSERVATION LEGACIES

The year 2020 was one of widespread social mobilization in protest of the killings of Ahmaud Arbery, Breonna Taylor, George Floyd, and numerous other African Americans. In the context of a growing critique of mainstream conservation in the US (see Chapter 3), the events of 2020 forced environmental and conservation advocates nationwide to confront the legacies of their founders and organizations—legacies that often included racist and exclusionary language, policies, and advocacy ranging from overt to covert. Leaders of the Sierra Club, Save the Redwoods League, Audubon Society, Wildlife Conservation Society, Bronx Zoo, and other groups penned public acknowledgments of their organizations' troubling origins: John James Audubon both owned and traded enslaved people; John Muir spoke in disparaging terms of the recently emancipated Black Americans he met on his thousand-mile walk through the South and wrote that there was "no right place in the landscape" for Native people in California's Sierra Nevada.

Numerous conservation leaders were adherents and promotors of eugenics, the pseudoscientific philosophy of human improvement through promotion of "superior" genetic lines that served as the central philosophical pillar of the Nazis. For example, Gifford Pinchot served on the advisory council of the American Eugenics Society and was a delegate to the first two International Eugenics Congresses. Leaders in the American eugenics movement such as Madison Grant, Henry Fairfield Osborn, and Charles Goethe were founding members of the Save the Redwoods League and Roosevelt's Boone and Crockett Club and were active in wildlife and national park protection campaigns.[74] Garrett Hardin continued to promote eugenics even after the horrors of the Nazis came to light; this perspective is reflected in his famous essay on "The Tragedy of the Commons."

The membership of conservation and environmental organizations has tended to be overwhelmingly White for most of their histories, and some organizations actively discouraged non-White membership and participation. As the discussion about conservation's past, present, and future continues, an important question is whether the connections between conservation and racial exclusion are coincidental (in other words, were Muir, Roosevelt, Pinchot, and others simply products of their time) or something more integral. Carolyn Finney, William Cronon, Laura Pulido, Carolyn Merchant, Dorceta Taylor, and Jake Kosek are just a few of the scholars who have provided thoughtful treatments of the relationships between race and conceptions of nature and the environment in the US.[75] The themes they examine have become increasingly salient as the nation's conservation agencies and organizations continue to confront their own histories.

## Conclusion

By the postwar period, the Forest Service was no longer a young upstart struggling for recognition—it had matured into a widely recognized and respected agency with a reputation for efficiency and quality. In what would become a seminal work in the field of public administration, Herbert Kaufman's 1960 book *The Forest Ranger* described its titular characters in terms of their strong cultural ethos and personal identification with the US Forest Service, creating a homogeneity of thought and action even within a decentralized agency spread across millions of acres of mostly rural work stations.[76] Kaufman conducted his study at what might have been the agency's zenith, the point at which it enjoyed its highest levels of public adoration, political deference, bureaucratic efficiency, and cultural cohesion.

Yet beneath the surface, tensions were starting to grow more visible. Many leaders of the wilderness movement, such as Aldo Leopold, were foresters steeped in the scientific management tradition who had come to question their own training and their own past management practices. Leopold is perhaps best known for his revolutionary vision of integrated wildlife management that made room for wolves, grizzly bears, and other species deemed commercially undesirable, but he was also an incisive critic of prevailing forest management approaches. He rejected what he called the "cabbage farming" model of forestry and called for a more holistic and integrated approach based around a complex idea of land health in contrast to the simplifying program of sustained yield that dominated federal forestry.[77] As the postwar Forest Service came to mirror in its forest practices the industry it had long sought to tame, an overriding focus on timber production created increasing conflicts with other forest values. At the same time, the agency's insistence on eliminating disturbances was quietly setting the stage for an ecological crisis.

The Wilderness Act represented a major incursion into the autonomy of the Forest Service, signaling that Congress did not trust the agency to protect the preservation and recreation values important to a growing segment of the American public. As the tumultuous 1960s gave way to the environmental era of the 1970s, the agency would find its autonomy further curtailed and its mission further complicated by countervailing demands and publics. The call of early conservationists for government intervention on behalf of forest values had been answered and the resulting institutions proved remarkably enduring. However, Progressive Era models of governance and turn-of-the-century understandings of forest dynamics would be forced to change as the complexities of both society and ecology became impossible to ignore.

## Questions for Discussion

1. Why was the federal government the entity that ultimately led many of the forest governance innovations of the late nineteenth and early twentieth centuries? What opportunities and limitations did state governments face in leading forest conservation efforts?
2. Why was the control of fire—including of the low-intensity fire favored by "light burning" proponents—so important to the Forest Service?
3. What are the advantages and disadvantages of a forest governance model built upon trained foresters managing government-owned land?
4. What interests or social groups benefited most from the establishment of the national forest system? What interests or social groups bore the greatest costs?
5. Why did the federal government ultimately not require compliance with forestry practice standards on private lands? What are the advantages and disadvantages of the cooperative model that was instituted as an alternative to federal regulation?
6. What comes to mind when you think of "wilderness"? Is it accurate to say that wilderness areas protect vestiges of "the world as it was in the beginning," in

the words of President Johnson? Does the recognition of forests as cultural landscapes undermine the argument for wilderness preservation?

## Notes

1 Tucker, 1981.
2 Langille, 1904.
3 Langston, 1995.
4 Muir, 1897.
5 Muir, 1916.
6 Taylor, 2016, pp. 192–208.
7 Carpenter, 2001; Taylor, 2016.
8 Taylor, 2016.
9 Robbins, 1985a, p. 4.
10 Jacoby, 2001, p. 42.
11 Taylor, 2016, p. 221.
12 Montrie, 2011, pp. 35–55; Taylor, 2016, pp. 211–221.
13 White (1991, p. 139) estimates that there were 375 different federal land laws passed between 1789 and 1834.
14 Copeland, 2013.
15 Finney, 2014, p. 60.
16 Ochiai, 2001.
17 Hine & Faragher, 2000; Shannon, 1945.
18 Dana & Fairfax, 1980.
19 Puter & Stevens, 1907, p. 21.
20 Miller, 2015; Taylor, 2016, p. 277.
21 Clayton, 2019.
22 Smith et al., 2009.
23 Blumm & Wigington, 2013.
24 Dodds, 1987.
25 Taylor, 2016, p. 265.
26 Lowood, 1990.
27 Catton, 2016, p. 36.
28 Miller, 2015.
29 Carpenter, 2001.
30 Carpenter, 2001.
31 Hays, 1959.
32 Searle, 1971.
33 Catton, 2016; Searle, 1971.
34 Catton, 2016, p. 42.
35 Wilkinson, 2015.
36 Robbins, 1985a; Wilkinson, 2015.
37 Johnson, 2007; Kennedy & Quigley, 1998; Winkel, 2014.
38 Robbins, 1985a; Schallau & Alston, 1987.
39 Twight, 1990.
40 Langston, 1995.
41 Pyne, 2021, p. 101; Scott, 1998.
42 Pyne, 1982, pp. 189–190.
43 Hudson, 2011.
44 Pyne, 1982, p. 195.
45 Egan, 2009.
46 Cumming, 1969; Little et al., 1948.

47 Carter et al., 2015, p. 69.
48 Huntsinger & McCaffrey, 1995; Norgaard, 2019; Pyne, 1982.
49 Leopold, 1920.
50 Pyne, 1982, pp. 112–119.
51 Bramwell, 2012, pp. 328–330.
52 Newfont, 2012, pp. 68–70.
53 Robbins, 1985b.
54 Morgan, 1978, p. 185.
55 Dana & Fairfax, 1980, pp. 169–170.
56 Carter et al., 2015, p. 29; Steen, 2004.
57 Zimmerman, 1976, p. 22.
58 Pyne, 1982, p. 115.
59 Pyne, 1982, p. 233.
60 Zimmerman, 1976, pp. 52–53.
61 Furniss, 2007.
62 Hirt, 1994.
63 Hirt, 1994.
64 Carter et al., 2015, p. 76.
65 Carter et al., 2015, pp. 60–62; Steen, 2004, pp. 280–281.
66 Rutkow, 2012, p. 284.
67 Carter et al., 2015, p. 74.
68 Lewis & Miller, 2018, pp. 4–5.
69 Leopold, 1925, p. 398.
70 Leopold, 1949; Marshall, 1930.
71 Dana & Fairfax, 1980, p. 157.
72 The four federal agencies that administer wilderness areas under the Act are the US Forest Service, National Park Service, Bureau of Land Management, and US Fish and Wildlife Service.
73 With subsequent additions to the National Wilderness Preservation System, a total of over 111 million acres within 762 distinct wilderness areas have been designated as of the time of this writing.
74 Allen, 2013; Spiro, 2009.
75 Cronon, 1995; Finney, 2014; Kosek, 2006; Merchant, 2003; Pulido, 1996; Taylor, 2016.
76 Kaufman, 1960.
77 Leopold, 1949.

## References


Allen, G. E. (2013). "Culling the herd": Eugenics and the conservation movement in the United States, 1900–1940. *Journal of the History of Biology*, *46*(1), 31–72.

Blumm, M. C., & Wigington, T. (2013). The Oregon & California railroad grants lands' sordid past, contentious present, and uncertain future: A century of conflict. *Boston College Environmental Affairs Law Review*, *40*(1–76), 1.

Bramwell, L. (2012). 1911 Weeks Act: The legislation that nationalised the US Forest Service. *Journal of Energy & Natural Resources Law*, *30*(3), 325–336.

Carpenter, D. P. (2001). *The forging of bureaucratic autonomy: Reputations, networks, and policy innovation in executive agencies, 1862–1928*. Princeton University Press.

Carter, M. C., Kellison, R. C., & Wallinger, R. S. (2015). *Forestry in the US South: A history*. Louisiana State University Press.

Catton, T. (2016). *American Indians and national forests*. University of Arizona Press.

Clayton, J. (2019). *Natural rivals: John Muir, Gifford Pinchot, and the creation of America's public lands*. Pegasus.

Copeland, R. W. (2013). In the beginning: Origins of African American real property ownership in the United States. *Journal of Black Studies*, 44(6), 646–664.

Cronon, W. (1995). The trouble with wilderness: Or, getting back to the wrong nature. In W. Cronon (Ed.), *Uncommon ground: Toward reinventing nature* (pp. 69–90). W. W. Norton & Company.

Cumming, J. A. (1969). Prescribed burning on recreation areas in New Jersey: History, objectives, influence, and technique. In 9th Tall Timbers Fire Ecology Conference, Tallahassee, FL (pp. 251–269).

Dana, S. T., & Fairfax, S. K. (1980). *Forest and range policy* (2nd ed.). McGraw-Hill.

Dodds, P. G. (1987). The Oregon and California lands: A peculiar history produces environmental problems. *Environmental Law*, 17(3), 739–766.

Egan, T. (2009). *The Big Burn: Teddy Roosevelt and the fire that saved America*. Houghton Mifflin Harcourt.

Finney, C. (2014). *Black faces, White spaces: Reimagining the relationship of African Americans to the great outdoors*. University of North Carolina Press.

Furniss, M. M. (2007). *A history of forest entomology in the Intermountain and Rocky Mountain areas, 1901 to 1982* (General Technical Report RMRS-GTR-195). US Department of Agriculture, Forest Service, Rocky Mountain Research Station.

Hays, S. P. (1959). *Conservation and the gospel of efficiency: The progressive conservation movement, 1890–1920*. Harvard University Press.

Hine, R. V. N., & Faragher, J. M. (2000). *The American West: A new interpretive history*. Yale University Press.

Hirt, P. W. (1994). *A conspiracy of optimism: Management of the national forests since World War Two*. University of Nebraska Press.

Hudson, M. (2011). *Fire management in the American West: Forest politics and the rise of megafires*. University Press of Colorado.

Huntsinger, L., & McCaffrey, S. (1995). A forest for the trees: Forest management and the Yurok environment, 1850 to 1994. *American Indian Culture and Research Journal*, 19(4), 155–192.

Jacoby, K. (2001). *Crimes against nature: Squatters, poachers, thieves, and the hidden history of American conservation*. University of California Press.

Johnson, K. N. (2007). Will linking science to policy lead to sustainable forestry? Lessons from the federal forests of the United States. In K. M. Reynolds, A. J. Thompson, M. Köhl, M. A. Shannon, D. Ray, & K. Rennolls (Eds.), *Sustainable forestry: From monitoring and modeling to knowledge management and policy science* (pp. 14–34). CABI.

Kaufman, H. (1960). *The forest ranger: A study in administrative behavior*. Johns Hopkins Press.

Kennedy, J. J., & Quigley, T. M. (1998). Evolution of USDA Forest Service organizational culture and adaptation issues in embracing an ecosystem management paradigm. *Landscape and Urban Planning*, 40(1–3), 113–122.

Kosek, J. (2006). *Understories: The political life of forests in New Mexico*. Duke University Press.

Langille, H. D. (1904). *Supplemental report on the proposed Chesnimnus Forest Reserve* Oregon. US Department of Agriculture, Bureau of Forestry.

Langston, N. (1995). *Forest dreams, forest nightmares: The paradox of old growth in the Inland West*. University of Washington Press.

Leopold, A. (1920). "Piute forestry" vs. forest fire prevention. *Southwestern Magazine*, *2*, 12–13.
Leopold, A. (1925). Wilderness as a form of land use. *The Journal of Land & Public Utility Economics*, *1*(4), 398–404.
Leopold, A. (1949). *A Sand County almanac, and sketches here and there*. Oxford University Press.
Lewis, J. G., & Miller, C. (2018). Vast, incredible damage: Herbicides and the US forest service. *Forest History Today*, Spring/Fall, 3–12.
Little, S., Allen, J. P., & Moore, E. B. (1948). Controlled burning as a dual-purpose tool of forest management in New Jersey's pine region. *Journal of Forestry*, *46*(11), 810–819.
Lowood, H. E. (1990). The calculating forester: Quantification, cameral science, and the emergence of scientific forestry management in Germany. In T. Frängsmyr, J. L. Heilbron, & R. E. Rider (Eds.), *The quantifying spirit in the 18th century* (pp. 315–342). University of California Press.
Marshall, R. (1930). The problem of the wilderness. *The Scientific Monthly*, *30*(2), 141–148.
Merchant, C. (2003). Shades of darkness: Race and environmental history. *Environmental History*, *8*(3), 380–394.
Miller, C. (2015). Measured impact: William B. McKinley and late nineteenth-century conservation. *Forest History Today*, Spring/Fall, 32–36.
Montrie, C. (2011). *A people's history of environmentalism in the United States*. Continuum.
Morgan, G. T. (1978). Conflagration as catalyst: Western lumbermen and American forest policy. *Pacific Historical Review*, *47*(2), 167–187.
Muir, J. (1897). The American forests. *Atlantic Monthly*, *80*, 145–157.
Muir, J. (1916). *A thousand-mile walk to the gulf* (W. F. Badé, Ed.). Houghton Mifflin.
Newfont, K. (2012). *Blue Ridge commons: Environmental activism and forest history in western North Carolina*. University of Georgia Press.
Norgaard, K. M. (2019). *Salmon and acorns feed our people: Colonialism, nature, and social action*. Rutgers University Press.
Ochiai, A. (2001). The Port Royal Experiment revisited: Northern visions of reconstruction and the land question. *The New England Quarterly*, *74*(1), 94–117.
Pulido, L. (1996). *Environmentalism and economic justice: Two Chicano struggles in the Southwest*. University of Arizona Press.
Puter, S. A. D., & Stevens, H. (1907). *Looters of the public domain*. Portland Printing House Publishers.
Pyne, S. J. (1982). *Fire in America: A cultural history of wildland and rural fire*. Princeton University Press.
Pyne, S. J. (2021). *The Pyrocene: How we created an age of fire, and what happens next*. University of California Press.
Robbins, W. G. (1985a). *American forestry: A history of national, state, and private cooperation*. University of Nebraska Press.
Robbins, W. G. (1985b). The social context of forestry: The Pacific Northwest in the twentieth century. *The Western Historical Quarterly*, *16*(4), 413–427.
Rutkow, E. (2012). *American canopy: Trees, forests, and the making of a nation* (1st Scribner hardcover ed.). Scribner.
Schallau, C. H., & Alston, R. M. (1987). Commitment to community stability: A policy or shibboleth? *Environmental Law*, *17*(3), 429–481.
Scott, J. C. (1998). *Seeing like a state: How certain schemes to improve the human condition have failed*. Yale University Press.

Searle, N. (1971). Minnesota National Forest: The politics of compromise, 1898–1908. *Minnesota History*, *42*(7), 242–257.

Shannon, F. A. (1945). A post mortem on the labor-safety-valve theory. *Agricultural History*, *19*(1), 31–37.

Smith, W. B., Miles, P. D., Perry, C. H., & Pugh, S. A. (2009). *Forest resources of the United States, 2007: A technical document supporting the forest service 2010 RPA Assessment*. (GTR-WO-78). USDA Forest Service.

Spiro, J. P. (2009). *Defending the master race: Conservation, eugenics, and the legacy of Madison Grant*. University Press of New England.

Steen, H. K. (2004). *The US Forest Service: A history* (Centennial ed.). Forest History Society in association with University of Washington Press.

Taylor, D. E. (2016). *The rise of the American conservation movement: Power, privilege, and environmental protection*. Duke University Press.

Tucker, G. J. (1981). *Historical sketches of Wallowa National Forest*. US Government Printing Office.

Twight, B. W. (1990). Bernhard Fernow and Prussian forestry in America. *Journal of Forestry*, *88*(2), 21–25.

White, R. (1991). *"It's your misfortune and none of my own": A new history of the American West*. University of Oklahoma Press.

Wilkinson, C. (2015). "Greatest good of the greatest number in the long run": TR, Pinchot, and the origins of sustainability in America. *Colorado Natural Resources Energy & Environmental Law Review*, *26*, 69–79.

Winkel, G. (2014). When the pendulum doesn't find its center: Environmental narratives, strategies, and forest policy change in the US Pacific Northwest. *Global Environmental Change*, *27*, 84–95.

Zimmerman, E. (1976). *A historical summary of state and private forestry in the US Forest Service*. US Department of Agriculture, Forest Service, State and Private Forestry.

# 3 Late 20th-Century Forest History

On September 30, 1994, the tarred and feathered effigies of two local environmentalists hanging from nooses appeared on Main Street in Joseph, Oregon. The act of symbolic violence came toward the end of a tumultuous year in Wallowa County: Joseph's Boise-Cascade sawmill shuttered in May and the R-Y Timber mill closed in October, eliminating over 100 jobs in a town whose population barely broke 1,000. Although Boise-Cascade and R-Y Timber both owned timberlands nearby, their mills were dependent upon logs cut from federal lands to make up part of their supplies. When timber sales and grazing practices on the Wallowa-Whitman National Forest were curtailed due to endangered species restrictions associated with Snake River salmon populations, local natural resource industries suffered—and economic pain reverberated throughout the county. A group of Wallowa County citizens took it upon themselves to send a message with the effigy, targeting the leaders of two environmental organizations they blamed for the county's sudden change in economic fortunes.

In Wallowa County and nationwide, the final decades of the 20th century brought dramatic changes to the forest policy and governance landscape. A suite of laws passed in the 1970s increased the federal role in private forest conservation, created new environmental and procedural obligations on public lands, and expanded the opportunities for individuals and interest groups to force compliance via the courts. By the 1990s, the new environmental policy framework was on a crash course with longstanding approaches to federal timber management, leading to substantial and often sudden changes in priorities and practices. At the same time, forests across ownerships experienced an increase in wildfires, insect outbreaks, and disease infestations while private forests in many urban fringe areas fell to expanding residential development. In this chapter, we trace changes since the 1960s, highlighting developments in both federal and nonfederal forest policy and governance.

## Policy Changes of the 1960s and 1970s

The 1960s and 1970s brought numerous changes to the country's environmental policy landscape amid growing public concern regarding the state of the nation's land, water, and air. In 1968, the National Park Service officially recognized the

DOI: 10.4324/9781003043669-3

potential benefits of both natural and prescribed fire in its policy handbooks and began instituting a prescribed fire program at some national parks, breaking with the fire exclusion policy that had been in place for decades.[1] These changes were inspired by the report of a 1963 national wildlife committee chaired by Starker Leopold (son of Aldo Leopold) that set a new vision for the conservation mission for national parks. It would be another decade before the Forest Service would quietly end its 10 a.m. policy by changing the guidance for wildfire response. National forest managers were now to take "appropriate suppression action" rather than ensuring that all fires were extinguished by the next morning.

Following the passage of the Multiple Use-Sustained Yield Act in 1960 and the Wilderness Act in 1964, national attention shifted to a broader suite of environmental concerns that included national forest management as well as a wide range of other issues: pesticides, toxic wastes, water and air pollution, species loss, and more. The late 1960s and early 1970s saw the founding of environmental groups such as the Natural Resources Defense Council and Environmental Defense Fund, organizations armed with both scientific and legal expertise deployed in an ongoing battle against industrial polluters. By this point, the Sierra Club had already demonstrated its potency as a national activist organization under the influence of firebrand environmentalist David Brower. Fighting alongside The Wilderness Society, Brower's Sierra Club notched victories against federal dam proposals in the 1950s and 1960s that would have irrevocably altered Dinosaur National Monument and the Grand Canyon. These campaigns showed that the environmental lobby was a force to be reckoned with. In the wake of Rachel Carson's publication of *Silent Spring* in 1962 and growing concern over the unintended consequences of novel chemical pesticides and herbicides, the legal and political infrastructure to regulate human impacts on the environment was assembled in short order. In the closing days of 1969, Congress passed the law that is now seen as the Magna Carta of American environmental policy, the National Environmental Policy Act—better known as NEPA (Box 3.1). This landmark policy requires the study and public disclosure of the environmental impacts associated with all proposed major federal actions, as well as the impacts of plausible alternatives. President Richard Nixon signed it into law on New Year's Day 1970, the opening stroke in a decade that would transform the nation's forest policy landscape.

The year 1970 ushered in the first Earth Day, organized to call attention to the impacts of industrial society on the earth's air, water, land, and biota. Nixon would go on to establish the Environmental Protection Agency (EPA) that same year to lead and coordinate efforts to protect the nation's air and water. Before the decade was through, Congress would pass much of what is now considered to be the foundational environmental legislation of the US: the Clean Air Act (1970), Clean Water Act (1972), Federal Insecticide, Fungicide, and Rodenticide Act (1972), Endangered Species Act (1973), Safe Drinking Water Act (1974), Resource Conservation and Recovery Act (1976), Toxic Substances Control Act (1976), and dozens of other environmental and conservation laws. Many of these represented substantial restructurings of previous, less comprehensive laws.

The new suite of legislation extended federal regulatory reach to nearly all lands and waters of the nation. Especially influential were NEPA, which would come to transform federal forest management, and the Clean Water Act and Endangered Species Act, which eventually changed the management of both private and public forests nationwide.

While the Clean Water Act's most aggressive provisions targeted "point sources"—the pipes, culverts, and other discrete conveyances that discharge waste materials into waters defined as navigable under the Act—its "nonpoint" regulations and wetland dredge-and-fill permit system drove the creation of both incentives and prohibitions designed to improve the water draining from forests, farms, and other extensive landscapes (see Chapter 4). Private forest owners who had long been accustomed to operating with little to no government oversight soon found themselves being encouraged—and in some cases required—to comply with a new suite of environmental rules. The Clean Water Act and associated state regulations influenced the design of logging roads and skid trails and complicated the harvesting of trees along waterways and in wetlands. New pesticide restrictions put in place under the 1972 Federal Insecticide, Fungicide, and Rodenticide Act meant changes in how private owners and public agencies responded to native and non-native pest outbreaks, and pesticide and fertilizer applicators now had to be licensed and operate under a written plan.[2] It was the Endangered Species Act, however, that implied the greatest changes to private forest management; this law is discussed in detail in Chapter 4. In 1978, Congress passed the Cooperative Forestry Assistance Act, which expanded the US Forest Service's role in providing technical assistance and financial support to private forest owners via partnerships with state forestry agencies. The Act also authorized the Forest Service to work with local governments in support of urban forest conservation (Chapter 10).

Legislation of the "environmental decade" also addressed federal land management head-on. The Forest Service received a substantial new policy framework in 1976 via the National Forest Management Act, technically a series of revisions to the Renewable Resources Planning Act passed just two years earlier. This second round of forest legislation came in response to a lawsuit brought by turkey hunters against the Monongahela National Forest in West Virginia. Upset with the aggressive clearcutting of some of their favorite hunting sites, the plaintiffs searched all the way back to the Forest Service's 1897 Organic Act to find the ideal legal lever to challenge the agency's timber practices: whereas clearcutting entailed marking the boundaries of a sale and harvesting all trees inside, the Organic Act required the Forest Service to cut only mature timber and to mark each tree prior to cutting. Coming on the heels of a damning assessment of timber management practices on the Bitterroot National Forest in Montana (the "Bolle Report"), and less than a decade after the passage of the Wilderness Act, the turkey hunters' legal victory signaled that the Forest Service's focus on aggressive timber production was starting to face substantial social and political headwinds.

The National Forest Management Act of 1976 authorized clearcutting under certain conditions but required that harvest practices not come at the cost of wildlife, watersheds, or future timber supplies.[3] Most importantly, the Act mandated that each national forest operate under a written plan detailing how the forest will be managed, including defining how different uses such as recreation, wildlife, and timber production will be provided for on the landscape. According to legislative language, the USFS must update these plans every ten to fifteen years and solicit public input in the process of writing and revising plans. Like NEPA, the National Forest Management Act was designed under a model of "comprehensive planning prior to action"[4] as well as on the principles of transparency and public involvement. In its requirements that the forest planning process be both interdisciplinary and open to public input and scrutiny, the National Forest Management Act effected a gradual but important change in the power to set direction for individual national forest units. Although foresters steeped in the ideals of sustained yield forestry still dominated the agency, they would eventually have to share decision-making power with other disciplines and with various publics and their diverse values for national forests.

The year 1976 also saw the passage of the Federal Lands Policy and Management Act, which for the first time established a comprehensive mandate and policy framework to guide the Bureau of Land Management.[5] Other legislation of the period created even greater openings for individuals and interest groups to actively participate in environmental governance—for example, the Endangered Species Act authorizes anyone to petition the federal government to list a species as threatened or endangered and allows anyone to sue to enforce the provisions of the Act. During this time, the Forest Service also expanded the ability of interested parties to administratively appeal agency decisions (such as forest plan decisions or project-level decisions regarding timber management), a process that halts implementation until the decision can be reviewed and resolved by higher authorities within the agency. In the 1970s, federal agencies thus entered what political scientist George Hoberg termed the "pluralist" period, in which they were granted an expanded scope of authority even as non-agency publics were given greater opportunities to enforce accountability through administrative and legal means.[6] As the 1970s gave way to the 1980s and 1990s, groups opposed to national forest management direction would come to use these opportunities to grind the agency's timber program to a virtual halt.

The federal agencies also began to grow more diverse during this time, although the pace of change was not always rapid. Whereas Herbert Kaufman in 1960 remarked on the ideological and philosophical homogeneity of the Forest Service workforce, in the 1970s the agency began to hire more women, people of color, and what became known as the "ologists": ecologists, wildlife biologists, hydrologists, archaeologists, and even some anthropologists and sociologists who introduced views on forest management that differed from the timber-centric perspective common among agency foresters at that time.[7] The 1970s, thus, brought both internal and external pressures to bear on the Forest Service and on the practice of forestry as a whole.

## BOX 3.1 THE NATIONAL ENVIRONMENTAL POLICY ACT

The National Environmental Policy Act, better known as NEPA, is considered one of the country's bedrock environmental laws despite the fact that its provisions are procedural rather than substantive (that is, it does not require the selection of the least environmentally damaging alternative). According to legislative language, an analysis of environmental impacts must be conducted and made public prior to initiating any "major federal action significantly affecting the quality of the human environment." This analysis is required to consider alternatives to the proposed federal action, including the alternative of doing nothing. The Act also established the Council on Environmental Quality, an agency whose role is to oversee the process of NEPA implementation among federal agencies as well as to provide other policy advice and development to the executive branch.

Two important questions related to NEPA implementation pertain to the definition of a "federal" action and the threshold for a "significant" impact on the environment. In general, any action with a clear federal nexus is subject to NEPA obligations, including actions directly taken by federal agencies as well as those that rely on federal funding or authorization. The threshold for a "significant" impact has been defined through decades of case law in concert with implementing regulations from the Council on Environmental Quality and agency-specific NEPA regulations; as a result, the precise definition of significance varies to some extent from agency to agency. A proposed action may rise to the level of "significance" depending upon its context and intensity. For example, a project that has large-scale or long-lasting impacts, that entails impacts on rare or unique resources, or that has impacts that are cumulative with the impacts of other projects may meet significance thresholds.

Once NEPA obligations are triggered by the presence of a federal nexus, planners have three broad procedural pathways from which to choose. *Categorical Exclusions* (CEs) exist for various classes of actions that are known to not have significant effects. These range from routine actions like campground maintenance to more substantial activities such as thinning or prescribed fires on federal forestlands below certain acreage limits. On the other end of the spectrum, proposed actions that cross the significance threshold are analyzed by way of an *Environmental Impact Statement* (EIS), which in the case of particularly complex projects can amount to multiple volumes and hundreds or even thousands of pages. In cases where it is unclear whether the threshold of significance will be met, an *Environmental Assessment* (EA) is used to make the determination. An EIS will be required if the EA concludes that significant impacts are likely; otherwise, the EA will result in a Finding of No Significant Impact (FONSI). A "mitigated

FONSI" is the result when mitigation actions are included to prevent the project from crossing the threshold of significance.

The US Forest Service is responsible for more NEPA decisions than any other federal agency, and the law has been the target of criticism from various groups who call for procedural requirements to be reduced or eliminated in order to expedite forest management. Common complaints are that NEPA interferes with the efficient management of federal forests for timber production, hazardous fuel reduction, or other management priorities and that its complex procedural requirements create opportunities for frivolous lawsuits that prevent projects from moving forward. Recent scholarship has provided insights to inform this debate. One analysis found that, of the 41,194 Forest Service NEPA decisions published between 2004 and 2020, 81.2% were CEs, 16.7% were EAs, and 2.1% were EISs.[8] The researchers reported that the median time for EIS completion was 2.8 years, for EAs it was 1.2 years, and for CEs it was 4 months. The small number of anomalously long NEPA processes in the dataset were largely explained by factors other than NEPA itself—factors such as inadequate staffing or funding, required compliance with other laws, or delays on the part of non-Forest Service personnel.[9] A similar study found that less than 1% of Forest Service NEPA decisions are ultimately litigated, representing 17% of EISs, 2% of EAs, and less than 1% of CEs.[10] The Forest Service prevails in these lawsuits 69% of the time. Although lawsuits are relatively uncommon, evidence suggests that substantial Forest Service staff time is often spent attempting to craft NEPA documents that can withstand possible legal or administrative challenges and that the fear of these challenges can influence the timing and structure of NEPA processes.[11]

## "Getting the Cut Out" on the National Forests

Despite the new suite of environmental legislation that emerged from the environmental decade—and in spite of continuing legal battles regarding the delineation of roadless areas eligible to be protected under the Wilderness Act—the Forest Service continued to give primacy to timber production on national forests. The agency had consistently sold over ten billion board feet of timber annually since the 1960s and this trend continued through the 1970s and into the 1980s. The agency's emphasis on "getting the cut out" was motivated by multiple factors: a forestry culture focused on maximizing wood fiber production;[12] political pressure from the highest levels of the executive branch to put the national forests to use producing marketable commodities and generating associated economic activity in the logging, milling, and transport sectors; the development of an "iron triangle" linking the Forest Service, forest business and rural community interests, and elected officials from timber states such as Oregon and Washington;[13] and a set of internal agency procedures and incentives that rewarded rangers based on the volume of timber they sold.

The incentives baked into the Forest Service's budgetary process encouraged managers to sell timber anywhere and everywhere—even in places where the costs of layout, sale administration, and building access roads exceeded the value of the timber harvested, a phenomenon that came to be known as the "below-cost timber sale." Economist Randall O'Toole provided an incisive analysis of Forest Service bureaucratic behavior in his 1988 book *Reforming the Forest Service*.[14] He detailed how the combination of a budget-maximization objective and the incentives created by various timber sale mechanisms led the agency to take many actions that were bureaucratically rational yet both economically and ecologically irrational—for example, building miles of new logging roads through previously roadless forestland to harvest low-value lodgepole pine near Yellowstone National Park. As environmental advocates ramped up challenges to the timber-oriented Forest Service throughout the 1970s and into the 1980s, they argued that the agency was degrading ecologically important habitats while also managing to waste taxpayer money—a lose-lose proposition. As evidence for this argument, environmental advocates could point to Southeast Alaska's Tongass National Forest, which annually received tens of millions of federal dollars throughout the 1980s to subsidize an economically marginal pulp and paper industry based on clearcutting old-growth forests.[15]

National forest leadership appeared impervious at times to the growing public calls for reform of the sustained-yield paradigm that continued to guide agency decision-making. The agency instead developed ever more sophisticated computational models for estimating, planning, and scheduling timber harvests, a process that culminated with the rollout of a tool called FORPLAN in 1979.[16] Like its predecessors, FORPLAN used linear programming to achieve a sustained yield of timber with maximum precision. Later revisions allowed forest managers to include other outputs beyond timber, but the overall approach was the same: maximize the production of outputs through a planning process so opaque that even FORPLAN's creator described it as "a formidable way for the national forests to insulate themselves from their critics."[17] For federal economists and others who saw national forests primarily as producers of quantifiable outputs, tools like FORPLAN achieved a level of mathematical efficiency that represented the realization of Pinchot's ideal of "the greatest good for the greatest number." For those who saw national forests as complex ecosystems, places for recreation, or lands on which the public ought to have a substantive voice, FORPLAN exemplified a technocratic approach to national forest management run amok.

## The Timber Wars

The 1980s and 1990s brought seemingly exponential growth in the number and visibility of organized environmental advocacy campaigns calling for forestry reforms. These took shape through numerous organizations that combined legal expertise with grassroots political organizing skills. Some of the common objectives of groups and coalitions like the Southern Appalachian Forest Coalition, Heartwood, Alliance for the Wild Rockies, Oregon Natural Resources Council,[18] and the Southeast Alaska Conservation Council included the protection and restoration of old-growth forests, a reduction in forest road density, conservation of wildlife habitat, and reform of prevailing forestry practices on private and especially on public land.[19] In 1989, current and former US Forest Service employees founded Forest Service Employees for Environmental Ethics, adding an insider voice to the growing calls for agency reform. In many cases, these grassroots efforts were formed because the large mainstream environmental organizations were, in the eyes of many activists, too willing to compromise on questions of old-growth forest harvesting.[20] Nowhere were the tensions between the Forest Service's quantitatively narrow approach to timber harvest scheduling and broader public values for forests more intense than in the temperate rainforests of the Pacific Northwest. These highly productive forests represented both the beating heart of the agency's timber program and some of the last remnants of centuries-old forests harboring species whose habitats had been progressively diminished under the influence of the Forest Service's sustained-yield paradigm.

Biologists had known since the 1970s that the northern spotted owl (Figure 3.1) depends on old-growth forest for its habitat and that its continued existence

*Figure 3.1* The northern spotted owl's old-growth forest habitat requirements became central to a legal strategy by environmental advocates to force changes to national forest and Bureau of Land Management timber programs in the Pacific Northwest. Image: National Archives.

was threatened by the aggressive timber program on northwestern federal forestlands.[21] The owl thrives in multilayered, multispecies forests dominated by Douglas-fir, western red cedar, hemlock, and other coniferous species that regularly live for centuries and grow hundreds of feet tall in the damp, cool conditions of western Oregon and Washington and northwestern California. Those same trees were the lifeblood of a timber industry that provided jobs, income, and cultural identity to numerous rural communities throughout the region, and powerful elected officials did their best to secure the continued flow of federal timber to those communities. Oregon Senator Mark Hatfield, for example, attached language to federal spending bills that exempted timber sales in the Northwest from restrictions related to the owl or to other environmental considerations.[22] Despite such efforts by friends in high places, timber-dependent communities suffered in the 1980s under the combined weight of an economic recession, weak domestic demand for wood products, competition from Canadian imports, the advent of a new generation of mills that required less human labor, and the overseas exportation of unprocessed logs harvested from private lands—all factors unrelated to environmental restrictions.[23]

By the end of the 1980s, regional and, later, national environmental campaigns were successfully using the Endangered Species Act and regulations implementing the National Forest Management Act to challenge the Forest Service's prevailing forestry practices in court. They had also learned to use the administrative appeals process to slow and even halt logging projects and forest plans. In the Northwest, where timber was still king, the collision between resource production and species preservation threatened a way of life and a regional economy. Rarely had forests been put under such intense national scrutiny, with the northern spotted owl appearing on the cover of *TIME* magazine and the evening news fixated on the conflicts between environmental advocates and the timber industry. The issue was widely framed in the media as one of "jobs versus owls," even though numerous policy actors and advocates cautioned that the conflict was not that simple.

Protests by environmental advocates included the use of "direct action" tactics in which activists would put themselves directly in harm's way, chaining themselves to logging equipment, constructing ramshackle blockades of logging roads, or building platforms in the canopies of trees slated for harvest where they would live sometimes for weeks or months on end. Some activists went so far as to "spike" trees, a practice that entails sinking nails or porcelain spikes into the trunk in order to sabotage tree felling and milling. The cutting and milling of spiked trees can damage sawblades and create extreme danger for workers in the woods and in the mills. In 1987, a 23-year-old worker named George Alexander was nearly killed when a spiked tree caused the bandsaw at a Louisiana-Pacific mill in northern California to unravel; a section of the broken saw sliced into his face and neck, breaking his jaw and several teeth and puncturing his jugular vein.[24] Despite the risks it posed to workers in the woods and in the mills, the controversial practice of tree spiking was surreptitiously implemented from California to Maine.[25]

In addition to the conflicts taking place in the woods and in the courts, the timber wars also included struggles over the framing of the dispute: environmental activists adopted a discourse centered around concepts such as "virgin," "primeval," and "cathedral" forests that were being defiled and violated by the incursions of industrial forestry.[26] Environmental leaders called their collective efforts to end old-growth forest harvesting the "Ancient Forest Campaign" as part of a strategy to garner public sympathy and support.[27] Imagery employed by the timber industry and its allies typically focused on the forest workers and the rural, blue-collar communities in which they lived. Environmental advocates were cast as clueless urban elites indifferent to the suffering of hard-working woodsmen and the families and rural communities they supported through their labor. Old-growth forests were described as "decadent," "overmature," and as "biological deserts" that would only benefit from being renewed through harvest. Signs affixed to the doors of timber-town businesses from restaurants to hair salons to sporting goods stores announced solidarity with the timber industry and the people it employed, while the spotted owl itself became an object of derision. One particularly widespread campaign encouraged residents to tie yellow ribbons around their fences and car antennas to demonstrate their support for a timber economy that suddenly found itself under siege.[28]

In lawsuits brought by the Seattle Audubon Society and allied environmental groups, the federal courts ruled time and again that the Forest Service and Bureau of Land Management could not continue their aggressive timber programs without violating federal environmental laws. The courts found that by progressively diminishing northern spotted owl habitat, the Forest Service was managing in violation of its own National Forest Management Act regulations requiring it to maintain "viable populations of existing native and desired non-native vertebrate species...well distributed in the planning area." The Endangered Species Act also came into play as various species including the marbled murrelet (a seagoing bird that nests in old-growth forest canopies), multiple populations of salmon and steelhead, and the northern spotted owl itself were listed as threatened or endangered in the first years of the 1990s. Subsequent rulings stipulated that the agencies must discontinue harvest activities until a plan was in place to protect the owl and other vanishing species. In rural towns from northern California to northern Washington, sawmills that had been fed a steady diet of federal timber went quiet as supplies were abruptly shut off.

Following the 1993 Forest Summit in Portland, convened by a freshly inaugurated President Clinton, Forest Service and Bureau of Land Management leaders (with the help of both federal and external scientists) went to work crafting a strategy that would meet the agencies' legal obligations to protect old growth-dependent species while restarting the timber economy. The result was the 1994 Northwest Forest Plan, built upon a model that included the following land use allocations: core conservation areas, including Late-Successional Reserves designed to protect and recover old-growth habitat; Matrix areas where timber production would still be a major focus; Riparian Reserves to help protect salmon and other aquatic species; and Adaptive Management Areas intended to provide living laboratories to experiment with new ways of reconciling resource production and species protection. These were complemented by an aquatic conservation strategy focused on maintaining and restoring watersheds to protect aquatic and riparian species.

The Northwest Forest Plan represented the Clinton administration's attempt to meet its legal obligations while continuing to send as much federal timber as possible to Northwest sawmills. Although Late-Successional Reserves were located in areas with large blocks of older forest, not all remaining old-growth stands were protected under the plan; those in the Matrix allocation were designated for eventual harvest unless they happened to be the immediate home to a listed species. The plan predicted that, going forward, the federal forests of the Pacific Northwest would produce around 1.2 billion board feet of timber annually, down from four times that level prior to the Audubon Society lawsuits.[29] Although some decried the inclusion of scattered older stands in the agencies' revised timber programs, environmental advocates had, in fact, won a resounding victory that far surpassed what many seasoned activists had thought possible just a few years earlier.

Thus, 24 years after the passage of NEPA, 21 years after the passage of the Endangered Species Act, and 18 years after the passage of the National Forest Management Act, the full weight of these laws was brought to bear on federal forest agencies, forcing them to shift to an entirely new management paradigm. The Northwest Forest Plan exemplified the concept of "ecosystem management" advanced by Jack Ward Thomas, President Clinton's hand-picked selection for Forest Service Chief and the first non-forester to hold the title (Thomas was a wildlife biologist by training). Under ecosystem management, the national forests were not to be managed primarily for timber production; rather, timber production would result from managing the forests for a suite of ecological objectives including restoration, habitat creation, and sensitive species protection. With the introduction of ecosystem management, the output-optimizing approach represented by FORPLAN was out, and landscape-scale planning informed by the field of conservation biology was in.

However, the Northwest timber wars were not ultimately resolved with the release of the Northwest Forest Plan. In 1995, a newly minted Republican congressional majority attached language to an emergency bill responding to the domestic terror attacks on the Murrah federal building in Oklahoma City; it waived all federal environmental laws in order to salvage federal trees killed or injured by fire. This innocuous-sounding "Salvage Rider" worked in practice to allow for one final round of old-growth harvest—including in stands identified for protection under the Northwest Forest Plan—and triggered a new wave of protests by environmental activists. Outrage over the rider convinced the membership of the Sierra Club to vote for the first time in favor of a "zero cut" policy that advocated for an end to all commercial logging on national forests.[30] Controversy over the last of the Salvage Rider sales, in the Eagle Creek area of the Mt. Hood National Forest, south of Portland, Oregon, lasted well into 2002 (Box 3.2). The Salvage Rider represented a new paradigm and new focal area for federal forest policy; it framed environmental restrictions as obstacles to both human safety and forest health in the face of the growing threat of wildfires and other disturbances.[31] As Chapter 5 describes, wildfire occurrence, severity, and suppression costs all trended sharply upward beginning in the 1990s and attention to fire, insect, and disease impacts on forests came to dominate the national forest policy debate in the wake of the Northwest Forest Plan.

The timber wars were not only fought on federal lands in the Northwest; they also included national forests in places like Colorado, Idaho, New Mexico, Arizona, Indiana, Illinois, and Pennsylvania along with some private lands—particularly in northern California where millennia-old redwoods on corporate timberlands were falling under the ax (see Chapter 9). Nor were the timber wars only fought via peaceful means; in 1990, environmental and labor organizers Judi Bari and Darryl Cherney were nearly killed when a bomb planted under their car detonated as they were preparing for activism in the redwood forest. A bombing targeted Forest Service facilities in New Mexico in 1996, following two failed attempts to assassinate a Forest Service employee in Nevada. A homemade bomb was found at the Detroit Ranger Station in Oregon and another nearby ranger station was destroyed in a suspected arson that same day. Environmentalists associated with the Earth Liberation Front took responsibility for arsons at timber industry offices in 1998 and 1999 along with a mill arson in 2001, all in the state of Oregon; they also claimed responsibility for arsons at various residential and commercial developments and tree biotechnology facilities nationwide. The latter acts of property damage came to be labeled using a new term: "eco-terrorism." Some pro-commodity production activists, including many affiliated with the Wise Use movement (a movement that appropriated Gifford Pinchot's terminology to attack environmental protections and advocate for continued development of natural resources on public lands), also embraced threats of violence as part of their efforts to realize political ends.[32]

## BOX 3.2 THE LAST SALVAGE RIDER SALE

A modern-day visitor to the Eagle Creek trail on the Mt. Hood National Forest south of Portland, Oregon, might take note of the deep shade provided by an overstory of 100-year-old western hemlocks, Douglas-firs, and western red cedars. What would be less obvious is that the surrounding forest was once the site of a pitched battle in the timber wars between environmental advocates and the US Forest Service. Eagle Creek was included in a list of projects designated under the 1995 Salvage Rider, which waived nearly all environmental laws to allow timber cutting in places affected by or susceptible to fire, insect outbreaks, or other disturbances. Environmental advocates contested the Eagle Creek timber sale over a period of several years starting in the late 1990s using tactics that had proven successful over the prior decade: activists chained themselves to forest gates and hoisted themselves into the canopy of trees, living on tiny platforms connected by rope bridges and ladders. Some used even more extreme tactics. In June 2001, three logging trucks parked near the site were set ablaze by improvised incendiary devices. The prime suspect in the arson—a charismatic activist who had previously brought attention to Eagle Creek by scaling the US Forest Service's regional office building in downtown Portland and living for nine days on a window ledge—fled to Canada where he was later arrested. In an even more tragic turn of events, 22-year-old Beth O'Brien fell to her death from an Eagle Creek canopy platform on April 12, 2002 as she and other activists awaited official confirmation of the sale's cancellation, announced publicly just days before. By some measures, Eagle Creek was of a piece with a series of other direct action campaigns by environmental advocates over the previous decade, entailing radical activism in defense of what advocates saw as some of the planet's last remaining old-growth forests. In other ways, however, Eagle Creek was unique—the forests slated for harvest were mature but did not meet old-growth criteria by most definitions, and the firebombing of logging trucks represented an escalation in direct action tactics. The events surrounding Eagle Creek were perhaps surprising given that, by 2001, the timber wars were largely over, with the environmental community already having claimed a victory beyond anything they might have envisioned when direct action protests and legal challenges first picked up steam in the late 1980s.

## The Aftermath

The transformative, if incomplete, resolution of the timber wars with the Northwest Forest Plan was significant for several reasons: it ended the sustained yield paradigm for national forest management in the Pacific Northwest and ushered in the ecosystem management period; it demonstrated the power of laws like the Endangered Species Act to force changes to forestry practices; it signaled that the courts and the executive branch—not Congress—were now largely calling the shots on federal forest policy; and it provided a template for environmental advocates to use in other federal forest-dominated landscapes nationwide. Threatened and endangered species such as the California spotted owl, Mexican spotted owl, Canada lynx, grizzly bear, wolverine, and various salmon and steelhead populations figured prominently in lawsuits challenging the Forest Service's roadbuilding and timber-cutting programs from Alaska to New Mexico. Many of these cases resulted in court-ordered logging moratoriums and requirements to manage for the preservation and recovery of listed species. The Northwest Forest Plan also appeared to signal the beginning of a new era of large landscape-scale plans informed by the science of conservation biology, but subsequent initiatives such as the Interior Columbia Ecosystem Basin Management Project faced stronger headwinds when it came time to formally adopt the recommendations of scientists, at least in part due to strong resistance by rural community members and elected officials who feared a loss of access to the benefits of forests.[33]

President Clinton continued to take advantage of his executive branch authority by directing the Forest Service to conserve the roadless condition of all remaining "inventoried roadless areas" (generally those eligible for wilderness status) on national forests and grasslands nationwide. The Roadless Area Conservation Rule, released just days before the end of Clinton's second term, had the practical effect of preventing roadbuilding except in emergency situations on all remaining roadless blocks of at least 5,000 acres. President Clinton also took advantage of authorities provided by the 1906 Antiquities Act to establish or expand 22 national monuments, with most of the designations—including the Giant Sequoia National Monument in California, the Cascade-Siskiyou National Monument in Oregon, and the Ironwood Forest National Monument in Arizona—coming during his final year in office.

By the year 2000, annual timber harvests on the national forests had fallen to less than two billion board feet—the lowest level since the agency's custodial period (Figure 3.2). Among other implications, this meant a steep drop in Forest Service employment, particularly on forests where timber receipts helped support staffing numbers. Across the agency, national forest units were consolidated,

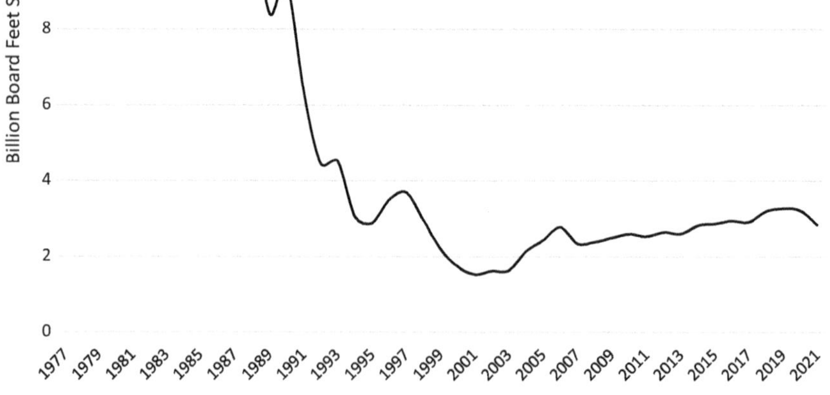


*Figure 3.2* Volume of timber sold across the National Forest System, 1977–2021 (fiscal year data). Source: US Forest Service Cut and Sold Reports.

field stations were shuttered, and forest managers struggled to fund even basic duties. For rural communities that had already suffered through a recession, layoffs, and the shutoff of federal timber supplies, the cut in Forest Service employment meant losing some of the only stable, well-paid jobs remaining. In addition to the economic impacts, there was also a heavy social cost as neighbors faced off against neighbors and as conflict over timber, economy, identity, employment, and preservation created divisions within rural towns and between those towns and regional urban centers. In mill towns across the West, the same Forest Service employees that had formerly been welcomed and even embraced as members of the community were now seen as unwelcome agents of an unpopular political agenda. Many federal employees found that local shop owners would refuse to serve them, and some faced threats of retaliatory violence.[34] The relationship between the Forest Service and many rural communities had been stretched to the point of breaking.

Out of the rubble of these difficult conditions emerged radical experiments in forest governance in which local leaders representing diverse viewpoints sought to identify solutions to combined ecological and economic dilemmas. Rather than waging winner-take-all battles in court, these efforts were founded in a collaborative rationality[35] that saw the potential for traditional adversaries to constructively work together. A pattern of increasingly large and destructive wildfires not only brought community members together to plan for fire protection but also led to new models of interagency partnerships for forest management and restoration across jurisdictional and property lines. Recent innovations in fire management are examined in Chapter 5, and the rise of community-based collaboration is discussed in Chapter 10.

## Beyond Western Federal Forests

Federal lands in the West remained at the center of US forest policy and governance deliberations in the late 20th century in light of the drama surrounding the northern spotted owl, Salvage Rider, and escalading wildfires. Yet much was happening elsewhere across the nation as well. Many of the timber companies active in the West had traditionally sourced the wood supplying their mills from a combination of their own timberlands and federal forests; the timber wars signaled that the federal side of that equation was going to be increasingly unpredictable and likely to come with reputational risks. Further, by the 1980s relatively little older forest remained on private lands in the Northwest, and timber firms were readying for a new round of disinvestment and relocation.[36] Even before the timber wars were fully underway, some of the largest timber companies were already selling parts of their northwestern properties and investing more heavily in the East and South as well as in international settings from Canada to Uruguay. Industry investments in southern forestlands were supported by inexpensive land prices, high productivity, short harvest rotations, the development of genetically improved pine stocks, and advances in technology that allowed for the creation of boards and panels from smaller trees.[37] Collectively, these factors led to the emergence of the South as the new center of forest production in the US.

In a sign that the crisis of legitimacy that buffeted national forestlands was also affecting other forests, owners and managers of millions of acres of private, state, county, and tribal forestlands nationwide began to adopt third-party forest certification in order to communicate their commitment to environmental sustainability principles to the public. By the end of the 1990s, a corporate restructuring process gave rise to new investor-driven forest industry models in which land and mills were largely separated (Chapter 9). As the timber industry restructured, one of its strategies was to take advantage of residential real estate markets by identifying and selling properties worth more as housing than as timberland. An extreme case emerged in 1988 when the timber company Diamond International put close to a million acres of timberland up for sale in New York and New England at residential—rather than timber—prices. Although much of the property ultimately went to other timberland companies and to conservation organizations, the various buyers announced plans to convert over 180,000 acres to residential developments. This transaction, along with other actual and threatened forestland conversion efforts, led Congress to push for a Northern Forest Lands Study in 1988, out of which grew the Northern Forest Lands Council two years later. The Council identified growing strains on the northern forest and its human communities stemming from a mix of residential development pressures, forest fragmentation, changing economic activities, economic disincentives to manage forests for non-residential uses, and the lack of a coordinated vision for conservation. Several policy mechanisms, primarily focused on conservation easements and other economic incentives to maintain forests in an undeveloped state as well as changes to tax policies, were recommended as means of addressing these challenges. In subsequent years, a similar mix of payments and changes to taxation were proposed and adopted in numerous states as means of promoting forest conservation; these efforts are discussed further in Chapters 6 and 8.

Ownership changes in private (especially corporate) timberland also spawned a new movement for community-owned forests in the 1990s and 2000s. Native American Tribes used both new strategies and new sources of revenue to repatriate lands of which they had long ago been dispossessed. At the same time that fire, insect, and disease outbreaks were stressing western landscapes, forests in the Midwest, South, and East were dealing with a complex mix of native and introduced diseases and pests—as well as substantial impacts from wildfires in states including Florida, Georgia, Arkansas, North Carolina, and Tennessee. The continued development of the wildland-urban interface nationwide continued to put additional pressure on forests, with consequent impacts on biodiversity, carbon storage, wildlife habitat, and water quality and quantity. All of these issues are examined more closely in the following chapters.

Another important development during this time was the emergence of a social movement that connected environmental management with the oppression of marginalized communities. What has come to be known as the Environmental Justice movement arose in the 1970s and 1980s as low-income communities and communities of color fought against the inequitable placement of toxic waste sites and other environmental hazards.[38] A key event took place in Warren County, North Carolina, in 1982 when residents of an overwhelmingly African American community adopted principles of nonviolent civil disobedience in their attempt to prevent the installation of a toxic waste dump in their backyard. Multiple studies in subsequent years documented the undue burden of toxic hazards experienced by Black and Latino communities nationwide.[39] Environmental Justice activists called out the mainstream environmental organizations for their failure to address these issues and for the overwhelmingly White composition of their boards and staff. In 1991, the First National People of Color Environmental Leadership Summit was held, and in 1994 President Clinton enacted an Executive Order on Environmental Justice. Although the Environmental Justice movement began with a focus on issues of waste and exposure to hazards, it quickly expanded to encompass issues such as the dispossession of Indigenous peoples (Chapter 7), the inequitable access to the benefits of forests in urban spaces (Chapters 10 and 11), Hispano access to timber and non-timber forest products in the Southwest (Chapter 11), working conditions for those in natural resource jobs (Chapter 11), and the burden placed on communities that rely on fish and game for subsistence or cultural practices. Beyond illuminating specific issues that were previously considered outside the purview of the environmental movement, Environmental Justice activists offered an alternative paradigm to the conventional dualism separating the category of "nature" (implicitly worthy of environmentalists' concern) and the category of "culture" (implicitly someone else's problem). The Environmental Justice paradigm instead frames environmental issues as inextricably bound up with questions of human equity, justice, and wellbeing.

## Conclusion

The ripples of discontent with the state of the American environment that appeared in the early postwar years became a tsunami by the 1970s. Reforms that began with the Wilderness Act and continued through the new suite of laws from the "environmental decade" and the subsequent Ancient Forest Campaign effected a revolution in forest policy and governance. Multiple influences combined synergistically during this time period: broad social concern with the environmental implications of industrial society; scientific findings that highlighted the loss of biodiversity and environmental quality as a result of changes to land use and chemical applications; policy language that provided openings for non-agency publics to participate in and contest agency decisions; new substantive obligations in these same policies for the protection of wildlife and water quality; the diversification of the government workforce; restructuring and reinvestment by the timber industry; the growing sophistication of varied interest groups that opposed the aggressive timber program on federal lands; and the ascendancy of the courts and executive branch in setting forest policy direction. The conflicts and policy dynamics of the 1990s set a template for the major trends that followed, from changes in the corporate structure of the timber industry to the tradeoffs between wildlife and economic development to policy disputes regarding the links between forest conservation and wildfires to the inequitable distribution of environmental benefits and burdens among social groups. These topics, examined in more detail in the following chapters, lie at the heart of some of the most important contemporary forest policy debates.

## Questions for Discussion

1. Why were so many of the country's major pieces of federal environmental legislation passed in a short period during the 1970s? What social and political forces were at play during that time that influenced this pattern of policymaking?
2. What features do these 1970s-era policies have in common? How are they different from policies that came before, such as the Weeks Act or Multiple Use-Sustained Yield Act?
3. At the time of the timber wars, media outlets often framed the conflict as "jobs versus owls." Is this an accurate framing? What factors made the conflict more complicated than this framing suggests?
4. Acts of violence and threats of violence were unfortunately common at the height of late 20th century conflicts over forests and their management. Why do you think some individuals and groups on both the pro-preservation and pro-timber sides came to see violence as appropriate or necessary? How could the conflict have been handled differently so as to alleviate the level of anger and resentment felt by some participants?
5. What is your understanding of the concept of "ecosystem management"? How does it differ from the concept of "sustained yield"? How would you expect the shift from sustained yield to ecosystem management to change the way decisions are made regarding the tradeoffs among the multiple values and uses for forests?

## Notes

1 Stephens & Ruth, 2005.
2 Carter et al., 2015, pp. 261–262.
3 Wilkinson, 1997.
4 Moseley & Charnley, 2014, p. 71.
5 Skillen, 2009.
6 Hoberg, 1992.
7 Brown & Harris, 1993.
8 Ruple et al., 2022.
9 Fleischman et al., 2020; Ruple et al., 2022.
10 Fleischman et al., 2022.
11 Mortimer et al., 2011; Ruple et al., 2022.
12 Clary, 1986.

13 Trosper, 2003.
14 O'Toole, 1988.
15 Nie, 2006.
16 Kent et al., 1991.
17 Johnson, 1987, p. 47.
18 The Oregon Natural Resources Council has since been renamed Oregon Wild.
19 Hays, 2007, pp. 26–50.
20 Bevington, 2009.
21 Yaffee, 1994, pp. 14–19.
22 Bevington, 2009, p. 120.
23 Cook, 1995; Loomis, 2016, pp. 195–198; Robbins, 1985.
24 Note that culpability for this case was never established, though there is some evidence that this particular spike was placed by a neighboring property owner and not by environmental activists (Bari, 1994).
25 Bari, 1994, pp. 264–282.
26 Davis, 1995.
27 Proctor, 1995.
28 Proctor, 1995.
29 Wilkinson, 1997, p. 678.
30 Bevington, 2009, pp. 134–138.
31 Vaughn & Cortner, 2005.
32 McCarthy, 2001.
33 Crow, 2006.
34 McCarthy, 2001.
35 Innes & Booher, 2010.
36 Robbins, 1985.
37 Carter et al., 2015, p. 263.
38 Taylor, 2011.
39 Bullard, 1990; Bullard et al., 2007; Commission for Racial Justice, United Church of Christ, 1987; General Accounting Office, 1983.

## References


Bari, J. (1994). *Timber wars*. Common Courage Press.

Bevington, D. (2009). *The rebirth of environmentalism grassroots activism: From the spotted owl to the polar bear*. Island Press.

Brown, G., & Harris, C. C. (1993). The implications of work force diversification in the US Forest Service. *Administration & Society*, 25(1), 85–113.

Bullard, R. D. (1990). *Dumping in Dixie: Race, class, and environmental quality*. Routledge.

Bullard, R. D., Mohai, P., Saha, R., & Wright, B. (2007). Racial and socioeconomic disparities in the distribution of environmental hazards: Assessing the evidence twenty years after Toxic Wastes and Race. In *Toxic wastes and race at twenty: 1987–2007*. United Church of Christ.

Carter, M. C., Kellison, R. C., & Wallinger, R. S. (2015). *Forestry in the US South: A history*. Louisiana State University Press.

Clary, D. A. (1986). *Timber and the Forest Service*. University Press of Kansas.

Commission for Racial Justice, United Church of Christ. (1987). *Toxic wastes and race in the United States: A national report on the racial and socio-economic characteristics of communities with hazardous waste sites*. Commission for Racial Justice, United Church of Christ.

Cook, A. K. (1995). Increasing poverty in timber-dependent areas in Western Washington. *Society and Natural Resources*, 8(2), 97–109.

Crow, T. R. (2006). Moving to the big picture: Applying knowledge from landscape ecology to managing US national forests. In A. H. Perera, L. J. Buse, & T. R. Crow (Eds.), *Forest landscape ecology* (pp. 157–180). Springer.

Davis, S. (1995). The role of communication and symbolism in interest group competition: The case of the Siskiyou National Forest, 1983–1992. *Political Communication*, *12*(1), 27–42.

Fleischman, F., Struthers, C., Arnold, G., Dockry, M., & Scott, T. (2020). US Forest Service implementation of the National Environmental Policy Act: Fast, variable, rarely litigated, and declining. *Journal of Forestry*, *118*(4), 403–418.

Fleischman, F., Struthers, C. L., Arnold, G., Dockry, M., & Scott, T. A. (2022). Corrigendum to: US forest service implementation of the National Environmental Policy Act: Fast, variable, rarely litigated, and declining. *Journal of Forestry*, fvac005.

General Accounting Office. (1983). *Siting of hazardous waste landfills and their correlation with racial and economic status of surrounding communities* (GAO/RCED-83-168). General Accounting Office.

Hays, S. P. (2007). *Wars in the woods: The rise of ecological forestry in America*. University of Pittsburgh Press.

Hoberg, G. (1992). *Pluralism by design: Environmental policy and the American regulatory state*. Praeger Publishers.

Innes, J. E., & Booher, D. E. (2010). *Planning with complexity: An introduction to collaborative rationality for public policy*. Routledge.

Johnson, K. N. (1987). Reflections on the development of FORPLAN. In T. W. Hoekstra, A. A. Dyer, & D. C. LeMaster (Eds.), *FORPLAN: An evaluation of a forest planning tool* (pp. 45–51). USDA Forest Service Rocky Mountain Forest and Range Experiment Station.

Kent, B., Bare, B. B., Field, R. C., & Bradley, G. A. (1991). Natural resource land management planning using large-scale linear programs: The USDA Forest Service experience with FORPLAN. *Operations Research*, *39*(1), 13–27.

Loomis, E. (2016). *Empire of timber: Labor unions and the Pacific Northwest forests*. Cambridge University Press.

McCarthy, J. (2001). States of nature and environmental enclosures in the American West. In N. L. Peluso & M. Watts (Eds.), *Violent Environments* (pp. 117–145). Cornell University Press.

Mortimer, M. J., Stern, M. J., Malmsheimer, R. W., Blahna, D. J., Cerveny, L. K., & Seesholtz, D. N. (2011). Environmental and social risks: Defensive National Environmental Policy Act in the US Forest Service. *Journal of Forestry*, *109*(1), 27–33.

Moseley, C., & Charnley, S. (2014). Understanding micro-processes of institutionalization: Stewardship contracting and national forest management. *Policy Sciences*, *47*(1), 69–98.

Nie, M. (2006). Governing the Tongass: National forest conflict and political decision making. *Environmental Law*, *36*, 385–480.

O'Toole, R. (1988). *Reforming the Forest Service*. Island Press.

Proctor, J. D. (1995). Whose nature? The contested moral terrain of ancient forests. In W. Cronon (Ed.), *Uncommon ground: Toward reinventing nature* (pp. 269–297). W. W. Norton & Company.

Robbins, W. G. (1985). The social context of forestry: The Pacific Northwest in the twentieth century. *The Western Historical Quarterly*, *16*(4), 413–427.

Ruple, J. C., Pleune, J., & Heiny, E. (2022). Evidence-based recommendations for improving National Environmental Policy Act implementation. *Columbia Journal of Environmental Law*, *47*(S), 273–358.

Skillen, J. R. (2009). *The nation's largest landlord: The Bureau of Land Management in the American West*. University Press of Kansas.

Stephens, S. L., & Ruth, L. W. (2005). Federal forest-fire policy in the United States. *Ecological Applications*, *15*(2), 532–542.

Taylor, D. E. (2011). Introduction: The evolution of environmental justice activism, research, and scholarship. *Environmental Practice*, *13*(4), 280–301.

Trosper, R. L. (2003). Policy transformations in the US forest sector, 1970–2000: Implications for sustainable use and resilience. In F. Berkes, J. Colding, & C. Folke (Eds.), *Navigating social-ecological systems: Building resilience for complexity and change* (pp. 328–351). Cambridge University Press.

Vaughn, J., & Cortner, H. (2005). *George W. Bush's healthy forests: Reframing the environmental debate*. University Press of Colorado.

Wilkinson, C. F. (1997). The National Forest Management Act: The twenty years behind, the twenty years ahead. *University of Colorado Law Review*, 68, 659–682.

Yaffee, S. L. (1994). *The wisdom of the spotted owl: Policy lessons for a new century*. Island Press.

# 4 Forests, Wildlife, and Water

In 1994, the timber sale program on the Wallowa-Whitman National Forest was enjoined (legally prohibited from moving forward) as the result of a lawsuit filed by a coalition of regional environmental advocacy organizations. Federal courts ruled that national forest staff could not plan or implement new timber sales until it completed a consultation process with the National Marine Fisheries Service (also called NOAA Fisheries) regarding the possible impacts of harvest on the Snake River chinook salmon populations listed as threatened under the Endangered Species Act. Northwest salmon and steelhead runs had declined precipitously over the previous century due to a combination of habitat alteration, fishing pressures, and especially the installation of hydropower infrastructure on the Columbia-Snake system. Timber harvesting and associated activities on federal lands in the Snake River watershed held at least the possibility of harming the species through the introduction of sediment, nutrients, and herbicides into the waterway or by removing streamside vegetation and thereby increasing water temperatures. Wallowa County leaders had already established a salmon recovery plan in partnership with the Nez Perce Tribe in 1993 in response to the Endangered Species Act listing.[1] Despite the presence of this (non-regulatory) plan, the Forest Service's consultation requirements under the Endangered Species Act were ruled to be non-negotiable. As detailed in Chapter 3, the sudden reduction in federal timber volume led to the shuttering of two sawmills in Joseph and generated an economic shock to the county's rural, resource-dependent communities.

The links between land use, water quality, and listed salmon runs meant greater scrutiny of forest activities like livestock grazing and herbicide application, implicating the county's other major economic activity: beef cattle production. The number of livestock allowed to graze on federal lands had undergone a slow, steady decline over several decades and some riparian zones on the national forests had been fenced off from livestock access. Public land grazing rights in the Hells Canyon area were revoked from sheep grazers in the interest of recovering native bighorn sheep populations.[2] In 2008, Wallowa County livestock producers confronted an additional threat with the establishment of the first breeding pair of gray wolves in the county since the species' eradication from Oregon in 1947. A previous wolf reintroduction program in neighboring Idaho, initiated by the

DOI: 10.4324/9781003043669-4

US Fish and Wildlife Service and supported by the Nez Perce Tribe, was successful enough that populations eventually expanded into Oregon by crossing the Snake River into Wallowa County. In addition to creating anxiety for livestock producers, the wolves' return generated concern that elk populations would be decimated, reducing the economic activity associated with hunting. Indeed, since the early twentieth century, Wallowa County could count on an economic bump in the fall as hunters poured in from across the region in pursuit of the county's abundant large game. As these examples illustrate, forest policy and governance dynamics are not always focused on trees. In this chapter we examine two issues that are often central to forest governance debates and decisions nationwide: forests as wildlife habitat and the nexus between forests and water.

## Forests and Wildlife

American conservation policy has long been driven by wildlife concerns. Wealthy sport hunters were a key constituency of the early conservation movement and the assembly of much of the federal conservation estate centered wildlife as a principal concern (Chapter 2). The vignette above from Wallowa County illustrates the power of wildlife concerns to drive contemporary forest management and governance changes. Other recent changes have revolved around habitat needs not only for wolves and salmon but also for species such as the grizzly bear, Canada lynx, Florida panther, Kirtland's warbler, and various subspecies of spotted owl, all animals listed as threatened or endangered under the federal Endangered Species Act. Yet wildlife policy and governance in the US reflect much more than this cornerstone piece of federal legislation. Common-law legal concepts from Britain, international treaties, patterns of state-federal cooperation, and constitutional separations of power all figure into the management of forests for wildlife.

Britain's monarchical system traditionally held that wildlife within the kingdom was the property of the Crown unless pursued by a hunter operating under a royal franchise. Licenses were afforded to certain members of the nobility for specific game species in designated areas. In British colonial America, on the other hand, hunting was widely practiced by colonists of all social classes. From a legal perspective, wildlife was generally treated by colonial law as belonging to no one until captured or killed.[3] Following the American Revolution, wildlife in the US came to be treated by the courts under the common-law doctrine of public trust, whereby the government manages wildlife "in trust" to the public, entailing governmental powers as well as obligations to serve public interests. Under this interpretation, states are primarily responsible for wildlife management and serve as the principal trustees in the relationship. Among other implications, the public trust doctrine in wildlife vests state governments with the responsibility for setting and enforcing hunting regulations, including on private lands.[4] The specific interpretation of public trust duties varies from state to state; some states explicitly acknowledge their role as trustee via constitutional or legislative language while others do not.[5] Some state courts have interpreted the trust obligation narrowly and others more expansively; for example, the California Supreme Court held that state agencies violated their public trust responsibilities when they allowed water diversions to lower water levels in Mono Lake to the point that non-game California gulls were exposed to predation.[6]

The federal government claims authority over wildlife in more limited situations: for example, where a species crosses state lines and is therefore part of the "interstate commerce" regulated by federal powers under the Constitution; where international conservation treaties are involved; where tribal rights to wildlife are at stake; or where wildlife is implicated in the management of federally owned land. Added to this list is the federal government's longstanding role protecting and recovering sensitive or vulnerable species. The Lacey Act of 1900 represents the federal government's earliest foray into nationwide wildlife regulation. The Act was passed to prohibit interstate trade in illegally taken wildlife—a problem that states acting alone were unable to solve. This was followed in 1918 by the Migratory Bird Treaty Act, which ratified a treaty with British Canada providing a framework for the conservation and protection of both game and nongame migratory bird species. The federal government protected the Bald Eagle from most hunting as early as 1940, and subsequent laws continued to expand the federal scope of sensitive species protection.

## The Endangered Species Act

The Convention on International Trade in Endangered Species of Wild Fauna and Flora (CITES) was finalized in 1973 via negotiations between 80 sovereign countries. The treaty provides a framework for preventing international trade from exacerbating the decline of dwindling plant and animal species. The Endangered Species Act was passed in that same year as part of the "environmental decade" policy reforms. The 1973 law overhauled its predecessors, the Endangered Species Preservation Act of 1966 and Endangered Species Conservation Act of 1969, and represented a watershed event in federal regulation of both wildlife and plant species at risk of extinction. Although the Endangered Species Act is principally implemented by the US Fish and Wildlife Service (for terrestrial and freshwater species) and NOAA Fisheries[7] (for marine and anadromous species), legislative language made it clear that threatened and endangered species conservation are to be part of the mission of all federal agencies.

The Endangered Species Act requires the US Fish and Wildlife Service and NOAA Fisheries to publish lists of species, or distinct populations of species, that are believed to be in danger of going extinct (endangered) and those that are likely to become endangered in the foreseeable future (threatened). Various legal obligations are triggered once a species is listed; among the most important is a prohibition on "take" of any animal species listed as endangered. The law defines "take" as "to harass, harm, pursue, hunt, shoot, wound, kill, trap, capture, or collect, or to attempt to engage in any such conduct," and courts have interpreted prohibited activities to include actions that harm or kill endangered species through habitat modification.[8] The US Fish and Wildlife Service automatically applies these same prohibitions on take to threatened species until alternative species-specific regulations are created. When a species is listed, it triggers the development of a federal plan detailing actions needed not only to avoid extinction but to recover populations to the point that protections under the Act are no longer necessary. Additionally, the US Fish and Wildlife Service and NOAA Fisheries can—but do not always—designate "critical habitat" for the listed species; this designation can include both public and private lands. Federal agencies are prohibited from destroying or adversely modifying critical habitat on any ownership category. Although critical habitat designation does not in itself place restrictions on what private landowners can do on their own properties, it can make some actions (such as permitting) more complicated due to the additional obligations imposed on federal agencies.

As of early 2022, 730 animals and 939 plants in the US were listed as threatened or endangered under the Act.[9] An estimated 90% of listed species are found partially or entirely on private lands,[10] and the Act operates somewhat differently on private land than it does on federal land. For example, federal agencies are required to consult with the US Fish and Wildlife Service or NOAA Fisheries before implementing activities that could affect listed species, whereas private landowners are not. Prohibitions against the take of listed plants apply on federal land but not on private land. Federal agencies are also required to work toward recovery of listed species, whereas private owners' obligations are largely limited to avoiding take.[11] The Department of Defense is subject to the same fundamental Endangered Species Act obligations as any other federal branch, although there are some special provisions that allow the military mission to take precedence even as recovery efforts advance.[12]

The US Fish and Wildlife Service and NOAA Fisheries can allow both private and public entities to carry out activities that result in take if conducted as part of an approved plan, such as a Habitat Conservation Plan or Safe Harbor Agreement; policy tools have also been developed to promote the conservation and recovery of candidate species—those that are under consideration for listing but not formally listed as threatened or endangered (Box 4.1). Projects that harm listed species can also be authorized by a high-level executive committee, dubbed the "God Squad," if certain criteria are met. This latter provision was added during 1978 amendments to the Act, motivated by a case in which a major federal dam project was enjoined by the courts due to its effects on the endangered snail darter, a three-inch-long fish.[13]

The Endangered Species Act has driven substantial changes to decision-making and management for forests nationwide. It makes its presence felt even in cases where species are not listed under the Act; the mere threat of possible future restrictions may be sufficient to change how forest owners make decisions. Private forest owners often dread the prospect of hosting a listed species due to the limitations on activities (economic or otherwise) they are allowed to pursue within or near the species' habitat. Indeed, fear of the implications of hosting a listed species can motivate landowners to take actions to prevent the creation of suitable habitat in the first place—or even quietly and illegally eliminate listed species upon their arrival. Because these kinds of activities work directly against the intended purpose of the Endangered Species Act, lawmakers and agencies have developed various incentive programs designed to encourage the recovery of dwindling species on private lands. The story of the red-cockaded woodpecker (or RCW as it is commonly known) illustrates the interplay of various regulatory and voluntary policy and governance mechanisms in support of the recovery of a forest-dependent species.

### *Longleaf Pine Forests and the Red-Cockaded Woodpecker*

The RCW (Figure 4.1) was once found in pine-dominated forests from Texas to Maryland and was especially abundant in the longleaf pine forests of the southeastern coastal plain. The species requires relatively large live pine trees that are susceptible to heart rot fungus to excavate its nests;[14] suitable trees are typically at least 60 years old. Trees of this age and condition became increasingly uncommon following the liquidation of southern pine forests in the early twentieth century and the more contemporary adoption of short-rotation forestry in which

*Figure 4.1* The red-cockaded woodpecker, listed as endangered as of the date of this writing, has lost the vast majority of the older pine forest habitat on which it depends. A suite of voluntary incentive programs and federal actions work in concert to help recover the species across its range. Photograph by Mark Pavlosky, Jr.

trees are harvested before they reach the age required by the bird. Researchers estimate that over 97 percent of the longleaf pine forests that once blanketed the coastal plan had been lost by the beginning of the twenty-first century.[15] The disappearance of fire-adapted, savannah-like forests put other species in jeopardy as well, including the gopher tortoise, eastern indigo snake, pine snake, and various plant species. The RCW received federal protection in 1970 under the Endangered Species Conservation Act of 1969 and was one of the first species listed as endangered following passage of the Endangered Species Act in 1973.[16] Given the predominance of private forestland across the South and the economic importance of the timber industry, potential restrictions on the take of RCWs threatened to conflict with the expectations of many forest landowners.

Nearly 50 years after the passage of the Endangered Species Act, recovery of the RCW is proceeding faster than some initial estimates had predicted.[17] This outcome is the result of numerous programs and policy tools applied throughout the South that serve to coordinate efforts across public, family, and corporate forest owners. Federally owned lands serve as the core territories of RCW recovery despite the relatively small federal footprint within the region. Federal recovery hotspots include not only national forests and national wildlife refuges but also various military installations.[18] Pine forests within the boundaries of installations like Fort Stewart and Fort Benning in Georgia, Eglin Air Force Base and Camp Blanding in Florida, and Fort Bragg and Camp Lejeune in North Carolina are typically older than nearby forests due to a lack of recent commercial harvesting, and additional buffer zones have been established around these sites. For example, the Army Compatible Use Buffer program and the Department of Defense's Readiness and Environmental Protection Initiative have been used to fund title purchases and conservation easements on pine forests outside the boundaries of Fort Bragg, Fort Benning, Camp Lejeune, and Fort Stewart. These properties prevent incompatible land uses (like residential or commercial development) from establishing near active training sites and serve to help recover listed species. A complementary program called Sentinel Landscapes coordinates conservation efforts among Department of Defense sites and nearby private lands and federal multiple-use lands. In 2017 the US Forest Service announced the "Million Acre Challenge," committing to putting one million additional national forest acres on a trajectory toward longleaf pine restoration by 2025.[19] Recovery efforts focused on federal properties have been successful enough in bolstering local RCW populations that they are now sources for translocation efforts, repopulating forests where the RCW is absent.

While RCW recovery efforts on federal lands require the creative use of acquisition and management tools both inside and outside of military installations and other federal properties, RCW recovery on the abundant privately owned forestlands within its historic range comes with a very different set of challenges. Pine forests managed for timber production are typically harvested well before trees reach a suitable age and condition for RCW nesting, and the bird's favored longleaf pine is generally less popular among timber producers than faster-growing species like slash and loblolly pine. Further, many landowners have strong incentives to avoid attracting RCWs given the restrictions on activities that could result in a potential take of the species. At least some southern landowners who had unwittingly created RCW habitat (by managing for longer rotations and applying prescribed fire) found themselves restricted from harvesting or implementing other ground-disturbing activities in areas surrounding nest sites.[20] As RCW-related restrictions started to take hold in the region, some landowners reportedly cleared their older pine forests out of fear of losing economic benefits and decision-making autonomy.[21]

Given these "perverse incentives" to avoid creating RCW habitat—and the predominance of private land within the bird's range—the US Fish and Wildlife Service and NGO partners have developed numerous incentive-based and cooperative programs to encourage recovery in private forests. A prime example is the Safe Harbor Agreement that was originally established in the North Carolina sandhills and later expanded to other states. Under this initiative, landowners are obligated only to protect the baseline number of RCWs on their property at the beginning of the agreement and are provided financial and technical support for activities such as prescribed fire application and cavity drilling that enhance RCW habitat. Each participating landowner receives an Enhancement of Survival Permit, allowing them to return to their baseline RCW population at any point without being legally liable for a violation of take prohibitions under the Endangered Species Act. State wildlife agencies are the primary entity responsible for administering the Safe Harbor Agreements within their jurisdictions.[22] The first fifteen years of the program saw the establishment of 265 RCW Safe Harbor Agreements covering over 640,000 acres.[23]

Other cooperative and incentive-based tools and initiatives supporting RCW recovery include Habitat Conservation Plans (HCPs) and cost-share and payment programs administered by various US Department of Agriculture agencies. An HCP provides an Incidental Take Permit (allowing legal take of listed species in the implementation of approved activities) in return for a commitment to offset or mitigate harm to the species' recovery. By detailing what activities within listed species habitat are allowed and how impacts will be minimized and compensated for, an HCP provides a framework for normal economic activities to proceed while continuing to support overall recovery. Statewide HCPs for the RCW were established as early as 1993, prior to the advent of the Safe Harbor Agreement option.[24] Payment and cost-share programs include programs funded under the Farm Bill such as the Environmental Quality Incentives Program administered by the Natural Resources Conservation Service and the Conservation Reserve Program administered by the Farm Service Agency, both of which can support longleaf pine establishment and maintenance (see Chapter 8). The Partners for Fish and Wildlife program and Working Lands for Wildlife program are additional cooperative, cost-share programs that have been used to support RCW recovery (Box 4.1).[25]

The success of RCW recovery to date can be credited to various factors, including the above-mentioned programs and the creative use of artificial structures (such as nest boxes and drilled cavity holes) across the bird's range. Another key element of success has been a shift over time from managing threats to the species' survival to engaging in landscape-scale ecosystem restoration.[26] This shift has been accomplished in large part through networks of private, public, and NGO partners working toward common goals over large areas. A central node in these networks is the Longleaf Alliance, a regional NGO established in 1995 at Auburn University's Solon Dixon Forestry Education Center in Andalusia, Alabama, with a focus on providing resources, information, and advice for longleaf pine restoration on private lands.[27] Ten years after its founding, the Longleaf Alliance formed a working group with numerous state, federal, private, NGO, and research entities that grew into the Longleaf Partnership Council; the Council's restoration activities are coordinated under the name America's Longleaf Restoration Initiative (ALRI).[28] ALRI receives financial support through both public and private sources and funds projects that are put into action through eighteen Local Implementation Teams in specific landscapes. Although some of ALRI's activities (like planting longleaf pine seedlings) will take decades to provide benefits to RCW populations, in other cases the benefits are more immediate—for example by implementing prescribed fire, thinning densely planted stands, and removing understory hardwood trees from longleaf pine forests. The voluntary and relationship-based approach embraced by the Longleaf Alliance and ALRI has succeeded in large part due to the trust that it has generated among private, public, and NGO partners and especially among family forest owners.[29]

### BOX 4.1 COOPERATIVE AND INCENTIVE-BASED TOOLS FOR THE RECOVERY OF RARE, THREATENED, AND ENDANGERED WILDLIFE SPECIES

*Partners for Fish and Wildlife* is a program administered by the US Fish and Wildlife Service focused on the recovery of Endangered Species Act-listed or at-risk species and migratory birds. Landowners enter into ten- to thirty-year commitments under which the program pays for specified restoration activities.

The *Working Lands for Wildlife* program, jointly administered by the Natural Resources Conservation Service and the US Fish and Wildlife Service, targets lands used for forestry, livestock, or agriculture production. It provides technical and financial assistance and regulatory assurances for the conservation or restoration of habitat for target species. The regulatory assurances come by way of tools listed below, such as Candidate Conservation Agreements with Assurances and Safe Harbor Agreements.

*Candidate Conservation Agreements* are formal agreements established between the agencies responsible for implementing the Endangered Species Act (the US Fish and Wildlife Service and NOAA Fisheries) and other federal, state, local, tribal, or non-governmental entities. These agreements specify actionable commitments to conserve candidate species (those that are not currently listed under the Endangered Species Act but could become listed pending agency review). Although the agreements do not come with specific permits—in contrast with Candidate Conservation Agreements with Assurances, described below—they may help to prevent a species from being listed.

*Candidate Conservation Agreements with Assurances* also pertain to candidate species. Nonfederal landowners entering into these agreements voluntarily commit to certain conservation actions that benefit the candidate species; in return, they receive an Enhancement of Survival Permit, which allows for incidental take and ensures that the landowner is not subject to additional conservation obligations if the species is eventually listed.

The *Safe Harbor Program* allows nonfederal owners of the habitat of a listed species to enter into agreements with the US Fish and Wildlife Service or NOAA Fisheries. By committing to actions that improve prospects for the species' recovery, the participating owner is assured that their conservation obligations will not increase along with expanding populations of the species. As part of the agreement, landowners receive an Enhancement of Survival Permit, which allows for the incidental take of any additional numbers of listed species above the baseline at the beginning of the agreement. Programmatic Safe Harbor Agreements allow multiple landowners within a state or region to participate in an umbrella agreement managed by a state or local agency.

*Habitat Conservation Plans* are binding agreements between nonfederal entities and either the US Fish and Wildlife Service or NOAA Fisheries that allow for economic activities to occur that could result in the unintended take of species listed under the Endangered Species Act. In exchange for agreeing to certain conservation measures and assessing the levels and impacts of anticipated incidental take, the nonfederal partner receives an Incidental Take Permit, allowing for unintentional take to occur under the conditions of the plan.

*Conservation Banking* is a market-like tool allowing "buyers" (entities whose actions will negatively affect Endangered Species Act-listed species) to purchase credits from "sellers" (owners of protected habitat for that species) to offset their impacts. To be eligible to sell credits, the participating seller must own nonfederal land that is permanently protected and managed for certain wildlife benefits. The US Fish and Wildlife Service and NOAA Fisheries determine the calculation of credits and debts associated with various activities.

### *Beyond the Red-Cockaded Woodpecker*

The story of RCW recovery and longleaf pine restoration is a complex one, entailing creative efforts on and around military installations and other federal lands combined with innovative tools and practices implemented across a patchwork of private, state, and county forestlands where ownership is dominated by thousands of individuals, companies, and agencies. Although the trajectory of RCW recovery has been positive amid broad support from various participants, not all forest-dependent wildlife species have benefited from the RCW's level of landowner and community support. For example, wolf reintroduction and population augmentation efforts have been underway for decades in various parts of the country, including gray wolf reintroduction in the northern Rocky Mountains and Southwest and red wolf recovery efforts in eastern North Carolina. Although these projects have managed to increase wolf populations in places where they were fully or nearly extirpated from their native ranges, they have not always been welcomed by local populations or elected officials, as the vignette from Wallowa County at the start of this chapter illustrates. After the success of some of these wolf recovery programs led the federal government to turn over wolf management to the states, elected officials in places like Idaho, Montana, Wyoming, and Wisconsin made it clear that their priority was to reduce wolf population numbers to the lowest levels possible.

In the Pacific Northwest, the northern spotted owl was the feathered catalyst for a complete overhaul of national forest and Bureau of Land Management timber practices as detailed in Chapter 3. Northern spotted owl populations continue to decline, however, despite the application of the conservation biology-focused Northwest Forest Plan to tens of millions of forested acres and the end to a decades-long program of old-growth forest conversion. In part this is the result of competition from the barred owl, a native of eastern forests that began to invade northwestern forests around the time of the timber wars, and in part it stems from the loss of complex forest habitat to wildfire, particularly on the eastern and southern fringes of the Northwest Forest Plan area. Twentieth-century fire suppression in places like the eastern slope of the Cascade Mountains and in the Klamath-Siskiyou region of southern Oregon and Northern California allowed for the development of forest stand structures to the northern spotted owl's liking. Wildfires in these same habitats have left many owls homeless[30] and recent fires have also consumed northern spotted owl habitat in the moist western Cascade forests that typically see fire far less frequently than the eastern slope and Klamath-Siskiyou. The Northwest Forest Plan emphasized protection of owl populations from timber harvesting but did little to protect them from fires burning under novel climatic and forest stand conditions.

Similar stories could be told about other wildlife species caught between the legacies of past land uses and emerging dynamics of global environmental change. For example, Canada lynx in southwestern Colorado and the Mt. Graham red squirrel in southern Arizona both saw their mountaintop Engelmann spruce forests devastated by insect outbreaks. The native spruce beetle was the major cause of mortality in both cases, although on Mt. Graham trees were also weakened by the introduced spruce aphid. Both locations had previously also seen substantial infrastructure development, including a major ski resort in Colorado and a world-class astronomical observatory on Mt. Graham. Among many factors stressing grizzly bear populations in the northern Rocky Mountains is the decline in whitebark pine, a major source of nutritious and oily seeds. Mountaintop whitebark pine populations have suffered due to complex, interacting factors that include a lack of fire, infestations by native mountain pine beetles active at higher elevations than their historic norm, and the spread of the invasive white pine blister rust. Various bat species, including three listed under the Endangered Species Act, have suffered sharp population declines due to white-nose syndrome, a deadly pathogen introduced from Europe. Decisions regarding management actions such as timber harvesting, prescribed fire application, and wildfire suppression within these bats' ranges must now consider both short-term and longer-term population impacts. For example, prescribed fire in the Appalachians may have short-term impacts on the endangered Indiana bat and the threatened northern long-eared bat; yet the absence of fire could degrade habitat for both species over time.[31]

Perhaps no wildlife management scenario is more complicated than that of anadromous salmonids: the steelhead and salmon species that are born in freshwater systems and live most of their lives in the ocean before returning to their natal waters to spawn. The management and recovery of these migratory species implicates dozens of international, federal, state, county, municipal, and tribal agencies and associated treaties and policies.[32] Relevant formal policies include not only the Endangered Species Act and Clean Water Act but also those such as the Clean Air Act, recreational and commercial fishing regulations, federal land grazing policy, federal and state energy policies, state forest practice acts, state water laws, local zoning and planning ordinances, municipal drinking water policies, tribal reserved rights, and many more. The complex causes of anadromous salmonids' more than century-long decline are often categorized in terms of the "four Hs": harvest, hydropower, hatcheries, and habitat. Of the four, only habitat directly implicates forests and their management. Forests slow and filter precipitation runoff, providing clean, clear water; streamside forests provide vital shade that maintains the low water temperatures needed by recently hatched and returning salmonids; trunks and large branches that fall into streams offer protection for juvenile salmon; infrequent landslides introduce the gravel and woody material needed for complex instream habitat. Conversely, ground-disturbing activities and forest roads can serve as sources of sediment into streams, smothering redds (salmonids' instream "nests" where fertilized eggs develop) and degrading habitat for juvenile salmonids. The removal of streamside forests can raise water temperatures and allow pesticides and other pollutants to enter waterways unfiltered. Culverts on forest roads can block salmon migration, cutting off access to spawning habitat. The case of salmonid dependence on forests illustrates the complexities of the topic to which we turn next—the inextricable connection between forests and water.

## The Water–Forest Connection

Water policy and governance in the United States involve many of the actors and interests profiled throughout this text: states, counties, municipalities, private firms, landowners, NGOs, tribes, and the federal government. What makes the water dimension of forest governance unique is the nature of water itself: unlike land, which (mostly) tends to stay in place, or trees that take root, grow, and die over decades or centuries, water is in constant motion. As it changes form and travels across landscapes, its boundaries are often impossible to demarcate; this creates challenges in applying the same kinds of property rights and responsibilities to water that are regularly attached to land or to other, less mobile, commodities. It also creates challenges in determining where the jurisdiction of one governmental body begins and the jurisdiction of another ends; this problem has vexed environmental managers, judges, and property owners since passage of the contemporary Clean Water Act in 1972.

Scholars going back at least as far as George Perkins Marsh recognized the connections between forest degradation and water degradation, and forest policy and governance processes in the US have long placed water concerns front and center. For example, many of the early twentieth century community-owned forests in New England were created with the specific objective of protecting town water supplies, and watershed protection was a primary goal animating New York's designation of Adirondack Park and Pennsylvania's establishment of over two million acres as state forests.[33] The Organic Act of 1897 included "securing favorable conditions of water flows" as one of three criteria for the establishment of federal forest reserves (later national forests), and many reserves were designated primarily with water protection in mind. The 1911 Weeks Act expanding national forests to the eastern states was justified as a measure to protect important watersheds. Forests of all ownerships play a critical role in municipal water supplies; it is estimated that 53% of the water supply nationwide—and 65% in the eleven western states—originates on forestlands.[34]

The relationship between forests and water is complex: forests can serve to increase water infiltration to the soil even as they reduce water yield due to increased evapotranspiration.[35] These relationships vary depending on multiple factors: the kind of forest, the age and size of trees, soil characteristics, and, especially, climatic conditions. From a water quality perspective, forests and forest soils serve to filter sediments, nutrients, and toxins.[36] Protecting these functions is the basis for forestry best management practices, described in Chapter 6. Forests also perform other vital watershed "services" such as lowering water temperature by shading streams, promoting groundwater recharge, moderating snowmelt, and contributing woody material to waterways to create fish, reptile, amphibian, and invertebrate habitat. When forest cover is lost as part of conversion to other land uses, many of the hydrologic benefits of forests are diminished—particularly if forests are replaced by impermeable surfaces like roads, parking lots, or residential structures.

The forest-water relationship becomes even more complicated when fire is introduced to the equation. Fire in forests can improve or degrade water quality and quantity depending on the specifics of the forest system and fire event.[37] Important variables include the severity of the fire, the affected soil type, slope, climatic conditions, and the timing and magnitude of post-fire precipitation events. Low-severity fires, including most prescribed fires, typically do little to alter forest hydrology.[38] On the other hand, a particularly dangerous combination is that of high severity fire and heavy post-fire precipitation—a scenario typical of many large wildfire events in the western states since the 1990s.[39] The 2010 Schultz Fire on the Coconino National Forest in Arizona, for example, was followed by a series of heavy rainfall events that sent water rushing into steep drainages and on into the city of Flagstaff, where flooding damaged or destroyed 85 homes, severed water supply pipelines, and took the life of a 12-year-old girl who was swept away by the torrent.[40] The total economic damage from the Schultz Fire and its aftermath was estimated to be at least $95 million, the vast majority from flood damage and flood mitigation costs rather than costs directly related to the fire event itself.[41] Similar stories could be told across the West where mountain terrain, high-severity wildfire, and heavy rainfall events have coincided.

## Policy and Governance Influences on the Forest-Water Nexus

The Clean Water Act is the nation's principal federal legislation for governing water quality nationwide; it is complemented by the Safe Drinking Water Act, which applies to water used for human consumption. Although the federal government regulated the dumping of refuse into navigable waterways as early as 1899 with the Rivers and Harbors Act, it was a series of amendments to the Federal Water Pollution Control Act in 1972 that produced the framework for the modern Clean Water Act. The 1972 legislation responded to a growing crisis triggered by use of the nation's waterways as dumping grounds for an increasingly toxic blend of industrial and municipal wastes. The Act included the goals of eliminating the discharge of pollutants into US waters by 1985 and making the nation's waters fishable and swimmable by 1983. These ambitious goals were not achieved, but the Clean Water Act has persisted through various subsequent amendments and is credited with substantially improving water quality nationwide.[42]

The Act underwent substantial revisions in 1977, 1981, and 1987. It remains a complex law with separate provisions targeting pollution discharges into waterways, alteration of wetlands, and the protection and restoration of water bodies according to their designated uses. The Act distinguishes between point sources—discrete conveyances of pollutions such as pipes, ditches, and culverts—and nonpoint sources such as farms, roadways, and forests that contribute pollutants to waterways via processes of infiltration and runoff across the landscape as a whole. Whereas discharge permits from the Environmental Protection Agency (EPA) are required for point sources, nonpoint sources are generally managed via programs led by states, tribes, territories, and local governments (see Chapter 6). Federal permits are also required under the Act for actions that add dredged or fill material to waterways, including wetlands; these "dredge-and-fill" permits (or "Section 404" permits) are administered by the US Army Corps of Engineers with oversight from the EPA. In all cases, the federal scope of the Clean Water Act is limited to waters defined as navigable and therefore subject to federal powers via the Commerce Clause of the US Constitution. However, defining which waters fall inside or outside of federal jurisdiction under the Act has long been unclear and remains contested as of the date of this writing (Box 4.2).

The influence of the Clean Water Act is felt on forestlands through various policy mechanisms. First, certain forestry activities related to road construction, log sorting, and site preparation can potentially trigger federal permit requirements, either because they are classified as point sources or because they result in the introduction of fill material into designated wetlands. Normal, ongoing silvicultural activities are generally exempt from wetland dredge-and-fill permits, subject to some limitations. Second, under the cooperative federalism of the Clean Water Act, states, tribes, and territories are empowered to create their own programs for reducing nonpoint source pollution. Sets of voluntary or mandatory rules known as forestry best management practices (or BMPs) are designed to minimize the role of forestry activities in contributing sediment, pesticides, and other pollutants into waterways. These rules are voluntary in many states, but there are often strong incentives for forest operators to comply (see Chapter 6). The provisions of the Clean Water Act are overlain on state-level laws and institutions governing the allocation of water for use in irrigation, industrial, municipal, or other uses.

The wetland dredge-and-fill component of the Clean Water Act makes it the primary federal law regulating the conversion of wetlands, including forested wetlands. The goal of "no net loss" of wetlands was formally adopted by President George H. W. Bush in 1989. Under this principle, dredge-and-fill permits are provided only under the condition that the permittee commit to a mitigation plan for wetland conservation. Permittees can meet this obligation through various means: directly creating, restoring, protecting, or enhancing wetlands; purchasing mitigation "credits" representing wetlands that have been restored or protected in offsite mitigation banks; or paying fees to a public or nonprofit organization to finance wetland compensation.[43] Although the avoidance of wetland loss is the stated first priority of the regulatory agencies (the EPA and US Army Corps of Engineers), the mitigation banking option has created the possibility for hundreds of for-profit ventures to sell credits representing wetlands conserved on private lands.[44]

## BOX 4.2 JUDICIAL AND REGULATORY INTERPRETATIONS OF THE CLEAN WATER ACT

As one of the federal environmental laws with the greatest impact on private landowners, the Clean Water Act has been the subject of extensive legal and political scrutiny—as well as important changes in its application—since its passage. The most vexing problem to resolve is defining exactly what waters are subject to the provisions of the Act. This is, perhaps, an inevitable outcome of any attempt to draw clear lines around bodies of water. The US Constitution designates the federal government as the entity responsible for regulating interstate commerce, a role that has long included regulating navigable waters used for transportation. The Clean Water Act itself unhelpfully defines navigable waters as "waters of the United States, including the territorial seas." Various US Supreme Court cases have done little to clarify exactly what waters are and are not subject to federal jurisdiction under the Act. A 1985 decision (*United States v. Riverside Bayview Homes*) held that wetlands adjacent to actually navigable waters do indeed fall under the scope of the Act, but a 2001 decision (*Solid Waste Agency of Northern Cook County v. US Army Corps of Engineers*) held that isolated wetlands do not—even if they are used by migratory birds that serve as objects of interstate commerce. A 2006 case (*Rapanos v. United States*) split the Supreme Court 4–1–4, with conservative justices ruling that the Act applies only to "relatively permanent, standing or flowing bodies of water" while the lone Justice Kennedy opined that even intermittent wetlands could be included if they "significantly affect the chemical, physical, and biological integrity of other covered waters." A 2015 Obama Administration definition of the Clean Water Act's scope, dubbed the Clean Water Rule, was mired in conflicting legal challenges until it was replaced by a narrower definition, the Trump Administration's 2019 Navigable Waters Rule. As of the time of this writing, the Biden Administration has promised to rewrite the definition of waters subject to federal Clean Water Act protections even as the Supreme Court has signaled that it is likely to narrow the definition of navigable waters subject to federal regulation.

### *State-Level Water Policy*

States, tribes, and territories play varied and important roles under the Clean Water Act beyond establishing and overseeing BMPs. For example, they define designated uses for each water body (such as for recreation, public drinking water supplies, fish and wildlife, or agricultural and industrial uses), and these uses affect how designated waters are managed under the Act. Under 1987 amendments to the Act, states, tribes, and territories are required to identify water bodies that do not meet water quality standards and to develop plans for recovering them. Every two years, they submit to the EPA a comprehensive report on the quality of all water within the state or relevant jurisdiction; this report includes a list of impaired water bodies, sometimes called 303(d)-listed waters after the relevant section of the Clean Water Act. For each impaired water body, the state, tribe, or territory details the source of degradation and is responsible for developing a recovery plan. A key element in recovery plans is the designation of a total maximum daily load (TMDL) for each pollutant; a TMDL represents the maximum amount of that pollutant that can enter the waterway from both point and nonpoint sources without impairing the designated use. TMDLs inform the allocation of point-source permits under the Clean Water Act and are also used to design nonpoint source pollution control programs. Lists of impaired water bodies and associated TMDLs and recovery plans are subject to federal oversight, and federal matching funds are available for implementing approved plans. Most states administer point source permits under the Act, and there are also mechanisms under which states can take over dredge-and-fill permitting responsibilities from the federal government. As of 2022, Michigan, New Jersey, and Florida had completed the required process to take over this permitting authority from the US Army Corps of Engineers.

States regulate water use through various mechanisms beyond the Clean Water Act. A key distinction in US water law is between states that historically employed the prior appropriations doctrine and those that use riparian water law.[45] Riparian water law is common in places that are relatively humid and where demand for water as an irrigation source is traditionally low. In these states, rights to withdraw and use water arise as a consequence of owning property that borders a water source. Rather than specifying a quantity of water the landowner is allowed to withdraw, riparian water law traditionally allows for "reasonable use" that does not interfere with the reasonable uses of other rights-holders. However, many riparian law states have instituted permit systems in light of increasing demand and now represent what are known as "regulated riparian" systems. Prior appropriations water law, conversely, prevails in places where water demand has historically outstripped supply—principally in the arid regions of the Intermountain West. Here, rights to use specified quantities of water are allocated on the principle of "first in time, first in right," meaning that the user with the most senior water right is given their full allocation of water before the user with the second-most senior right is allowed to use any of their own right, and so on down the line of seniority. Continuous and beneficial use of water is required to maintain water rights under this doctrine, but ownership of land abutting the water source is not.

Irrigation, navigation, industry, and mining demands drove the development of much of the nation's water law, and the rules of access to and use of water have been modified over time as a more diverse set of values developed regarding waterways. These changes have not always come easily, particularly given the deference to senior water rights (almost always used for consumptive purposes) in prior appropriations systems. However, courts have ruled that the most senior water rights holders in many western watersheds are the federal government and Native American tribes. Tribal water rights in the West arise from two separate principles: first, that water rights attach to tribal communities, including reservations and other communities sanctioned by the federal government, and second, that water rights are necessary to fulfill longstanding practices predating Euro-American colonization. Such practices include the retained rights (see Chapter 7) associated with activities like fishing in the "usual and accustomed places."[46] The federal government's senior water rights come from its long-term presence as a landowner and the implied rights of water use associated with its extensive land holdings in the West. Combined with various state laws that recognize (and provide mechanisms to protect) instream uses as legitimate and beneficial water rights, the establishment of federal and tribal seniority has allowed for the legal protection of some non-consumptive uses in the West.

The federal Wild and Scenic Rivers Act of 1968 is another policy tool oriented toward the protection of instream water flows. Although passed just four years after the Wilderness Act and motivated by similar concerns regarding the need to protect undeveloped natural systems, the Act's very different policy mechanisms reflect the complicated legal status of water. Under this law, rivers with "outstandingly remarkable values" can be designated for protection in their free-flowing state. The three categories of protected status include "wild," "scenic," and "recreational," in ascending order of road and infrastructure development. Designation comes either through an act of Congress or through a separate process involving a petition to the Secretary of the Interior from the governor of the state in which the river is located. Designated river segments are protected from federally funded or permitted hydropower projects, yet designation by itself does not affect the distribution of water rights or impose any restrictions on the use of private lands bordering the designated river. The Act authorizes the public acquisition of land along designated river corridors and encourages—but does not mandate—local governments to use zoning and land use planning processes to support the outstandingly remarkable values for designated rivers.[47]

### *Watershed Conservation and Restoration*

The spatial arrangement of forests and of forestland ownership is another important consideration in understanding water policy and governance. National forests contain headwater streams in many watersheds, making them especially important elements of the landscape for protecting water quality and quantity.[48] Numerous municipal watersheds are owned at least partially by the federal government as national forests, national parks, Bureau of Land Management lands, or under other designations. The national forests alone contain "over 400,000 miles of streams and over 3.5 million acres of lakes and wetlands…and furnish drinking water to about 60 million Americans living in about 3,400 communities."[49] State forestlands are also frequently represented in the ownership matrix of municipal watersheds. Where public lands comprise major components of municipal watersheds, protecting water quality is often an overarching objective; in some cases, this means closing the watershed to most or all public access and restricting certain land uses to protect water quality and reduce the chances of human-caused forest disturbances. In fire-prone regions, protecting watersheds from high-severity wildfire events may take precedence on public lands. Yet achieving this objective can be limited by the same factors that frustrate public land management more broadly: inadequate planning and implementation capacity within land management agencies and persistent conflict among sectors of the public regarding appropriate management direction.

Governing forests for water quality in places where private forestlands predominate implies a different set of complicating factors. Many watersheds in areas dominated by private lands have highly fragmented ownership patterns, with numerous smaller parcels belonging to numerous individual property owners. This makes watershed protection difficult given the high costs involved in communicating and coordinating among owners. And because private property rights tend to be given deference, it can be difficult to control the flow of sediment, nutrients, and residential or agricultural chemicals into the watershed from these myriad private parcels. BMPs for forestry, agriculture, construction, and other activities play an important role in maintaining water quality. However, in places characterized by widespread forest conversion to housing, industrial uses, or agriculture, these rules and guidelines may only serve to blunt the worst of the watershed impacts.

Recent decades have seen the development of governance innovations meant to improve the management of both municipal watersheds and watersheds for which non-consumptive uses predominate. One innovation that has taken hold across much of the country takes advantage of what has been called "natural infrastructure" for source water protection. The basic principle is that targeted investments in conservation across a municipal watershed can yield water quantity and quality benefits, often (but not always) at a lower cost than treating degraded water through built infrastructure.[50] One of the earliest and best-known examples of this approach within the US was piloted in New York City's watershed in the Catskill Mountains, where approximately $1.5 billion in payments to private landowners to support forest conservation, improved livestock management, and installation of modern septic tanks successfully preempted the need to spend $8 to $10 billion on a new water filtration plant. Subsequent projects in other watersheds have used some of these same measures in concert with payments for riparian forest restoration, road decommissioning, and improved forest management as tools for improving water quality and reducing or eliminating the need for certain kinds of water treatment facilities. Incremental changes such as these, implemented over the scale of an entire watershed, hold the potential to reduce sedimentation, water temperature, and inputs of undesirable bacteria and other pollutants to the point that less intensive treatment is required to make the water safe for human consumption.[51] Investing in natural infrastructure typically delivers additional benefits beyond water treatment cost savings, such as improving wildlife habitat, conserving open space, generating ecotourism opportunities, and improving watershed resilience in the face of climate variability—including resilience to extreme precipitation events and droughts.[52] Four examples of source water protection programs focused on forestland are profiled in Box 4.3.

Source water protection was recognized as an optional tool in the 1996 revisions to the Safe Drinking Water Act, and states are eligible to fund these activities through various federal programs including the Drinking Water State Revolving Fund and Clean Water State Revolving Fund, both administered by the EPA. Another mechanism for funding source water protection is related to the states' and EPA's regulation of point-source pollution under the Clean

Water Act. Some states have created programs whereby the pollution reduction required to bring a point source into compliance is achieved through improvements elsewhere in the watershed rather than (or in addition to) improvements at the point-source facility itself. Known as "water quality trading," this mechanism is designed to achieve watershed benefits at a lower cost through efficiencies in reducing total pollutant inputs into the watershed. For example, the city of Medford's wastewater treatment plant in southern Oregon discharges water into the Rogue River in exceedance of temperature limits under the Clean Water Act. Rather than installing mechanical cooling technology, the city has financed reductions in water temperature by purchasing credits from the NGO The Freshwater Trust associated with riparian reforestation activities on private lands within the Rogue watershed. Other commonly employed funding mechanisms for natural infrastructure come from user fees (paid via drinking water or stormwater bills), municipal bonds, general state or municipal budgets, and many of the private land conservation programs profiled in Chapter 8.

## BOX 4.3 EXAMPLES OF SOURCE WATER PROTECTION PROGRAMS IN ACTION

*From Forests to Faucets, Colorado*: In the wake of the 1996 Buffalo Creek Fire and the 2002 Hayman Fire, both of which triggered massive post-fire sedimentation events, Denver Water signed a Memorandum of Understanding with the US Forest Service to improve forest conditions within municipal watersheds. Under the partnership, Denver Water committed to matching the Forest Service's $16.5 million investment in forest restoration and watershed improvement activities over five years, all of which took place on national forestlands in the watersheds of the Upper South Platte, South Platte, St. Vrain, Colorado, and Blue Rivers. The investment was funded through a small fee increase for water users that amounted to less than a 1% change to their monthly bill. Additional partners include the Colorado State Forest Service and the Natural Resources Conservation Service, which have helped to expand the scope of the initiative to include state and private lands.

*Rio Grande Water Fund, New Mexico*: Paralleling the situation in Colorado, New Mexico forests have experienced multiple uncharacteristically severe wildfire events that compromised water quality for downstream users. The Rio Grande Water Fund represents a network of over 90 private, public, and non-profit organizations working toward the common goal of restoring forests and reducing the risk of high-severity wildfire across varied ownerships in the Upper Rio Grande Basin. The objective is to restore 600,000 acres over a 20-year period. The Fund began with private donations and leveraged these with state and federal contributions for a combined investment of over $50 million as of 2019. An executive committee with broad representation works to identify treatment priorities and activities, and the network is facilitated by the NGO The Nature Conservancy.

*Upper Neuse Clean Water Initiative, North Carolina*: Water from the Upper Neuse Basin supplies the fast-growing "Research Triangle" of Raleigh, Durham, and Chapel Hill; the 770 square mile watershed is largely privately owned. Falls Lake, a key drinking water reservoir, was listed as an impaired waterway under section 303(d) of the Clean Water Act due to elevated levels of chlorophyll-*a*, indicating an excess of algae due to high nutrient loads. Beginning in 2005, the City of Raleigh and a coalition of land trusts began identifying priority land parcels for conservation. Conservation easements, land acquisitions, best management practice implementation, reforestation, and landowner outreach and education efforts are some of the key tools funded via utility hookup fees and small increases to water bills. The City of Durham also increased water rates to fund land acquisition near its drinking water reservoirs. To date, the Initiative has protected over 7,500 acres and established over 80 miles of stream buffers.

*Savannah River Clean Water Fund, Georgia/South Carolina*: This bistate initiative has a goal of conserving 60% of forest cover in the watershed of the lower Savannah River, a supplier of municipal water to over half a million residential customers. In 2014 the Fund was developed as an independent non-profit organization by The Nature Conservancy and has since established partnerships with utilities, landowners, and state and federal forest managers. The Fund focuses on conservation easements and best management practice adoption to promote forest conservation, with a goal of protecting at least 8,000 acres per year. The Longleaf Alliance, an established nonprofit organization with a long record of working with private landowners in the South, plays a lead role in landowner outreach for the Fund.

## Conclusion

Water runs clear and cold in the rivers and streams of Wallowa County, providing what should be an ideal place for salmon even alongside competing uses for irrigation and municipal consumption. Yet Wallowa County is connected to a much larger system: it ultimately drains into the Snake River, which itself later joins the mighty Columbia River. A salmon fry born in the crystalline waters of the Wallowa Mountains will have to navigate eight major federal dams before making it to the Pacific Ocean—and then pass these same eight dams again on its return journey to spawn. In between, it will have to avoid being eaten by predators, caught in fishing nets, or succumbing to a variety of other hazards. The listing of Snake River chinook under the Endangered Species Act set in motion a series of changes that upended Wallowa County's forest management and associated economic activities, but the ultimate fate of this far-ranging species will depend on decisions well outside of the county's control. The species has suffered from transformations throughout a massive river basin, and only a basin-wide approach holds any chance of achieving a successful recovery.

Both water and wildlife are complex objects for governance because both are prone to crossing lines of ownership and jurisdiction. The governance of wildlife and water entails complex and varied interactions among policies, public entities, user groups, and property owners, complicating simplistic notions of property and boundaries. The "ownership" of wildlife and water is always contextual and often contested; states typically play a leading role in managing them, but are under obligations to conserve them for the public rather than in a position of possessory ownership. The federal government overrides these state responsibilities in special situations, and tribes hold rights that cut across federal and state authorities. Recent experiences provide ample evidence of water quality and individual wildlife species suffering not only from local land use decisions but from global environmental change dynamics, the topic to which we turn in Chapter 5.

## Questions for Discussion

1. Why did the British colonies in America adopt wildlife laws that were so different from the rules prevailing in England at the time? How might that colonial experience have influenced the later development of US wildlife law?
2. Why did the federal government first become involved in regulating wildlife given that states are the primary governments charged with wildlife responsibilities under the public trust doctrine? Why do you think the federal role in wildlife conservation expanded over the course of the twentieth century?
3. Recovery efforts for the red-cockaded woodpecker appear to be succeeding due to a combination of federal commitments and voluntary incentive programs for private landowners. Could those same ingredients work for other species like spotted owls, wolves, and salmon, or are there reasons why recovering those species is more complicated?
4. Why do you think the Clean Water Act was designed to retain federal control over point source and dredge-and-fill permitting while devolving more authority to states, tribes, and territories regarding nonpoint source pollution control?
5. What are some of the advantages and potential drawbacks of market-based wildlife and water conservation programs such as wetland mitigation banking, water quality trading, and conservation banking?

## Notes

1 Waage, 2001.
2 The grazing allotments were closed to prevent transmission of disease from domestic to wild sheep.
3 Blumm & Ritchie, 2005.

4 Freyfogle et al., 2019, pp. 59–60.
5 Blumm & Paulsen, 2013.
6 Blumm & Ritchie, 2005.
7 NOAA stands for the National Oceanic and Atmospheric Administration.
8 *Babbitt v. Sweet Home Chapter of Communities for a Greater Oregon*, 515 US 687, 115 S. Ct. 2407 (1995); *Defenders of Wildlife v. Bernal*, 204 F.3d 920 (9th Cir. 1999); *Arizona Cattle Growers' Association v. Fish and Wildlife Service*, 273 F.3d 1229 (9th Cir. 2001).
9 This includes species as well as distinct population segments.
10 United States General Accounting Office, 1994.
11 Evansen et al., 2021.
12 Under the Endangered Species Act, the Secretary of Defense is authorized to exempt any action by the implementing agencies if considered necessary for national security. Additionally, Department of Defense lands are exempt from critical habitat designations if the applicable integrated natural resources management plan required by the Sikes Act already contains conservation measures for that species.
13 des Rosiers, 1991.
14 Jusino et al., 2015.
15 Jose et al., 2006.
16 Weiss et al., 2019.
17 Stephens et al., 2019.
18 Hoctor et al., 2006.
19 Riley et al., 2019.
20 Alavalapati et al., 2006.
21 Costa & Gooch, 2011.
22 Costa & Gooch, 2011; Zhang & Mehmood, 2002.
23 Costa & Gooch, 2011.
24 Costa & Gooch, 2011.
25 Riley et al., 2019.
26 Weiss et al., 2019.
27 Johnson & Gjerstad, 1998.
28 Riley et al., 2019.
29 Riley et al., 2019.
30 Stephens et al., 2019.
31 Loeb & O'Keefe, 2014.
32 Wilkinson & Conner, 1983.
33 Clayton, 2019; Hays, 2007, pp. 119–121; Lyman et al., 2013, p. 91.
34 Brown et al., 2008.
35 Keleş, 2019; Pramova et al., 2012.
36 Gartner, McGray, et al., 2014; Stuart & Edwards, 2006.
37 Hallema et al., 2018.
38 Hallema et al., 2018; Neary, 2019.
39 McCarthy, 2014.
40 Neary, 2019.
41 Colavito et al., 2021.
42 see, e.g., Andreen, 2013.
43 Hough & Robertson, 2009.
44 Hough & Robertson, 2009.
45 Several states use both systems, particularly states that contain distinct humid and arid zones.
46 Royster, 2000.
47 Paveglio et al., 2022; Perry, 2017.
48 Brown et al., 2008.
49 Tidwell, 2016, p. 216.
50 Price & Heberling, 2018.

51 Gartner, Mehan, et al., 2014.
52 Price & Heberling, 2018.

## References

Alavalapati, J. R. R., Stainback, G. A., & Matta, J. R. (2006). Longleaf pine restoration. In S. Jose, E. J. Jokela, & D. L. Miller (Eds.), *The longleaf pine ecosystem: Ecology, silviculture, and restoration* (pp. 403–412). Springer.

Andreen, W. L. (2013). Success and backlash: The remarkable (continuing) story of the Clean Water Act. *George Washington Journal of Energy & Environmental Law, 4*(1), 25–37.

Blumm, M. C., & Paulsen, A. (2013). The public trust in wildlife. *Utah Law Review, 2013*(6), 1437–1504.

Blumm, M. C., & Ritchie, L. (2005). The pioneer spirit and the public trust: The American rule of capture and state ownership of wildlife. *Environmental Law, 35*, 673–720.

Brown, T. C., Hobbins, M. T., & Ramirez, J. A. (2008). Spatial distribution of water supply in the coterminous United States. *Journal of the American Water Resources Association, 44*(6), 1474–1487.

Clayton, J. (2019). *Natural rivals: John Muir, Gifford Pinchot, and the creation of America's public lands*. Pegasus.

Colavito, M., Combrink, T., Hjerpe, E., Edgeley, C., Burnett, J., & Meador, A. S. (2021). *Full-cost accounting remeasurement of the 2010 Schultz Fire: Understanding the long-term socio-economic implications of high-severity wildfire and post-wildfire flooding* (ERI White Paper–Issues in Forest Restoration). Ecological Restoration Institute, Northern Arizona University.

Costa, R., & Gooch, R. (2011). The red-cockaded woodpecker: Conservation through statewide Safe Harbor Agreements. In K. Arha & B. H. Jr. Thompson (Eds.), *The Endangered Species Act and federalism: Effective conservation through greater state commitment* (pp. 165–186). RFF Press.

des Rosiers, J. (1991). The exemption process under the Endangered Species Act: How the "God Squad" works and why. *Notre Dame Law Review*, 66(3), 825–862.

Evansen, M., Harl, H., Carter, A., & Malcom, J. (2021). Status of the recovery mandate under section 7(a)(1) of the US Endangered Species Act. *Frontiers in Conservation Science, 2*, 768628.

Freyfogle, E. T., Goble, D. D., & Wildermuth, T. A. (2019). *Wildlife law: A primer* (2nd ed.). Island Press.

Gartner, T., McGray, H., Mulligan, J., Epstein, J., & Dinshaw, A. (2014a). Adaptation: Forests as water infrastructure in a changing climate. In V. A. Sample & R. P. Bixler (Eds.), *Forest conservation and management in the Anthropocene: Conference proceedings* (pp. 313–337). US Department of Agriculture, Forest Service, Rocky Mountain Research Station.

Gartner, T., Mehan, G. T. I., Mulligan, J., Roberson, J. A., Stangel, P., & Qin, Y. (2014b). Protecting forested watersheds is smart economics for water utilities. *Journal American Water Works Association, 106*(9), 54–64.

Hallema, D. W., Sun, G., Caldwell, P. V., Norman, S. P., Cohen, E. C., Liu, Y., Bladon, K. D., & McNulty, S. G. (2018). Burned forests impact water supplies. *Nature Communications*, 9(1), 1307.

Hays, S. P. (2007). *Wars in the woods: The rise of ecological forestry in America*. University of Pittsburgh Press.

Hoctor, T. S., Noss, R. F., Harris, L. D., & Whitney, K. A. (2006). Spatial ecology and restoration of the longleaf pine ecosystem. In S. Jose, E. J. Jokela, & D. L. Miller (Eds.), *The longleaf pine ecosystem* (pp. 377–402). Springer.

Hough, P., & Robertson, M. (2009). Mitigation under Section 404 of the clean water act: Where it comes from, what it means. *Wetlands Ecology and Management*, *17*(1), 15–33.

Johnson, R., & Gjerstad, D. (1998). Landscape-scale restoration of the longleaf pine ecosystem. *Restoration & Management Notes*, *16*(1), 41–45.

Jose, S., Jokela, E. J., & Miller, D. L. (2006). The longleaf pine ecosystem. In S. Jose, E. J. Jokela, & D. L. Miller (Eds.), *The longleaf pine ecosystem: Ecology, silviculture, and restoration* (pp. 3–8). Springer.

Jusino, M. A., Lindner, D. L., Banik, M. T., & Walters, J. R. (2015). Heart rot hotel: Fungal communities in red-cockaded woodpecker excavations. *Fungal Ecology*, *14*, 33–43.

Keleş, S. (2019). An assessment of hydrological functions of forest ecosystems to support sustainable forest management. *Journal of Sustainable Forestry*, *38*(4), 305–326.

Loeb, S. C., & O'Keefe, J. M. (2014). Indiana bats, northern long-eared bats, and prescribed fire in the Appalachians: Challenges and considerations. In T. A. Waldrop (Ed.), *Proceedings, wildland fire in the Appalachians: Discussions among managers and scientists* (pp. 73–81). US Department of Agriculture, Forest Service, Southern Research Station.

Lyman, M. W., Danks, C., & McDonough, M. (2013). New England's community forests: Comparing a regional model to ICCAs. *Conservation and Society*, *11*(1), 46–59.

McCarthy, L. F. (2014). Water source protection funds as a tool to address climate adaptation and resiliency in southwestern forests. In V. A. Sample & R. P. Bixler (Eds.), *Forest conservation and management in the Anthropocene: Conference proceedings* (pp. 329–343). US Department of Agriculture, Forest Service, Rocky Mountain Research Station.

Neary, D. G. (2019). Forest soil disturbance: Implications of factors contributing to the wildland fire nexus. *Soil Science Society of America Journal*, *83*, S228–S243.

Paveglio, T. B., McGown, B., Wilson, P. I., & Krumpe, E. E. (2022). The Wild and Scenic Rivers Act at 50: Managers' views of actions, barriers and partnerships. *Journal of Outdoor Recreation and Tourism*, *37*, 100459.

Perry, D. M. (2017). *The uneven geography of river conservation in the US: Insights from the application of the Wild and Scenic Rivers Act* [PhD Thesis]. University of Oregon.

Pramova, E., Locatelli, B., Djoudi, H., & Somorin, O. A. (2012). Forests and trees for social adaptation to climate variability and change. *WIREs Climate Change*, *3*(6), 581–596.

Price, J. I., & Heberling, M. T. (2018). The effects of source water quality on drinking water treatment costs: A review and synthesis of empirical literature. *Ecological Economics*, *151*, 195–209.

Riley, S. P., Abernethy, R., & McKenzie, D. F. (2019). Case studies of scalable wildlife conservation in working landscapes. *Wildlife Society Bulletin*, *43*(3), 363–371.

Royster, J. V. (2000). Winters in the East: Tribal reserved rights to water in riparian states. *William and Mary Environmental Law and Policy Review*, *25*(1), 169–201.

Stephens, S. L., Kobziar, L. N., Collins, B. M., Davis, R., Fulé, P. Z., Gaines, W., Ganey, J., Guldin, J. M., Hessburg, P. F., & Hiers, K. (2019). Is fire "for the birds"? How two rare species influence fire management across the US. *Frontiers in Ecology and the Environment*, *17*(7), 391–399.

Stuart, G. W., & Edwards, P. J. (2006). Concepts about forests and water. *Northern Journal of Applied Forestry*, *23*(1), 11–19.

Tidwell, T. L. (2016). Nexus between food, energy, water, and forest ecosystems in the USA. *Journal of Environmental Studies and Sciences*, 6(1), 214–224.

United States General Accounting Office. (1994). *Endangered Species Act: Information on species protection on nonfederal lands* (GAO/RCED-95-16). General Accounting Office.

Waage, S. A. (2001). (Re)claiming space and place through collaborative planning in rural Oregon. *Political Geography*, *20*(7), 839–857.

Weiss, S. A., Toman, E. L., & Corace, R. G. (2019). Aligning endangered species management with fire-dependent ecosystem restoration: Manager perspectives on red-cockaded woodpecker and longleaf pine management actions. *Fire Ecology*, *15*(1), 1–14.

Wilkinson, C. F., & Conner, D. K. (1983). The law of the Pacific salmon fishery: Conservation and allocation of a transboundary common property resource. *University of Kansas Law Review*, *32*, 17–110.

Zhang, D., & Mehmood, S. R. (2002). Safe harbor for the red-cockaded woodpecker: Private forest landowners share their views. *Journal of Forestry*, *100*(5), 24–29.

# 5 Climate Change, Disturbances, and Adaptation

The Joseph Creek watershed has been a focal area for Wallowa County forest restoration since the late 1990s. The creek itself supports populations of salmon, including the Snake River chinook that were listed as threatened under the Endangered Species Act. By the time of the listing decision in 1992, the forests of the Joseph Creek watershed were already showing the effects of past fire exclusion and land use practices. Many of the stands were overstocked, with elevated rates of disease and insect infestation as too many trees competed for too little available water. In some stands, as many as eight of every ten trees had been killed by bark beetles. Tree mortality on the forest outpaced forest growth. While much of the watershed historically supported frequent, low-intensity fires, decades without fire left these semiarid forests overcrowded with trees and at risk of the kind of stand-replacing wildfire that could degrade water quality and salmon habitat (Figure 5.1). Additionally, invasive plants—most of them introduced during the twentieth century from Europe or Asia—were steadily increasing their presence within the watershed, displacing native species and reducing forage quality for both wildlife and livestock. Community organizations and a local collaborative group helped to lead a collaborative assessment and restoration effort in the upper part of the watershed and followed this with a second effort focused on the lower watershed.

Prior to the shutdown of timber harvesting in 1994, typical management projects on the Wallowa-Whitman National Forest were relatively straightforward: national forest managers would prepare mature stands to be harvested and sell the harvest rights to the highest bidder using a timber sale contract. If the national forest needed bridges, roads, or recreation infrastructure to be built or maintained, it would pay a contractor via a service contract. The Upper Joseph Creek project envisioned by community leaders was a departure from the previous model of forest management; it included elements such as culvert replacement, road repair, and removal of unmerchantable small-diameter trees along with the harvest of some commercial-grade timber. The value of this timber did not offset other project costs, meaning that funding was needed to supplement the costs of implementation. In part due to its status as one of the nation's early efforts in collaborative, landscape-scale forest restoration, the Upper Joseph Creek project was able to attract investments from federal, state, county, foundation, and

DOI: 10.4324/9781003043669-5

*Figure 5.1* A 1908 photograph shows Special Agent J. T. Jardine employing low technology to extinguish a fire burning through an open ponderosa pine stand in the Wallowa National Forest. Image: US Forest Service.

non-profit sources. The Lower Joseph Creek project, implemented nearly two decades later, was not able to garner the same level of private and nonprofit economic support. Instead, project leaders relied on various new tools to help implement restoration: funds provided by the State of Oregon under its Federal Forest Restoration Program and federal support from a new initiative called the Collaborative Forest Landscape Restoration Program.

The Upper and Lower Joseph Creek projects exemplify contemporary approaches to forest restoration on federal multiple-use lands. Rather than focusing on singular objectives, both projects integrated multiple concerns. They did so by taking a landscape perspective; indeed, both projects benefited from complementary restoration work conducted on neighboring, nonfederal forestlands. Both projects attempted to remedy the effects of decades of fire exclusion and other activities that left forests susceptible to historically uncharacteristic insect outbreaks and wildfires. To plan and implement these projects, community partners had to look not only to past conditions and disturbance patterns but also to the future, where a changing climate and spreading populations of introduced species would further complicate an already complex management scenario. In this chapter, we consider America's forests under the multiple pressures of environmental change. We examine policies and governance models that have emerged to address these pressures and explore their outcomes. We focus especially, but not exclusively, on fire given its important historic role in American forests and its current status as both a threat and a potential tool for restoration.

## Introduction

Signs of stress on the nation's forests are evident in diverse geographies and ecosystems. In lower elevations of the Rocky Mountains, some forests are failing to recover following large fire events; although established trees are typically able to persist under a changing climate, seedlings in the post-fire environment are desiccating under conditions that are hotter and drier than those in which previous stands established.[1] At higher elevations in the Rockies, subalpine forests are burning more frequently than at any time in the past two millennia.[2] In California's Sierra Nevada range, giant sequoia trees that have endured for hundreds and even thousands of years with regular, recurring fires are dying at unprecedented rates due to recent high-intensity wildfires.[3] Along the Atlantic and Gulf coasts, so-called "ghost forests" are visible in low-lying coastal zones where sea-level rise has allowed salt water to infiltrate the soil and kill standing forests.[4] Although patterns of sea-level rise and forest retreat have been ongoing for centuries, current rates of change along the mid-Atlantic coast have been estimated as being two to fourteen times greater than during the preindustrial era.[5]

In other forest systems, many tree species are succumbing to novel pests and pathogens. The emerald ash borer was first discovered near Detroit, Michigan, in 2002 and has since spread to over half of all US states, killing hundreds of millions of ash trees in the process. Sudden oak death was first discovered in northern California and southern Oregon forests in the mid-1990s and is now found in at least seven other states, with potential for spread to many more. The agent in sudden oak death, an algae-like water mold, infects over 100 tree and forest shrub species and can provoke widespread mortality in some systems.[6] An introduced fungal complex known as rapid ‘Ōhi‘a death is spreading across the Hawaiian islands, threatening the state's most abundant native tree species.

As we take stock of these worrying trends in forest loss, it is important to emphasize that environmental change is nothing new: fires, diseases, insect outbreaks, windstorms, floods, droughts, and other events classified as "disturbances" have been part of forest ecology for as long as there have been forests. Numerous plant, wildlife, and fungal species have co-evolved with disturbances and benefit from the physical and ecological upheaval induced by events like fires and windstorms. However, when we look at contemporary patterns of environmental change, some important differences stand out. The first difference has to do with the sheer scale and scope of human impacts on America's forests: the US population is approximately 330 million at the time of this writing, and the energy, transportation, and housing infrastructure supporting those 330 million people has transformed forests in ways that have no historical precedent. Among other important patterns is the vast proliferation of houses, outbuildings, utility corridors, and other elements of human infrastructure in flammable areas away from urban or suburban cores. The rapid expansion of this so-called "wildland-urban interface" brings with it an abundance of new ignition sources that can allow fires to establish in places and times that they would normally be absent.[7] When fire does occur in these landscapes, there is an expectation that firefighting agencies will spare no expense in protecting the built infrastructure that people value (as well as the surrounding scenery), making fire management more complex, more expensive, and ultimately more dangerous for the firefighters themselves.

Second, forest conditions in many places have been greatly altered following past harvesting activities, fragmentation, conversion to non-forest uses, the introduction of domestic livestock, and especially the suppression of lightning-ignited and Indigenous burning. These varied influences have changed both the kinds of disturbances experienced by forests and the capacity of forests to withstand or adapt to disturbances. Research in many forest types shows profound changes in forest structures, processes, and species composition that can often be traced back to the land-use changes instituted in the wake of Euro-American settlement; these changes are especially apparent in forests that were historically characterized by frequent or moderately frequent fires.[8] Whereas recurring fires in these systems once consumed both dead and live woody material, maintained relatively open forest conditions, and created a diversity of forest patches across space, fire exclusion has allowed living and dead wood to accumulate. This has increased fuel loads, the continuity of flammable material, and the homogeneity of forest patches.[9] When fires eventually occur under such conditions, they tend to burn much hotter than fires burning under historically characteristic conditions and may reach into the canopy of mature trees where they spread as crown fires, leading to widespread tree mortality. These same overstocked and homogenous conditions can also set up some kinds of forests for uncharacteristic insect and disease outbreaks.[10]

Third, the growth of global trade networks over the past two centuries has allowed for the widespread movement of plant, animal, insect, fungal, and microbe species across their traditional borders and outside of the environments in which they evolved. Not all of these introduced species are able to survive in their new environment, but some of those that do survive are able to spread unchecked in the absence of coevolved predators or other limits on population growth.[11] Examples of devastating introduced diseases include the chestnut blight, white pine blister rust, and Dutch elm disease, which respectively decimated populations of the American chestnut (once one of the most abundant and productive species in eastern forests), multiple species of five-needle pines, and several species of elms that were important both in undeveloped and urban forests. Introduced plants, insects, and animals like the Brazilian pepper tree, hemlock woody adelgid, spongy moth, feral hog, and brown tree snake have already fundamentally transformed many forest systems. Other introduced species including the emerald ash borer, spotted lanternfly, and Asian long-horned beetle are continuing to expand their ranges at the time of this writing.

Fourth, the global climate is changing rapidly due to increased greenhouse gas concentrations in the atmosphere. Specific patterns vary across the US, but an overall warming trend has led to earlier snowmelt, longer fire seasons, more frequent and severe droughts and heatwaves, and changes to precipitation patterns in many places.[12] Indeed, climate is a key driver of forest fire occurrence and severity; although local forest conditions matter for individual fires, larger climatic trends account for most of the variation from year to year in fire occurrence and behavior.[13] These same climate drivers also allow for extreme weather events to coincide in space and time with much greater frequency than in the past.[14] Combinations of extreme heat and extreme drought that were once exceptional are increasingly commonplace. Combined with changes in forest structure, warmer temperatures and more frequent droughts allow many insect species to reproduce and spread more quickly and can give them the upper hand over their tree hosts.[15] Collectively, such changes set the stage for some of the unusually destructive wildfire, insect outbreak, and post-fire flooding events that have become nearly annual events in recent decades.

## Fire Returns with a Vengeance

Chapter 3 describes the US fire policy changes in the 1960s and 1970s that represented a break with the full-suppression paradigm established decades before. General climate patterns across the US also underwent a shift around this time: after decades characterized by a relatively moist, cool pattern in western forests that contributed to limited fire activity, large fires began to reappear in the warmer and drier 1980s. The year 1988 brought a massive wildfire in and near Yellowstone National Park that drew critical scrutiny to the National Park Service's "let burn" policy.[16] The fire largely behaved in ways consistent with the fire ecology of lodgepole pine and other high elevation forests; still, the shock of seeing blackened forests in one of the country's most beloved national parks led some elected officials to push for more aggressive fire control measures.[17] Fire policy was again thrust into the center of national political debate less than a decade later with the 1995 "Salvage Rider," focused on the harvest of fire-damaged trees (see Chapter 3). Designed to undo the old-growth forest protections established by the Clinton administration through the Northwest Forest Plan, this legislation took shape in the wake of the 1994 fire year in which over four million acres burned and 38 wildland firefighters tragically lost their lives. Fourteen of these fatalities occurred in a single incident on Storm King Mountain, Colorado, during the South Canyon Fire. Overall federal fire suppression costs approached one billion dollars that year.

As described in detail below, a series of national policies beginning in the 1990s recognized the dilemma created by decades of fire exclusion and called for the restoration of fire as an ecological process. However, a combination of climatic changes and overstocked conditions in forests that had been deprived of natural fire for decades came to vastly complicate the effort to implement this vision in practice. A key event occurred in 2000 when a prescribed fire in northern New Mexico escaped control. By the time the Cerro Grande wildfire was extinguished, it had burned 48,000 acres (the largest wildfire ever recorded in New Mexico at that time), resulted in the evacuation of the city of Los Alamos, and destroyed 255 homes as well as several structures on the Los Alamos National Laboratory; fortunately, it did not lead to the release of radioactive contamination from laboratory facilities.[18] Extensive wildfires burned elsewhere throughout the West in 2000; major fires occurred in Idaho, southern Colorado, eastern Oregon, eastern Washington, and western Montana, claiming the lives of fifteen wildland firefighters. US military and international crews were brought in to help combat fires in places like Montana's Bitterroot Valley, where 70 homes and 170 other structures were lost to the flames.

The year 2002 brought three western wildfires that dwarfed Cerro Grande's footprint: the 138,000-acre Hayman Fire in Colorado, the 467,000-acre Rodeo-Chediski Fire in Arizona (composed of two separate fires that ultimately merged, both ignited by humans), and the nearly 500,000-acre Biscuit Fire in southern Oregon and northern California. Federal firefighting costs in the first decade of the new century regularly exceeded one billion dollars annually on their way to two and even three billion by the second decade. Even larger fires broke records set just a few years before: in 2011 Arizona experienced the 538,000-acre Wallow Fire, followed by Colorado's 208,000-acre Cameron Peak Fire in 2020 and New Mexico's Hermit's Peak/Calf Canyon fires in 2022, a complex of over 340,000 acres started by escaped prescribed burns. California saw its largest wildfire on record in 2020, eclipsed by an even larger fire just a year later that burned nearly a million acres. A US Forest Service report in 2015 stated that fire spending was consuming over half of the agency's annual budget, up from just 16% two decades before, with core forest management, planning, and operations budgets being reduced to make up the difference.[19]

As discussed in Chapter 1, forests throughout much of the present-day US co-evolved with fire—both human- and lightning-caused—and a growing incidence of wildfire is not by itself an indicator that forests are ecologically unhealthy. Indeed, fire ecologists have argued that many of the nation's forests need more—not less—of the right kind of fire.[20] Even during some of the massive events named above, fires behaved more or less within historic parameters in at least part of the burn area. In other cases, however, wildfires were clearly behaving in ways far outside the historic norm. This was especially true in forests with historically low-severity fire regimes such as ponderosa pine and dry mixed-conifer forests. These forest types experienced high levels of tree mortality in events such as Arizona's Rodeo-Chediski and Wallow fires, Montana's Bitterroot complex of 2000, New Mexico's 2011 Las Conchas fire, and many others.

Several of these large and high intensity fires also overwhelmed emergency response and evacuation capabilities and led to the loss of human infrastructure and lives. An especially tragic example was northern California's Camp Fire of 2018, the deadliest wildfire disaster in California history. The fast-moving firestorm destroyed 19,000 structures and killed 85 residents as it nearly wiped the town of Paradise off the map. Even the normally humid forests of the Appalachians and western Cascade Mountains have not been spared: the Chimney Tops fire in 2016 killed fourteen people and destroyed 2,500 homes near Gatlinburg, Tennessee; a complex of wildfires burning under conditions of extreme drought in western Oregon in 2020 took the lives of eleven people and incinerated thousands of homes and other structures.[21]

Fires behaving in ways inconsistent with their historic patterns and burning through human communities were not the only sign of trouble in the nation's forests around the turn of the twenty-first century. Outbreaks of native insects and diseases also began to occur in historically unprecedented patterns, in some cases killing host trees across millions of contiguous acres. The southern pine beetle, so named because its range was largely restricted to the southern pine belt, began to appear as far north as New York and Connecticut. It is believed that the beetle has been able to expand its range as a result of the milder winters associated with global climate change.[22] Outbreaks of native bark beetles starting around the turn of the twenty-first century were responsible for forest mortality over an area of at least 13 million acres in the interior West, spanning the range from low elevation piñon-juniper woodlands to the highest subalpine forests.[23] In many cases, the same interacting patterns of past forest management practices, fire exclusion, and climate change that created the contemporary wildfire dilemma were also drivers of these slower moving "forest health" crises.

## Fire and Forest Policy Developments

National forest policy change in the twenty-first century has largely been oriented toward addressing the growing dilemmas associated with wildfire, insect, and disease outbreaks.[24] New policies and programs have been advanced through a mix of federal agency (and interagency) plans, executive-level regulations, and laws passed through the traditional congressional channels. Although the details are new, the overall approach continues the model of federal leadership and state cooperation that was forged in the earliest days of the US Forest Service as an agency.

### *Executive Agency and Interagency Policies*

Interagency policy changes regarding fire have largely been led by the Departments of Agriculture and Interior. An early effort was the 1995 Federal Wildland Fire Management Policy and Program Review, which stressed the need for coordination among agencies, local collaboration, and sustained support for a program of fuel reduction. In a sign that federal entities were ready to fully break from the full-suppression paradigm of the past, the Review stated that "wildland fire as an ecological process and natural change agent will be incorporated into the planning process." This was followed by the 2000 National Fire Plan, the ten-year Comprehensive Strategy and implementation plan crafted in partnership with the Western Governors' Association, a 2001 update of the Wildland Fire Policy and Program Review, and the 2009 Guidance for Implementation of Federal Wildland Fire Management Policy, which provided explicit support for using naturally ignited fires to achieve forest restoration and fuel reduction objectives. The National Cohesive Wildland Fire Management Strategy was finalized in 2014; its three overarching goals continue to guide federal fire management as of the date of this writing. The "Resilient Landscapes" goal focuses on the land itself, continuing the difficult task of safely reintroducing fire and restoring forest conditions that will respond well to fire. The "Fire Adapted Communities" goal focuses on the built environment, using primarily cooperative and incentive-based programs and tools to encourage homeowners and communities in flammable landscapes to better prepare for the inevitability of future fires. The "Safe and Effective Wildfire Response" goal focuses on wildfire suppression activities, acknowledging that many fire starts will continue to be suppressed due to their potential impacts on both people and forests.

Other important changes to fire and forest policy have been instituted through executive-level programs and initiatives. An early effort of the George W. Bush administration was to make administrative rule changes to environmental laws such as NEPA and the Endangered Species Act as well as to laws allowing citizens and organizations to appeal and litigate federal agency management decisions. The administration argued that policies such as these prevented Forest Service and Bureau of Land Management thinning and prescribed fire projects from moving forward efficiently and therefore allowed wildfires to ravage forests and wildland-urban interface communities (see Chapter 3); academic analyses did not always support this narrative, however.[25] The Healthy Forests Initiative, released in 2002, consisted of a series of executive-level actions that reduced NEPA obligations for certain kinds of Forest Service and Bureau of Land Management projects (thinning, prescribed fire, and post-fire salvage harvests), limited administrative appeals for some agency decisions, changed consultation requirements for threatened and endangered species, and eliminated some elements of the Northwest Forest Plan. Not all of these changes ultimately survived judicial review.[26]

The Obama administration revised administrative regulations of the National Forest Management Act to emphasize restoration, ecological integrity, consideration of climate impacts, and resilience to disturbances.[27] These changes complemented other executive-level actions that required federal agencies to better plan for resilience under a changing climate and encouraged an "All Lands" approach to addressing wildfire and other risks across property boundaries.[28] In 2014, the Chiefs of the Forest Service and Natural Resources Conservation Service announced the creation of the Joint Chiefs' Landscape Restoration Partnership, a competitive funding program for restoration projects of up to three years in duration that span the boundary separating national forests and private lands.[29] Many of the projects funded under this initiative focus on fuel reduction and forest restoration but have also included issues such as post-hurricane recovery, watershed improvement, and endangered species conservation. The initiative was later passed into law as part of the 2021 Infrastructure Investment and Jobs Act. A key executive policy in the Trump administration was the Shared Stewardship initiative, described below.

***Legislative Changes***

In 2003, Congress passed into law the Healthy Forests Restoration Act (HFRA) as companion legislation to the Healthy Forests Initiative. Among other changes, HFRA created expedited NEPA procedures and Endangered Species Act consultation obligations as well as revisions to the administrative appeals process for national forest and Bureau of Land Management fuel reduction projects.[30] It also created an expedited NEPA process for projects in forests experiencing—or at risk of experiencing—severe disease or insect damage, and offered new opportunities for communities at risk to craft Community Wildfire Protection Plans (See Box 5.1).[31] Other policy developments since the passage of HFRA include passage of the 2009 Federal Land Assistance, Management and Enhancement (FLAME) Act, which required the Departments of Interior and Agriculture to develop a cohesive strategy for wildfire management, and the 2018 "Fire Funding Fix," which created an emergency fund to pay the costs of large wildfire suppression needs. The fix responded to a recurring pattern in which fire-related cost overruns led the US Forest Service and Department of Interior to raid non-fire funds (including funds for forest restoration treatments) or seek emergency appropriations.

Congress has also passed a series of laws meant to encourage forest restoration, fire reintroduction, and the protection of communities at risk of fire through new contracting and funding tools. Stewardship end result contracting authorities were developed in large part to facilitate restoration projects on national forest and Bureau of Land Management lands that include the removal of both non-merchantable material (such as small trees and brush) and commercially valuable trees—projects like the Upper and Lower Joseph Creek restoration initiatives described at the beginning of the chapter. Stewardship contracts allow timber receipts to be used to offset the costs of non-commercial work or to be reinvested in future restoration projects. They also allow treatments to be designated by end-results descriptions rather than by marking trees to be harvested or retained. In order to attract the private investments in processing capacity that add value to restoration by-products, stewardship contracting policy initially allowed for contracts of up to ten years in duration; legislative language has since been approved authorizing the extension of contracts for up to twenty years in certain areas. Since nationwide authorization in 2003, thousands of stewardship contracts have been used to implement fuel reduction and restoration projects on national forest and Bureau of Land Management lands—in many cases implementing integrated plans developed through collaborative efforts involving federal land managers, local community representatives, and other key partners (see Chapter 10).

The Collaborative Forest Landscape Restoration Program (CFLRP) was authorized in 2009 to support landscape-scale forest restoration projects planned in collaboration among Forest Service managers and relevant communities. Through a competitive proposal review and selection process, the CFLRP funds implementation on national forests for periods of up to ten years; inclusion of adjacent non-national forest land is encouraged but not directly funded by the program.[32] For example, the Lower Joseph Creek project described at the beginning of this chapter was included in the Northern Blues CFLRP proposal, selected for funding in 2021. The project entails management on up to 900,000 acres across the Malheur, Umatilla, and Wallowa-Whitman National Forests with the goal of leaving treated areas more resilient to fire, drought, insects, diseases, and varied climate-induced disturbances. In the first ten years of the CFLRP national program, nearly four million acres nationwide were treated with fire or mechanical thinning, and funded projects attracted nearly half a billion dollars of partner investments. Building upon the vision of long-term and large-scale investments in forest restoration embodied in stewardship contracting policy, policies like CFLRP and the Joint Chiefs' Landscape Restoration Partnership represent a model of federal support for collaboratively designed forest restoration over large landscapes.[33] The 2021 Infrastructure Investment and Jobs Act increased funding for the CFLRP and Joint Chiefs' Landscape Restoration Partnership and included over $5 billion in additional funding for forest restoration treatments; the Biden administration announced that the allocation of these funds would be prioritized to areas considered to be at greatest risk of undesirable wildfire impacts.

### BOX 5.1 THE HEALTHY FORESTS RESTORATION ACT

The Healthy Forests Restoration Act (HFRA) is considered to be one of the most significant changes to federal forest policy since the National Forest Management Act of 1976. The passage of HFRA in 2003, along with its administrative counterpart the Healthy Forests Initiative, helped to set a new paradigm for federal forest policy, one that combines the themes of forest restoration, participation of communities in federal forest project planning, and limitations on both environmental analysis and opportunities to challenge proposed projects on federal land. These policy mechanisms were all supported by a narrative emphasizing the need to act expeditiously to treat forest conditions and avert future catastrophes.[34] Some of HFRA's key elements include:

- An emphasis on fuel reduction through thinning and the application of prescribed fire, especially in the wildland-urban interface;
- Expedited NEPA and administrative appeals/objections processes for projects focused on fire, insect, or disease hazard risk reduction;
- The promotion of partnerships, both between varied federal agencies (such as the Forest Service and Bureau of Land Management) and between federal, state, tribal, and other entities; and
- The empowerment of local communities to participate in wildfire planning.

An important component of HFRA is the authorization of Community Wildfire Protection Plans—documents created by community leaders and members themselves that define the boundaries of the wildland-urban interface and identify priorities, practices, and responsible parties for reducing the threat of wildfire to people and human infrastructure. HFRA provides reduced NEPA obligations for federal land fuel reduction projects within the wildland-urban interface zones defined in these plans and commits substantial federal funding to projects within designated interface zones.

Another characteristic of HFRA, mirrored in subsequent policies, is its creation of optional policy tools that local-level forest managers can choose to employ. This is a contrast to the mandatory provisions associated with earlier laws such as NEPA, the National Forest Management Act, and the Endangered Species Act. The 2014 Farm Bill amended Title VI of HFRA to create new authorities for governors to declare "Insect and Disease Emergency Areas," triggering obligations for the US Forest Service to respond. Continuing a common policy trend in recent years, the 2014 Farm Bill also created new categorical exclusions from environmental analysis under NEPA for certain projects related to reducing insect or disease risk.

## Implementing Restoration and Fire Reintroduction Policies

Wildfire management in the US is confronted with what has been called the "fire paradox": faced with the risk of unacceptable wildfire impacts, managers often choose to suppress fires soon after they ignite; however, doing so allows for the continued accumulation of vegetative fuel, thereby setting the stage for future fires that burn even hotter and are harder to control.[35] As fire historian Stephen Pyne summarized, "the more we try to remove fire from places that have coevolved with it, the more violently fire will return."[36] Following decades of research, practice, and collaborative deliberations in forests nationwide, forest managers and community partners in many places have come to a consensus on a broad definition of the dilemma and on the kinds of actions needed to address it. In general, they agree that many historically fire-adapted forests need remedial work to change the way fire behaves when it inevitably returns. Although there are often disagreements over the details of individual projects, there is a broadly recognized need to manage forest conditions to reduce the likelihood of extreme fire events that threaten human communities, degrade watersheds, and burn so hot as to leave little standing in their wake. There is also broad agreement that relevant communities of place and communities of interest should be involved in the site-specific decision-making processes that determine how to move forward with projects to reduce fuel loads and allow for the safe restoration of fire.

The primary management tools that are consistently applied in projects meant to improve forest resilience to fire are mechanical thinning (using manual labor and equipment to remove some proportion of the living and dead trees that contribute to heavy fuel loads and high fire severity) and prescribed fire (the intentional reintroduction of fire under conditions conducive to both control and the achievement of resource objectives).[37] Both of these tools hold the potential to change the volume and arrangement of combustible material in forests and allow fire to behave in ways considered more socially and ecologically appropriate. Both tools can be challenging to apply in practice, however. Mechanical thinning is often expensive to conduct over large areas because it tends to focus on the removal of small-diameter and low-value material, meaning that removal costs are in many cases not offset by the sale of resulting forest products. Many forest owners and agencies lack the funding to implement thinning over large areas, especially if there is weak demand for the harvested material. It is important to note that the category of "thinning" as a management activity is broad, encompassing a range of prescriptions and methods; the effect of any given thinning treatment on fuel loads depends on the details. While some kinds of thinning in some kinds of forests may be effective in reducing the risks of uncharacteristic fire,[38] other approaches in other kinds of forest may not have this effect.[39] In most cases, prescribed burning following a thinning treatment is recommended as part of a long-term program of fostering forest resilience.[40] However, the application of prescribed fire can be challenged by limitations on adequate weather conditions, limited burn crew capacity, considerations of smoke impacts to nearby communities and transportation corridors, and even restrictions on particulate emissions under the Clean Air Act.[41] As the Cerro Grande, Hermit's Peak/Calf Canyon, and other escaped prescribed fires have illustrated, putting fire back on the ground can also come with substantial risks to people and infrastructure.

Another tool managers can use is to allow unplanned, naturally ignited fires to burn; this tool has been referred to by various names, including "prescribed natural fire," "wildland fire use," "wildfire for resource benefit," and "managed wildfire." Some naturally ignited fires were allowed to burn in national parks as early as 1968 and in remote wilderness areas on national forests starting in 1972.[42] The use of managed wildfire has been endorsed in national policies like the 2009 Guidance for Implementation of Federal Wildland Fire Management Policy, and some national forests have made changes to their forest plans to allow for its use.[43] Although this approach is supported by many scholars and conservation advocates, it can be difficult for fire managers to implement in practice—particularly in places where human lives and infrastructure could be at risk if a fire were to burn out of control. Managers often report being under pressure from elected officials and local communities to suppress fire starts.[44] US Forest Service performance metrics, cultural beliefs, and the ambiguous policy direction regarding wildfire management also tend to encourage the short-term risk management approach associated with fire suppression, even if it leads to further entrenchment of the wildfire dilemma over the long term.[45] In the absence of advanced planning and both political and public support, the professional risk to forest managers of allowing a fire to burn often outweighs the risk of suppression.[46]

## State and County Involvement in Federal Forest Restoration

As detailed in Chapter 6, states and counties have long been important partners in the administration of federal forest conservation programs on private lands. Recent federal policy reforms have created limited but important opportunities for states and counties to play a role in managing federal lands as well. Policy language in the 2001 Interior Department appropriations bill created a pilot program authorizing the Colorado State Forest Service to conduct restoration activities on national forestlands as part of work on adjacent state or private lands. Known as the Good Neighbor Authority, the policy was designed to improve federal-state coordination and to allow forest managers to realize efficiencies by conducting restoration projects across ownership boundaries. The pilot Good Neighbor Authority was later expanded, and by 2014 it was authorized nationwide. Under the authority, states, counties, and federally recognized tribes can enter into agreements with the US Forest Service or Bureau of Land Management to conduct restoration, fuels treatments, pest and disease control, planning, and conduct timber sales and road work on federal lands; unlike in the pilot program, these activities are not limited to areas adjacent to lands where the nonfederal partner is already working. In using this authority, states may be able to retain some project proceeds (for example, timber sale revenues) to reinvest in future Good Neighbor Authority projects and states can also use their own—rather than federal—contracting processes when hiring planners, loggers, restoration practitioners, and others to perform work in the woods.[47] As detailed in Chapter 7, tribes can also take advantage of other authorities to implement restoration programs on federal lands.

Some states have established programs in which state funds are spent to address capacity limitations in federal land management. These programs have been especially common in western states where federal lands constitute large proportions of the landscape and where federal forest management potentially affects the backyards, watersheds, and workplaces of large numbers of residents. For example, the state of Montana introduced the Forests in Focus program in 2014; a major element of this initiative entails investing state funds to accelerate restoration and management on the 17 million acres of national forestland within the state. The first round of funding targeted 25 projects over approximately 300,000 acres, supporting both planning and implementation of projects in or adjacent to areas designated as priority landscapes in the State Forest Action Plan. Along similar lines, Oregon's Federal Forest Restoration Program (originally called the Federal Forest Health Program) supported analysis, site preparation, and project implementation on national forests statewide to achieve priority restoration, fuel reduction, and economic development projects—including on the Lower Joseph Creek project profiled at the beginning of this chapter. Through 2022, California has spent over $1.5 billion of carbon cap-and-trade revenue (see Chapter 6) on various fire risk reduction-related programs, including on federal lands.[48] In other cases, states and counties have developed ad-hoc funding mechanisms to respond to emerging issues; an example is the financial and capacity contributions to cross-boundary forest management by states and counties in the Black

Hills region in the face of the massive mountain pine beetle outbreak that lasted from roughly 2000–2015 (Box 5.2).[49]

The most recent advancement in state–federal partnership evolution came in 2018, when the Trump administration announced the Shared Stewardship initiative. The initiative built upon developments such as those mentioned above as well as the "All Lands" concept embraced by the Obama administration and policies like the Good Neighbor Authority, CFLRP, and Joint Chiefs' Landscape Restoration Partnership. Although Shared Stewardship did not come with any new policy tools or dedicated project funding, it signaled that recent movements by states to take on stronger roles managing federal forestlands as part of broader landscapes would be welcomed and encouraged. As of the date of this writing, 29 states and 3 regional alliances have signed formal memoranda of understanding with the US Forest Service that detail their goals, priorities, and strategies under Shared Stewardship. Early research shows that the strategy has triggered relatively little substantive change in some states, while other states have hired coordinators and created new mechanisms for increasing their role in federal forest management. For example, Utah hired new positions focused on Shared Stewardship topics and adapted an existing competitive project review framework to evaluate proposals for funding work on national forests within the state.[50]

Counties have also increased their engagement in federal forest activities in many parts of the country (Box 5.2). One way for county governments and other nonfederal entities to increase their involvement in federal forest management is through the use of stewardship agreements, in which state, county, tribal, or NGO partners commit to working with the US Forest Service on mutually beneficial programs of restoration for community benefit.[51] For example, in 2017, Tuolumne County, California, entered into a ten-year master stewardship agreement with the Stanislaus National Forest. Among other things, this allows the county to serve as the contracting entity for federal land management projects such as the 94,000-acre Social and Ecological Resilience Across the Landscape (SERAL) project intended to reduce stand densities, reintroduce prescribed fire, and protect both human communities and Endangered Species Act-listed species like the California spotted owl from uncharacteristic wildfire.

**BOX 5.2 RESPONDING TO A MOUNTAIN PINE BEETLE OUTBREAK IN THE BLACK HILLS**

Beginning in the late 1990s, tens of millions of acres of forest from Colorado to northern British Columbia were affected by the mountain pine beetle, a native insect prone to periodic outbreaks that can kill trees over large areas. This most recent outbreak is believed to have been driven by a combination of factors related to forest conditions (an abundance of mature lodgepole pine and ponderosa pine trees, in some cases growing at high density) and climate drivers (warmer temperatures that allow beetles to survive and reproduce more easily and drought conditions that stress trees' defense systems).[52] The Black Hills area of South Dakota and Wyoming was one of the hardest-hit regions in the US; the abundant ponderosa pine trees on the Black Hills National Forest and adjacent state and private lands serve as the basis for both a regional timber industry and a major part of the area's tourism economy. In the wake of legal challenges to proposed thinning projects, then-Senate Majority Leader Tom Daschle attached a policy rider to an appropriations bill in 2002 that allowed the projects to proceed without further analysis or review. In exchange, the bill expanded wilderness protections in the national forest. The action was justified as an emergency measure needed to confront the threat of wildfire and beetle outbreaks. At around this same time, the Black Hills National Forest supervisor convened the National Forest Advisory Board, a multi-stakeholder group representing various local interests, as a means of resolving conflicts regarding the forest plan revision; this was followed by the creation of a larger collaborative entity known as the Conservation Leaders. These groups helped to coordinate landscape-scale plans to respond to the mountain pine beetle outbreaks, including thinning and hazard tree removal projects across federal, state, and private property boundaries.

The State of South Dakota eventually contributed more than $13 million over five years to the response effort, and county governments in both South Dakota and Wyoming collectively contributed over $10 million. Federal, state, county, and other partners coordinated treatments over hundreds of thousands of acres of forests. Traditional timber sale contracts were used to harvest commercially valuable stands at risk of beetle infestation while tools including stewardship contracts and intergovernmental cooperative agreements were used for non-commercial activities like the removal of hazard trees from travel corridors.[53] Rural communities in the Black Hills responded in their own way, creating an annual arts and culture festival that culminates in the ceremonial burning of a giant wooden bark beetle (Figure 5.2). Despite being lauded as a success,[54] the response was characterized by some observers as too aggressive; they point to a US Forest Service Research Station report that recommends reductions in the Black Hills National Forest timber program to allow forests to recover from years of heavy harvest pressure.[55] Area elected officials and timber industry representatives have pushed back against the suggestion of a reduction in harvest, citing the dependence of local mills on federal timber. In a possible sign of things to come, the Rushmore Forest Products mill in Hill City shuttered in 2021, with the owners citing a lack of supply as the principal cause. The closure led to the loss of 120 jobs in this rural area with few alternative income streams outside of the tourist high season.

*Figure 5.2* Communities in the Black Hills have organized the "Burning Beetle" festival annually since 2012. The climax of the midwinter arts and culture festival is the ceremonial burning of a wooden sculpture representing the mountain pine beetle that has taken a toll on area forests. Photograph by the author.

## Private Forests

Federal lands have been a major focus of wildfire and native insect and disease policies because such a large proportion of western forests—where these issues are most prevalent—is owned and managed by the federal government. Yet nonfederal forestlands are also under growing stress from the combined effects of environmental change, and various policies and governance approaches are targeted at these forests. The presence of large numbers of small-acreage owners in places dominated by private forests can present a major challenge to restoring forests and reintroducing fire at appropriately large scales. Private forest owner participation in Shared Stewardship or Joint Chiefs' Landscape Restoration Partnership projects is one way to "scale up" the impact of fuel reduction activities in places like these. Watershed-oriented initiatives such as those profiled in Chapter 4 are also promising approaches to coordinating across ownership boundaries. In some places, state and NGO partners have worked with private landowners to establish prescribed burn associations, which provide training, equipment, and coordination on a voluntary basis to help accomplish fire reintroduction on private lands. As an alternative to participation in coordinated landscape-scale programs, the State and Private Forestry branch of the US Forest Service administers programs targeted at individual forest owners designed to encourage stand density reduction and other activities that can impart resilience to disturbances,[56] and some states have supplemented these programs with additional funding and outreach support (see Chapter 6).

Many private forest owners need assistance with objectives other than, or in addition to, creating conditions that are resilient to fire and native insect and disease species. Controlling the spread and impacts of introduced insects, diseases, plants, and animals is a top priority in many parts of the country. Yet successful invasive species management, like fire management, is only possible through the coordinated action of multiple landowners; landscapes with highly fragmented ownership and use patterns are especially difficult settings for controlling invasions.[57] The US Department of Agriculture helps to coordinate responses by conducting invasive species research, surveillance, and outreach across property boundaries. Key invasive species management programs are administered by the Forest Service, Animal and Plant Health Inspection Service, and Agricultural Research Service.[58] Federal and state quarantines, in which the transportation of untreated wood (such as firewood) outside of infested zones is restricted, have been used as a policy tool to slow the spread of species like the emerald ash borer and Asian long-horned beetle. These practices can be used in conjunction with eradication programs in which infested trees outside of containment areas are felled, chipped, and burned.[59]

## Tribal Lands and Resources

Indigenous populations are especially vulnerable to climate change impacts due to high levels of direct cultural and nutritional dependence on natural resources (see Chapter 11 for an example) as well as disproportionate rates of poverty and poor public health outcomes.[60] The transmission of cultural knowledge, including Indigenous traditional ecological knowledge (detailed in Chapter 7), may be tied to particular ecosystems, processes, or species that are threatened by changes to climatic patterns, fire regimes, or forest conditions.[61] Along similar lines, introduced insect, plant, animal, and disease species can create disproportionate impacts on Indigenous populations. For example, the black ash tree, currently threatened by the invasive emerald ash borer, is of paramount cultural and economic importance to the Wabanaki Nations in Maine (the Penobscot Nation, Passamaquoddy Tribe, Arrostook Band of Mi'kmaq, and Houlton Band of Maliseet). The black ash is a central species in the Wabanaki creation story, and baskets woven from this species represent an important product both for personal use and commercial sale. The Wabanaki have been working with the State of Maine and the research community to prevent the spread of the emerald ash borer and develop control mechanisms.[62]

Tribal members may face unique barriers to participation in decision-making for environmental management, but supportive policies and institutions can address at least some of these barriers. Potawatomi tribal member and scholar Kyle Powys Whyte suggests four policies that are foundational for the consideration of tribal rights and that support environmental justice in the face of climate change (see also Chapter 7).[63] First, the government-to-government relationship between tribes and the federal government is important because it serves as the basis for tribal sovereignty and establishes a unique role for tribal governments in climate change adaptation. Second, the federal government's trust responsibility creates obligations for the government to "collaborate with tribes in their efforts to work with or resist corporations and private citizens, expand their political authority off-reservation, and pursue ecological outcomes favored by both sovereigns."[64] Third, various policy documents (including, most recently, a 2021 White House memorandum[65]) promote the integration of climate and environmental science with Indigenous traditional ecological knowledge and collaboration between tribal and non-tribal knowledge holders. Finally, multiparty governance is important due to the cross-boundary nature of climate change impacts and is reflected in various organizations and initiatives such as the Columbia River Inter-Tribal Fish Commission, Great Lakes Tribal Climate Change Summit, Landscape Conservation Cooperatives, and various climate science assessments and strategies.[66]

Chapter 7 discusses the many and varied approaches by tribal forest managers to conserve and restore forests both on land currently held by tribes and those within traditional territories that are under the ownership of other entities. The reintroduction of fire to landscapes where it has been excluded is a central tribal restoration strategy in many landscapes and is often guided by principles and knowledges that differ in important ways from those guiding non-tribal fire use.[67] Although the application of traditional practices to forests both on and off reservations was opposed by government foresters for much of the twentieth century, recent policies and programs have created opportunities to bring Indigenous knowledges and practices back into the woods. Indigenous fire knowledge holds the potential to inform planning and implementation for forest recovery in places where fire has been excluded and can both complement and serve as an alternative to fire reintroduction programs informed by western science.[68]

## Fire-Adapted Communities

Since the onset of the contemporary wildfire era in the late twentieth century, substantial funding and policy attention have been directed at reducing the risks to people and their valuable property in and near fire-prone forests. The ultimate goal, as articulated in the 2014 Cohesive Strategy, is to promote fire-adapted communities, defined as "human communities consisting of informed and prepared citizens collaboratively planning and taking action to safely co-exist with wildland fire."[69] Although federal land managers often prioritize forest treatments near wildland-urban interface areas, the forests directly surrounding homes and human infrastructure are typically not federally owned—meaning that residents, property owners, or other entities are primarily responsible for their management. Importantly, research has shown that characteristics of the immediate home environment (within roughly 30 feet of the structure) such as roofing materials, vent screens, the location of flammable material, and the arrangement of vegetation are the most important variables in determining whether a structure survives a wildfire event.[70]

As important as modifications to the immediate home environment are, they make up only a part of a fire-adapted community. There is ultimately no one singular approach to making a community "fire adapted,"[71] but common elements include: firebreaks and defensible space in and around the community; local fire response capacity; adequate ingress and egress to allow for safe community evacuation and for entry of emergency vehicles; communications systems for coordinating action among residents; local plans, codes, and ordinances for reducing vulnerability to wildfire and prioritizing action; and awareness and recognition of risks among residents.[72] Some of these activities may be mandatory in communities that have active homeowners' associations, that are located in states or counties with strong ordinances, or where insurance policies are especially demanding, but often their implementation will require voluntary action on the part of residents—including, in many cases, collective action. State and federal grants, voluntary programs like FireWise USA, and communication and outreach initiatives like the Fire-Adapted Communities Network can all encourage activities that are considered to promote community fire adaptation. However, research shows that even communities that have been directly affected by wildfire events may not show signs of adaptation; for example, some have failed to take steps to reduce their exposure to future fires.[73]

Wildland-urban interface communities are a diverse mix of places that range from remote rural outposts to gated suburban neighborhoods.[74] Not all wildland-urban interface residents and communities are equally vulnerable to the negative effects of uncontrolled fire and not all receive the same level of attention and public investment. A 2018 analysis estimated that 12.4 million people in the US lived in census tracts with high social vulnerability and exposure to wildfire risk; majority Black, Hispanic, and Native American census tracts had a roughly 50% greater vulnerability to wildfire compared to other tracts.[75] The statistical relationship between wildfire exposure and social vulnerability has been observed in other studies as well,[76] and this pattern may be exacerbated by the limited ability of low-income households to invest in actions that reduce their vulnerability.[77] Existing public support may not always reach socially vulnerable households, even where risk exposure is high.[78] Individual fire events have been characterized by poor support, outreach, fire response, and communication to marginalized groups such as undocumented immigrants and residents that do not speak English.[79] Together, this research points toward the need to consider both the diversity of wildland–urban interface communities and the diversity of community members in terms of their access to resources and actions to improve their safety in the face of growing wildfire risks.

## Conclusion

Fire historian Stephen Pyne has termed our modern era the "Pyrocene": a period in which humans have gained control of combustion for powering an industrial and post-industrial society even as we have lost control of fires on the landscape—many of them supercharged by the climatic changes resulting from our industrial combustion processes. Concerns regarding uncharacteristic insect, disease, and wildfire events under climate change and altered forest conditions have driven many of the recent forest policy and governance changes in the US. In contrast to the model of stand-level forest planning and management that held sway for most of the twentieth century, there is now a recognized need to think and act over long periods and at a landscape scale, reflected in initiatives such as the CFLRP and the use of stewardship contracting. A landscape-scale approach often means coordinating across multiple landowners and natural resource agencies and reconciling issues as diverse as forests, water, wildlife, recreation, natural resource businesses, and human communities within a single program of work as envisioned by policies like the Joint Chiefs' Landscape Restoration Initiative, the Good Neighbor Authority, prescribed burn associations, and the Shared Stewardship initiative. Consequently, decision-making processes are increasingly incorporating a broad array of perspectives and a diversity of landowners and decision-makers. In some cases, this means working to integrate different kinds of knowledge: the scientific knowledge of researchers, the professional knowledge of resource managers, the local knowledge of forest users and residents, and the Indigenous knowledge of those with cultural ties to forests that extend millennia into the past.

This new approach represents a break with many of the foundational principles of scientific forestry in the US. Pinchot and other scientific forestry proponents argued that fire should be eliminated to the extent possible, forests should be managed for a narrow range of marketable commodities, and trained experts should make forest management decisions with minimal outside influence. Contemporary forest governance, by contrast, is forced to reckon with complexity: the ecological complexity of both native and introduced disturbance agents acting under altered forest and climatic conditions; the complexity of risk tradeoffs between planned and unplanned fire; and especially the complexity of diverse landowners, diverse management agencies, and a diverse society with multiple and varied understandings of forests as well as varied patterns of vulnerability to forest fires and other disturbance events. An additional dimension of complexity comes from the vast diversity of the nation's forests and the diversity of disturbance regimes: although many forests have changed substantially as a result of decades or even centuries of fire exclusion, this model does not apply equally to all forests. A high-intensity stand-replacing fire may be outside the historic pattern for many ponderosa or longleaf pine forests but not for most lodgepole pine forests. Treatments like thinning and prescribed fire are supported by scientists and stakeholders in some forest types, but these same tools may not provide broadly desirable or acceptable outcomes in other forest types. Separating "good" from "bad" fire, understanding the influences of climate and past land uses on contemporary disturbance patterns, and designing forest treatments that are socially and ecologically sensible are complex problems. Addressing them will require substantive participation from perspectives across a wide spectrum of social categories.

## Questions for Discussion

1. What are the differences between contemporary patterns of environmental change and patterns typical of earlier periods in human history?
2. How has national policy evolved to address wildfire risks? What policy tools are used, what entities are provided authority, and what are the guiding principles?
3. What was significant about passage of the Healthy Forests Restoration Act of 2003? How do this law and the Healthy Forests Initiative attempt to address dilemmas associated with wildfires and insect and disease outbreaks?
4. Why might the growing attention to fires, insects, and diseases create new opportunities for states, counties, and other nonfederal entities to play a more active role in federal forest governance and management?
5. What does it mean for a community to be "fire adapted"? Why might some communities face more obstacles than others in preparing to live with fire?

## Notes

1 K. T. Davis et al., 2019; Stevens-Rumann et al., 2018.
2 Higuera et al., 2021.
3 Shive et al., 2022.
4 Kirwan & Gedan, 2019.
5 Schieder & Kirwan, 2019.
6 Grünwald et al., 2019; Rizzo & Garbelotto, 2003.
7 Balch et al., 2017; Downing et al., 2022; Mietkiewicz et al., 2020.
8 Covington & Moore, 1994; Hessburg & Agee, 2003; Langston, 1995; Parsons & DeBenedetti, 1979; Stephens et al., 2015.
9 Hessburg & Agee, 2003; Keane et al., 2002.
10 Langston, 1995.
11 Lovett et al., 2016.
12 Westerling, 2016.
13 E.g., Abatzoglou & Kolden, 2013; Higuera et al., 2015; Miller et al., 2009; Sherriff & Veblen, 2008; van Mantgem et al., 2013; Westerling, 2016; Williams et al., 2019.
14 Alizadeh et al., 2020; Khorshidi et al., 2020.
15 Dukes et al., 2009; Raffa et al., 2008.
16 Lichtman, 1998.
17 van Wagtendonk, 2007.
18 van Wagtendonk, 2007.
19 USDA Forest Service, 2015.
20 M. D. Abrams, 1992, 2005; Falk, 2006; Frost, 2020; North et al., 2012.
21 Higuera & Abatzoglou, 2021.
22 Lesk et al., 2017.
23 Hicke et al., 2016; Raffa et al., 2008.
24 Schultz et al., 2021.

25 Vaughn & Cortner, 2005.
26 J. B. Davis, 2004.
27 J. Abrams et al., 2021.
28 Charnley et al., 2017; T. Timberlake et al., 2017.
29 Cyphers & Schultz, 2019.
30 J. B. Davis, 2004.
31 Gandhi et al., 2019.
32 Butler & Schultz, 2019.
33 Schultz, Timberlake, et al., 2019c; Schultz & Moseley, 2019.
34 Vaughn & Cortner, 2005.
35 Calkin et al., 2015; Reinhardt et al., 2008.
36 Pyne, 2021, p. 3.
37 Agee & Skinner, 2005; Brown et al., 2004.
38 Fulé et al., 2012; Johnston et al., 2021.
39 Agee & Lolley, 2006; Graham et al., 1999.
40 Hudak et al., 2020; Ryan et al., 2013; Stephens & Moghaddas, 2005.
41 Ryan et al., 2013; Schultz, McCaffrey, et al., 2019a.
42 van Wagtendonk, 2007.
43 J. Abrams et al., 2021; T. J. Timberlake et al., 2020; Young et al., 2020.
44 Canton-Thompson et al., 2008; Donovan et al., 2011; Greiner et al., 2021.
45 Schultz, Thompson, et al., 2019b; Thompson et al., 2018.
46 Seielstad, 2015.
47 Bertone-Riggs et al., 2018.
48 https://subscriber.politicopro.com/article/eenews/2022/05/13/wildfires-eat-up-1-9b-of-calif-cap-and-trade-revenue-00030578.
49 Steen-Adams et al., 2022.
50 Kooistra et al., 2022.
51 Cowan et al., 2022.
52 Bentz et al., 2010; Fettig et al., 2013; Jenkins et al., 2014.
53 Bobzien & Van Alstyne, 2014; Steen-Adams et al., 2022.
54 Bobzien & Van Alstyne, 2014.
55 Steen-Adams et al., 2022.
56 Gandhi et al., 2019.
57 Epanchin-Niell et al., 2010.
58 Gandhi et al., 2019.
59 Poland & McCullough, 2006.
60 Chief et al., 2014; Daigle & Putnam, 2009.
61 Whyte, 2013.
62 Chief et al., 2014; Ranco et al., 2012.
63 Whyte, 2013.
64 Whyte, 2013, p. 527.
65 https://www.whitehouse.gov/wp-content/uploads/2021/11/111521-OSTP-CEQ-ITEK-Memo.pdf.
66 Voggesser et al., 2013.
67 Long et al., 2020.
68 Lake et al., 2017; Long et al., 2020; Steen-Adams et al., 2019.
69 US Department of Agriculture & US Department of the Interior, 2014, p. 91.
70 Calkin et al., 2014; Cohen, 2000.
71 Paveglio & Edgeley, 2020.
72 Frank et al., 2015.
73 J. Abrams et al., 2015; Mockrin et al., 2018; Paveglio et al., 2016.
74 Paveglio et al., 2015.
75 Davies et al., 2018.
76 Lynn & Gerlitz, 2006.

77 Collins, 2005, 2008.
78 Adams & Charnley, 2020; Ojerio et al., 2011.
79 Davies et al., 2018; Méndez et al., 2020.

## References

Abatzoglou, J. T., & Kolden, C. A. (2013). Relationships between climate and macroscale area burned in the western United States. *International Journal of Wildland Fire, 22*(7), 1003–1020.

Abrams, J., Greiner, M., Schultz, C. A., Evans, A., & Huber-Stearns, H. (2021). Can forest managers plan for resilient landscapes? Lessons from the United States national forest plan revision process. *Environmental Management, 67*(4), 574–588.

Abrams, J., Knapp, M., Paveglio, T. B., Ellison, A., Moseley, C., Nielsen-Pincus, M., & Carroll, M. S. (2015). Re-envisioning community-wildfire relations in the US West as adaptive governance. *Ecology and Society, 20*(3), 34.

Abrams, M. D. (1992). Fire and the development of oak forests. *BioScience, 42*(5), 346–353.

Abrams, M. D. (2005). Prescribing fire in eastern oak forests: Is time running out? *Northern Journal of Applied Forestry, 22*(3), 190–196.

Adams, M. D. O., & Charnley, S. (2020). The environmental justice implications of managing hazardous fuels on federal forest lands. *Annals of the American Association of Geographers, 110*(6), 1907–1935.

Agee, J. K., & Lolley, M. R. (2006). Thinning and prescribed fire effects on fuels and potential fire behavior in an Eastern Cascades Forest, Washington, USA. *Fire Ecology, 2*(2), 3–19.

Agee, J. K., & Skinner, C. N. (2005). Basic principles of forest fuel reduction treatments. *Forest Ecology and Management, 211*(1), 83–96.

Alizadeh, M. R., Adamowski, J., Nikoo, M. R., AghaKouchak, A., Dennison, P., & Sadegh, M. (2020). A century of observations reveals increasing likelihood of continental-scale compound dry-hot extremes. *Science Advances*, 6(39), eaaz4571.

Balch, J. K., Bradley, B. A., Abatzoglou, J. T., Nagy, R. C., Fusco, E. J., & Mahood, A. L. (2017). Human-started wildfires expand the fire niche across the United States. *Proceedings of the National Academy of Sciences, 114*(11), 2946–2951.

Bentz, B. J., Régnière, J., Fettig, C. J., Hansen, E. M., Hayes, J. L., Hicke, J. A., Kelsey, R. G., Negrón, J. F., & Seybold, S. J. (2010). Climate change and bark beetles of the western United States and Canada: Direct and indirect effects. *BioScience, 60*(8), 602–613.

Bertone-Riggs, T., Cyphers, L., Davis, E. J., & Hardigg, K. (2018). *Understanding good neighbor authority: Case studies from across the West*. Rural Voices for Conservation Coalition.

Bobzien, C., & Van Alstyne, K. (2014). Silviculture across large landscapes: Back to the future. *Journal of Forestry, 112*(5), 467–473.

Brown, R. T., Agee, J. K., & Franklin, J. F. (2004). Forest restoration and fire: Principles in the context of place. *Conservation Biology, 18*(4), 903–912.

Butler, W. H., & Schultz, C. A. (2019). *A new era for collaborative forest management: Policy and practice insights from the Collaborative Forest Landscape Restoration Program*. Routledge.

Calkin, D. E., Cohen, J. D., Finney, M. A., & Thompson, M. P. (2014). How risk management can prevent future wildfire disasters in the wildland-urban interface. *Proceedings of the National Academy of Sciences, 111*(2), 746–751.

Calkin, D. E., Thompson, M. P., & Finney, M. A. (2015). Negative consequences of positive feedbacks in US wildfire management. *Forest Ecosystems*, *2*(1), 1–10.

Canton-Thompson, J., Gebert, K. M., Thompson, B., Jones, G., Calkin, D., & Donovan, G. (2008). External human factors in incident management team decisionmaking and their effect on large fire suppression expenditures. *Journal of Forestry*, *106*(8), 416–424.

Charnley, S., Kelly, E. C., & Wendel, K. L. (2017). All lands approaches to fire management in the Pacific West: A typology. *Journal of Forestry*, *115*(1), 16–25.

Chief, K., Daigle, J. J., Lynn, K., & Whyte, K. P. (2014). Indigenous experiences in the US with climate change and environmental stewardship in the Anthropocene. In V. A. Sample & R. P. Bixler (Eds.), Forest conservation and management in the Anthropocene: Conference proceedings. Proceedings. (Vol. 71, pp. 161–176). US Department of Agriculture, Forest Service, Rocky Mountain Research Station.

Cohen, J. D. (2000). Preventing disaster: Home ignitability in the wildland-urban interface. *Journal of Forestry*, 98(3), 15–21.

Collins, T. W. (2005). Households, forests, and fire hazard vulnerability in the American West: A case study of a California community. *Global Environmental Change Part B: Environmental Hazards*, 6(1), 23–37.

Collins, T. W. (2008). The political ecology of hazard vulnerability: Marginalization, facilitation and the production of differential risk to urban wildfires in Arizona's White Mountains. *Journal of Political Ecology*, *15*(1), 21–43.

Covington, W. W., & Moore, M. M. (1994). Postsettlement changes in natural fire regimes and forest structure: Ecological restoration of old-growth ponderosa pine forests. *Journal of Sustainable Forestry*, 2(1–2), 153–181.

Cowan, E. R., Grimm, K. E., Davis, E. J., Nielsen, E. A., & Waltz, A. E. M. (2022). New hands in US public lands management: The role and influence of nonagency partners in Forest Service stewardship agreements. *Journal of Forestry*, *120*(3), 302–315.

Cyphers, L. A., & Schultz, C. A. (2019). Policy design to support cross-boundary land management: The example of the Joint Chiefs Landscape Restoration Partnership. *Land Use Policy*, *80*, 362–369.

Daigle, J. J., & Putnam, D. (2009). The meaning of a changed environment: Initial assessment of climate change impacts in Maine–Indigenous peoples. In G. L. Jacobson, I. J. Fernandez, P. A. Mayewski, & C. V. Schmitt (Eds.), *Maine's climate future: An initial assessment* (pp. 35–38). University of Maine.

Davies, I. P., Haugo, R. D., Robertson, J. C., & Levin, P. S. (2018). The unequal vulnerability of communities of color to wildfire. *PLOS ONE*, *13*(11), e0205825.

Davis, J. B. (2004). The Healthy Forests Initiative: Unhealthy policy choices in forest and fire management. *Environmental Law*, *34*(4), 1209–1245.

Davis, K. T., Dobrowski, S. Z., Higuera, P. E., Holden, Z. A., Veblen, T. T., Rother, M. T., Parks, S. A., Sala, A., & Maneta, M. P. (2019). Wildfires and climate change push low-elevation forests across a critical climate threshold for tree regeneration. *Proceedings of the National Academy of Sciences*, *116*(13), 6193–6198.

Donovan, G. H., Prestemon, J. P., & Gebert, K. (2011). The effect of newspaper coverage and political pressure on wildfire suppression costs. *Society & Natural Resources*, *24*(8), 785–798.

Downing, W. M., Dunn, C. J., Thompson, M. P., Caggiano, M. D., & Short, K. C. (2022). Human ignitions on private lands drive USFS cross-boundary wildfire transmission and community impacts in the western US. *Scientific Reports*, *12*(1), 1–14.

Dukes, J. S., Pontius, J., Orwig, D., Garnas, J. R., Rodgers, V. L., Brazee, N., Cooke, B., Theoharides, K. A., Stange, E. E., & Harrington, R. (2009). Responses of insect pests, pathogens, and invasive plant species to climate change in the forests of northeastern North America: What can we predict? *Canadian Journal of Forest Research*, *39*(2), 231–248.

Epanchin-Niell, R. S., Hufford, M. B., Aslan, C. E., Sexton, J. P., Port, J. D., & Waring, T. M. (2010). Controlling invasive species in complex social landscapes. *Frontiers in Ecology and the Environment*, 8(4), 210–216.

Falk, D. A. (2006). Process-centred restoration in a fire-adapted ponderosa pine forest. *Journal for Nature Conservation*, *14*(3–4), 140–151.

Fettig, C. J., Reid, M. L., Bentz, B. J., Sevanto, S., Spittlehouse, D. L., & Wang, T. (2013). Changing climates, changing forests: A western North American perspective. *Journal of Forestry*, *111*(3), 214–228.

Frank, M., Bowyer, D. J., Bratkovich, S., Erickson, G., Fernholz, K., Groot, H., & Howe, D. J. (2015). *Fire adapted communities: Nationwide efforts and community-based tools for living with wildfire* (p. 14). Dovetail Partners.

Frost, C. (2020). Restoring the fire continuum: Restoring fire integrity. In S. M. Hood, S. Drury, T. Steelman, & R. Steffens (Eds.), Proceedings of the Fire Continuum Conference-Preparing for the Future of Wildland Fire: Vol. RMRS-P-78 (pp. 84–94). US Department of Agriculture, Forest Service, Rocky Mountain Research Station.

Fulé, P. Z., Crouse, J. E., Roccaforte, J. P., & Kalies, E. L. (2012). Do thinning and/or burning treatments in western USA ponderosa or Jeffrey pine-dominated forests help restore natural fire behavior? *Forest Ecology and Management*, *269*, 68–81.

Gandhi, K. J., Campbell, F., & Abrams, J. (2019). Current status of forest health policy in the United States. *Insects*, *10*(4), 106.

Graham, R. T., Harvey, A. E., Jain, T. B., & Tonn, J. R. (1999). *The effects of thinning and similar stand treatments on fire behavior in Western forests* (PNW-GTR-463). USDA Forest Service, Pacific Northwest Research Station.

Greiner, S. M., Schultz, C. A., & Kooistra, C. (2021). Pre-season fire management planning: The use of Potential Operational Delineations to prepare for wildland fire events. *International Journal of Wildland Fire*, *30*(3), 170–178.

Grünwald, N. J., LeBoldus, J. M., & Hamelin, R. C. (2019). Ecology and evolution of the sudden oak death pathogen Phytophthora ramorum. *Annual Review of Phytopathology*, *57*, 301–321.

Hessburg, P. F., & Agee, J. K. (2003). An environmental narrative of Inland Northwest United States forests, 1800–2000. *Forest Ecology and Management*, *178*(1), 23–59.

Hicke, J. A., Meddens, A. J., & Kolden, C. A. (2016). Recent tree mortality in the western United States from bark beetles and forest fires. *Forest Science*, *62*(2), 141–153.

Higuera, P. E., & Abatzoglou, J. T. (2021). Record-setting climate enabled the extraordinary 2020 fire season in the western United States. *Global Change Biology*, *27*(1), 1–2.

Higuera, P. E., Abatzoglou, J. T., Littell, J. S., & Morgan, P. (2015). The changing strength and nature of fire-climate relationships in the northern Rocky Mountains, USA., 1902–2008. *PLOS ONE*, *10*(6), e0127563.

Higuera, P. E., Shuman, B. N., & Wolf, K. D. (2021). Rocky Mountain subalpine forests now burning more than any time in recent millennia. *Proceedings of the National Academy of Sciences*, *118*(25), e2103135118.

Hudak, A. T., Fekety, P. A., Kane, V. R., Kennedy, R. E., Filippelli, S. K., Falkowski, M. J., Tinkham, W. T., Smith, A. M., Crookston, N. L., & Domke, G. M. (2020). A carbon monitoring system for mapping regional, annual aboveground biomass across the northwestern USA. *Environmental Research Letters*, *15*(9), 095003.

Jenkins, M. J., Runyon, J. B., Fettig, C. J., Page, W. G., & Bentz, B. J. (2014). Interactions among the mountain pine beetle, fires, and fuels. *Forest Science*, 60(3), 489–501.

Johnston, J. D., Olszewski, J. H., Miller, B. A., Schmidt, M. R., Vernon, M. J., & Ellsworth, L. M. (2021). Mechanical thinning without prescribed fire moderates wildfire behavior in an Eastern Oregon, USA ponderosa pine forest. *Forest Ecology and Management*, *501*, 119674.

Keane, R. E., Ryan, K. C., Veblen, T. T., Allen, C. D., Logan, J. A., & Hawkes, B. (2002). *Cascading effects of fire exclusion in Rocky Mountain ecosystems: A literature review* (RMRS-GTR-91). US Department of Agriculture, Forest Service, Rocky Mountain Research Station.

Khorshidi, M. S., Dennison, P. E., Nikoo, M. R., AghaKouchak, A., Luce, C. H., & Sadegh, M. (2020). Increasing concurrence of wildfire drivers tripled megafire critical danger days in Southern California between 1982 and 2018. *Environmental Research Letters*, *15*(10), 104002.

Kirwan, M. L., & Gedan, K. B. (2019). Sea-level driven land conversion and the formation of ghost forests. *Nature Climate Change*, 9(6), 450–457.

Kooistra, C., Schultz, C. A., Abrams, J., & Huber-Stearns, H. (2022). Institutionalizing the United States Forest Service's Shared Stewardship strategy in the western United States. *Journal of Forestry*, fvac010. https://doi.org/10.1093/jofore/fvac010.

Lake, F. K., Wright, V., Morgan, P., McFadzen, M., McWethy, D., & Stevens-Rumann, C. (2017). Returning fire to the land: Celebrating traditional knowledge and fire. *Journal of Forestry*, *115*(5), 343–353.

Langston, N. (1995). *Forest dreams, forest nightmares: The paradox of old growth in the Inland West*. University of Washington Press.

Lesk, C., Coffel, E., D'Amato, A. W., Dodds, K., & Horton, R. (2017). Threats to North American forests from southern pine beetle with warming winters. *Nature Climate Change*, *7*(10), 713–717.

Lichtman, P. (1998). The politics of wildfire: Lessons from Yellowstone. *Journal of Forestry*, 96(5), 4–9.

Long, J. W., Lake, F. K., Goode, R. W., & Burnette, B. M. (2020). How traditional tribal perspectives influence ecosystem restoration. *Ecopsychology*, *12*(2), 71–82.

Lovett, G. M., Weiss, M., Liebhold, A. M., Holmes, T. P., Leung, B., Lambert, K. F., Orwig, D. A., Campbell, F. T., Rosenthal, J., & McCullough, D. G. (2016). Nonnative forest insects and pathogens in the United States: Impacts and policy options. *Ecological Applications*, 26(5), 1437–1455.

Lynn, K., & Gerlitz, W. (2006). Mapping the relationship between wildfire and poverty. In P. L. Andrews & B. W. Butler (Eds.), *Fuels management-How to measure success* (pp. 401–415). USDA Forest Service Rocky Mountain Research Station.

Méndez, M., Flores-Haro, G., & Zucker, L. (2020). The (in)visible victims of disaster: Understanding the vulnerability of undocumented Latino/a and indigenous immigrants. *Geoforum*, *116*, 50–62.

Mietkiewicz, N., Balch, J. K., Schoennagel, T., Leyk, S., St. Denis, L. A., & Bradley, B. A. (2020). In the line of fire: Consequences of human-ignited wildfires to homes in the US (1992–2015). *Fire*, *3*(3), 50.

Miller, J. D., Safford, H. D., Crimmins, M., & Thode, A. E. (2009). Quantitative evidence for increasing forest fire severity in the Sierra Nevada and southern Cascade mountains, California and Nevada, USA. *Ecosystems*, *12*(1), 16–32.

Mockrin, M. H., Fishler, H. K., & Stewart, S. I. (2018). Does wildfire open a policy window? Local government and community adaptation after fire in the United States. *Environmental Management*, 62(2), 210–228.

North, M., Collins, B. M., & Stephens, S. (2012). Using fire to increase the scale, benefits, and future maintenance of fuels treatments. *Journal of Forestry*, *110*(7), 392–401.

Ojerio, R., Moseley, C., Lynn, K., & Bania, N. (2011). Limited involvement of socially vulnerable populations in federal programs to mitigate wildfire risk in Arizona. *Natural Hazards Review*, *12*(1), 28–36.

Parsons, D. J., & DeBenedetti, S. H. (1979). Impact of fire suppression on a mixed-conifer forest. *Forest Ecology and Management*, *2*, 21–33.

Paveglio, T. B., Abrams, J., & Ellison, A. (2016). Developing fire adapted communities: The importance of interactions among elements of local context. *Society & Natural Resources*, *29*(10), 1246–1261.

Paveglio, T. B., & Edgeley, C. (2020). Fire adapted community. In S. L. Manzello (Ed.), *Encyclopedia of wildfires and wildland-urban interface (WUI) fires* (pp. 1–9). Springer Nature Switzerland AG.

Paveglio, T. B., Moseley, C., Carroll, M. S., Williams, D. R., Davis, E. J., & Fischer, A. P. (2015). Categorizing the social context of the wildland urban interface: Adaptive capacity for wildfire and community "archetypes." *Forest Science*, *61*(2), 298–310.

Poland, T. M., & McCullough, D. G. (2006). Emerald ash borer: Invasion of the urban forest and the threat to North America's ash resource. *Journal of Forestry*, *104*(3), 118–124.

Pyne, S. J. (2021). *The Pyrocene: How we created an age of fire, and what happens next*. University of California Press.

Raffa, K. F., Aukema, B. H., Bentz, B. J., Carroll, A. L., Hicke, J. A., Turner, M. G., & Romme, W. H. (2008). Cross-scale drivers of natural disturbances prone to anthropogenic amplification: The dynamics of bark beetle eruptions. *Bioscience*, *58*(6), 501–517.

Ranco, D., Arnett, A., Latty, E., Remsburg, A., Dunckel, K., Quigley, E., Lillieholm, R., Daigle, J., Livingston, B., & Neptune, J. (2012). Two Maine forest pests: A comparison of approaches to understanding threats to hemlock and ash trees in Maine. *Maine Policy Review*, *21*(1), 76–89.

Reinhardt, E. D., Keane, R. E., Calkin, D. E., & Cohen, J. D. (2008). Objectives and considerations for wildland fuel treatment in forested ecosystems of the interior western United States. *Forest Ecology and Management*, *256*(12), 1997–2006.

Rizzo, D. M., & Garbelotto, M. (2003). Sudden oak death: Endangering California and Oregon forest ecosystems. *Frontiers in Ecology and the Environment*, *1*(4), 197–204.

Ryan, K. C., Knapp, E. E., & Varner, J. M. (2013). Prescribed fire in North American forests and woodlands: History, current practice, and challenges. *Frontiers in Ecology and the Environment*, *11*(s1), e15–e24.

Schieder, N. W., & Kirwan, M. L. (2019). Sea-level driven acceleration in coastal forest retreat. *Geology*, *47*(12), 1151–1155.

Schultz, C. A., Abrams, J. B., Davis, E. J., Cheng, A. S., Huber-Stearns, H. R., & Moseley, C. (2021). Disturbance shapes the US forest governance frontier: A review and conceptual framework for understanding governance change. *Ambio*, *50*(12), 2168–2182.

Schultz, C. A., McCaffrey, S. M., & Huber-Stearns, H. R. (2019a). Policy barriers and opportunities for prescribed fire application in the western United States. *International Journal of Wildland Fire*, *28*(11), 874–884.

Schultz, C. A., & Moseley, C. (2019). Collaborations and capacities to transform fire management. *Science*, *366*(6461), 38–40.

Schultz, C. A., Thompson, M. P., & McCaffrey, S. M. (2019b). Forest Service fire management and the elusiveness of change. *Fire Ecology, 15*(1), 13.

Schultz, C. A., Timberlake, T., Wurtzebach, Z., McIntyre, K., Moseley, C., & Huber-Stearns, H. (2019c). Policy tools to address scale mismatches: Insights from US forest governance. *Ecology and Society, 24*(1), 21.

Seielstad, C. (2015). Reconsidering wildland fire use: Perspectives from the Northern Rockies. In R. E. Keane, M. Jolly, R. Parsons, & K. Riley (Eds.), Proceedings of the Large Wildland Fires Conference (Vol. 73, pp. 207–212). US Department of Agriculture, Forest Service, Rocky Mountain Research Station, Missoula, MT.

Sherriff, R. L., & Veblen, T. T. (2008). Variability in fire–climate relationships in ponderosa pine forests in the Colorado Front Range. *International Journal of Wildland Fire, 17*(1), 50–59.

Shive, K. L., Wuenschel, A., Hardlund, L. J., Morris, S., Meyer, M. D., & Hood, S. M. (2022). Ancient trees and modern wildfires: Declining resilience to wildfire in the highly fire-adapted giant sequoia. *Forest Ecology and Management, 511*, 120110.

Steen-Adams, M. M., Abrams, J. B., Huber-Stearns, H. R., Bone, C., & Moseley, C. (2022). Leveraging administrative capacity to manage landscape-scale, cross-boundary disturbance in the Black Hills: What roles for federal, state, local, and nongovernmental partners? *Journal of Forestry, 120*(1), 86–105.

Steen-Adams, M. M., Charnley, S., McLain, R. J., Adams, M. D. O., & Wendel, K. L. (2019). Traditional knowledge of fire use by the Confederated Tribes of Warm Springs in the eastside Cascades of Oregon. *Forest Ecology and Management, 450*, 117405.

Stephens, S. L., Lydersen, J. M., Collins, B. M., Fry, D. L., & Meyer, M. D. (2015). Historical and current landscape-scale ponderosa pine and mixed conifer forest structure in the Southern Sierra Nevada. *Ecosphere, 6*(5), 79.

Stephens, S. L., & Moghaddas, J. J. (2005). Experimental fuel treatment impacts on forest structure, potential fire behavior, and predicted tree mortality in a California mixed conifer forest. *Forest Ecology and Management, 215*(1), 21–36.

Stevens-Rumann, C. S., Kemp, K. B., Higuera, P. E., Harvey, B. J., Rother, M. T., Donato, D. C., Morgan, P., & Veblen, T. T. (2018). Evidence for declining forest resilience to wildfires under climate change. *Ecology Letters, 21*(2), 243–252.

Thompson, M. P., MacGregor, D. G., Dunn, C. J., Calkin, D. E., & Phipps, J. (2018). Rethinking the wildland fire management system. *Journal of Forestry, 116*(4), 382–390.

Timberlake, T. J., Schultz, C. A., & Abrams, J. (2017). *Resilience in land management planning: Policy mandates, approaches, and resources* (Working Paper 77). Ecosystem Workforce Program, Institute for a Sustainable Environment.

Timberlake, T. J., Schultz, C. A., Evans, A., & Abrams, J. B. (2021). Working on institutions while planning for forest resilience: A case study of public land management in the United States. *Journal of Environmental Planning and Management, 64*(7), 1291–1311.

US Department of Agriculture & US Department of the Interior. (2014). *The national strategy: The final phase in the development of the National Cohesive Wildland Fire Management Strategy*. US Department of Agriculture, Forest Service, and Department of the Interior.

USDA Forest Service. (2015). *The rising cost of wildfire operations: Effects on the Forest Service's non-fire work*. USDA Forest Service.

van Mantgem, P. J., Nesmith, J. C., Keifer, M., Knapp, E. E., Flint, A., & Flint, L. (2013). Climatic stress increases forest fire severity across the western United States. *Ecology Letters, 16*(9), 1151–1156.

van Wagtendonk, J. W. (2007). The history and evolution of wildland fire use. *Fire Ecology*, *3*(2), 3–17.

Vaughn, J., & Cortner, H. (2005). *George W. Bush's healthy forests: Reframing the environmental debate*. University Press of Colorado.

Voggesser, G., Lynn, K., Daigle, J., Lake, F. K., & Ranco, D. (2013). Cultural impacts to tribes from climate change influences on forests. *Climatic Change*, *120*(3), 615–626.

Westerling, A. L. (2016). Increasing western US forest wildfire activity: Sensitivity to changes in the timing of spring. *Philosophical Transactions of the Royal Society B: Biological Sciences*, *371*(1696), 20150178.

Whyte, K. P. (2013). Justice forward: Tribes, climate adaptation and responsibility. *Climatic Change*, *120*(3), 517–530.

Williams, A. P., Abatzoglou, J. T., Gershunov, A., Guzman-Morales, J., Bishop, D. A., Balch, J. K., & Lettenmaier, D. P. (2019). Observed impacts of anthropogenic climate change on wildfire in California. *Earth's Future*, *7*(8), 892–910.

Young, J. D., Evans, A. M., Iniguez, J. M., Thode, A., Meyer, M. D., Hedwall, S. J., McCaffrey, S., Shin, P., & Huang, C.-H. (2020). Effects of policy change on wildland fire management strategies: Evidence for a paradigm shift in the western US? *International Journal of Wildland Fire*, *29*(10), 857–877.

# 6 State and County Government Roles

On October 10, 2009, then-Governor Ted Kulongoski and representatives of the Nez Perce Tribe and Confederated Tribes of the Umatilla Indian Reservation held a small public ceremony near Wallowa Lake and the Old Chief Joseph gravesite to dedicate the newest addition to the Oregon State Park system. The ceremony, held in a gravel parking lot on the edge of the 62-acre parcel of rocky grassland dotted with ponderosa pines and Douglas-firs (Figure 6.1), represented a peaceful end to a land-use battle that had raged for the previous fifteen years. Long used for livestock grazing, the property in question was purchased in 1991 by a developer and businessman who soon made plans to subdivide it for sale as scenic homesites. Although the parcel was situated within the "urban growth boundary" of the city of Joseph—the area where residential development was expected to occur under Oregon land-use law—development proposals raised concerns among locals who used the property as a kind of informal commons for hiking, fishing, hunting, and other activities. Even more concerned were the Nez Perce and allied Plateau Tribes, who recognized the site for its historical and cultural importance as a traditional sockeye salmon fishing location and for its proximity to the final resting place of Old Chief Joseph.

Wallowa County approved a 2003 proposal by the landowner to develop eleven upscale homesites, but the decision was appealed to the state land-use court by the City of Joseph, a group of local residents, and three Plateau Tribes (the Nez Perce, Confederated Tribes of the Colville Reservation, and Confederated Tribes of the Umatilla Indian Reservation). The court reversed the county's decision, citing a failure to follow archaeological and other land-use planning requirements. By 2005, Oregon voters had approved a ballot measure that eviscerated much of the state's land-use planning framework; the landowners began building roads on the site under the new voter-approved rules. This action sparked dueling protests on and near the property, with Plateau Tribes and their allies emphasizing the sanctity of the land and the property owner and allies emphasizing the sanctity of private property rights. Before development proceeded further, the landowner accepted a buyout offer of 4.1 million dollars, of which 3.2 million came from a state conservation fund and the balance from the three Plateau Tribes that had appealed the earlier decision. The new state heritage site was named Iwetemlaykin, roughly translating to "at the edge of the lake."[1]

DOI: 10.4324/9781003043669-6

*Figure 6.1* The Iwetemlaykin State Heritage Site was dedicated in 2009 following a protracted social and political battle over preservation, residential development, and recognition of the site's cultural history. Photograph by the author.

The land-use conflict at the edge of Wallowa Lake—and its resolution—exemplify the roles of states and counties in governing the tensions among the varied private and public values in land. Through prescriptive regulations, voluntary incentives, participatory planning, and other means, state and county governments define the rights, privileges, and obligations associated with nonfederal-forestlands, all within the limits of federal constitutional principles and treaty protections. This chapter examines the many influences of states and counties in owning, managing, and governing nonfederal forests and participating in the governance of federal forests.

## State and County Forestlands

The story of Iwetemlaykin illustrates one means by which states and counties participate in forest governance: through the title ownership of forestland managed by state agencies. As discussed in Chapter 2, the first state forestry agencies were founded in the late nineteenth century to confront issues of uncontrolled wildfire and poor logging practices. State-led forest conservation efforts received federal support via policies such as the Weeks Act and Clarke-McNary Act in the first decades of the twentieth century. This led to the development of state capacities in fire protection and, eventually, to a more comprehensive set of roles and duties carried out by state foresters and forest agencies. At the same time that the federal government was amassing a conservation estate to be managed by federal government foresters, states in the Northeast, Upper Midwest, and Northwest took ownership of extensive areas of burned-over and cutover land. Other states held on to properties that had been in their possession since statehood (or properties exchanged for state land grants) and purchased additional sites of recreational or conservation value. These retained and purchased lands became the state forests, parks, preserves, and similar designations that currently account for approximately 7% of the nation's forestland.

Somewhat less common are county-level forestlands and forest management agencies. Minnesota and Wisconsin account for the vast majority of the 5.4 million acres of county-owned forestlands in the US, in both cases acquired largely on lands that had been stripped of timber and abandoned—sometimes following failed attempts at farming. Oregon, Michigan, New York, Washington, and Pennsylvania also contain substantial county-owned forest acreage.[2] States and counties have largely adopted the federal government's distinction between properties designated as "parks" (recreation-oriented lands such as state and county parks) and those designated as "forests" (multiple use-oriented lands such as state and county forests). State and county forests are frequently managed primarily for timber revenues, with conservation, recreation, and other uses subordinate—roughly paralleling the approach adopted by the US Forest Service in the decades prior to the timber wars.

State and county forestlands are required to be financially self-sustaining in many states; in others, they have obligations to return the greatest value to taxpayers. Timber harvest revenues directly support public schools in many states and counties. Compared to federal forest managers, state forest managers tend to operate under simpler policy mandates, generally with limited requirements for public involvement, and they work within less complex bureaucracies. Due to higher harvest levels and lower obligations for environmental protection and public process, state forestlands typically produce more timber on a per-acre basis than do similarly situated national forests.[3] County forest managers may have even greater latitude than do state managers to set and implement their own management practices due to the relative lack of complex legislation or multiple-use mandates.[4] As political scientist Steven Davis observed,

> The great irony, then, is that Gifford Pinchot's Progressive Era creation, the USFS, with its considerable lore and proud agency culture of bureaucratic professionalism and scientific expertise, is actually the forest agency that must act most often as referee, conciliator, honest broker, and juggler of diverse public needs, goals, and preferences, while tiny little county forest departments operate as they see fit according to the tenets of professional forestry.[5]

This observation does not apply across all county and state forest agencies, however; some have found themselves grappling with the same social and political dilemmas surrounding forest management that federal forest agencies have historically had to face. For example, managers of the Natchez Trace State Forest in western Tennessee were forced to implement a moratorium on clearcutting—the state's preferred harvest system—after public outcry mounted in the 1980s. A similar controversy struck the Massachusetts Department of Conservation and Recreation in 2010 when the agency implemented a series of smaller clearcuts on state forestland, including within Boston's municipal watershed. Conflict flared again in 2018; activists chained themselves to trees in the Wendell State Forest in an attempt to block a pending timber sale. In Oregon, Washington, and California—the epicenter of the timber wars of the 1990s—state forestlands have become embroiled in controversies over the seeming incompatibility of timber management and endangered species protection, and some have been the sites of direct-action protests (Box 6.1).

A growing trend in recent decades is the purchase by states and counties of full title to land or of conservation easements for open space preservation and habitat conservation. Funds for these purchases come from federal sources such as the Forest Legacy Program (see Chapter 8), from state and county general funds, or from dedicated public conservation funds. Several states maintain programs to acquire or protect environmentally important lands or lands crucial for water supply protection. Florida has been a leader in land acquisition since 1964 with the creation of the Land Acquisition Trust Fund, followed by the Land Conservation Act (1972), the Conservation and Recreational Lands program (1979), Preservation 2000 (1990), and, most recently, Florida Forever (1998). Through these combined programs the state has purchased title to over six million acres of land.[6] Voters in numerous states have approved ballot initiatives dedicating money specifically to forest and other open space protection. For example, in 2018 Georgia voters overwhelmingly approved a constitutional amendment that dedicates up to 80% of outdoor sporting goods sales tax funds to land and open space conservation. According to the Trust for Public Land, 2,863 land conservation ballot measures at the state or local level were presented to voters between 1988 and 2021; 76% of them passed, generating a total of over $83 billion in conservation funding.[7]

## BOX 6.1 OREGON'S ELLIOTT STATE FOREST

The Elliott State Forest, located in the coast range near Coos Bay, was established in 1930 through a land exchange with the federal government; subsequent lands added as private forests fell into tax delinquency and passed from the counties to the state. The Elliott was originally envisioned as a resource that would generate revenue for public education, a legacy of its origin in the "school lands" deeded to Oregon upon statehood. At the time of the exchange, much of the forest was comprised of younger stands regrowing in the wake of the 1868 Coos Bay Fire. The first state timber harvest on the Elliott occurred in 1945, and a regular program of timber sales began in earnest in the 1950s. By the late 1990s, harvests of the forest's extensive stands of Douglas-fir, western red cedar, and other valuable species were regularly bringing in over ten million dollars annually—but they were also diminishing habitat for species such as the marbled murrelet, a seabird that was listed as threatened under the Endangered Species Act in 1992. Following activist blockades of planned timber sales between 2009 and 2011, three conservation organizations filed a lawsuit against the Oregon Department of Forestry over its plans to accelerate harvest levels, including through harvesting older forests used by the marbled murrelet. The State of Oregon settled the lawsuit, dropping numerous planned timber sales and agreeing to new forest management considerations. The years 2013 and 2014 saw the forest operating at a net loss, and this revenue drain from what was meant to be an asset for public schools convinced state leaders to put the property up for sale. The highest bidder was a partnership between a timber management company, a conservation finance organization, and the Cow Creek Band of Umpqua Tribe of Indians, but the state canceled the deal following widespread public outcry. Oregon leaders finally settled on a plan to retain the forest in state ownership, decouple it from the school fund, and rededicate it as a research forest to be managed by Oregon State University. As part of this arrangement, a Habitat Conservation Plan to protect threatened and endangered species will be established, and the university will convene a stakeholder advisory committee to help guide research and management activities.

## State and County Governance of Private Forestland

States and counties play numerous roles in forest governance beyond their roles as forest owners and managers. A nationwide study conducted in 2003 identified a total of 276 separate state-level entities with some authority regarding nonfederal forests.[8] Air and water pollution regulation, forest management oversight, and fish and wildlife regulation were the most common activities for which state agencies were responsible. The same study reviewed 2001 government data and identified a total of 187 active federal programs potentially affecting nonfederal-forests, with annual program budgets exceeding 1.6 billion dollars. Many, though not all, of these programs operated through federal coordination with state or county entities via grants, loans, technical advice and information, service provision, and other mechanisms.[9] As of 2020, there were nearly 8,000 foresters employed full-time by state agencies nationwide.[10]

States and counties have at their disposal a broad array of policy tools for influencing the disposition and management of privately owned forests. These include policy carrots like tax incentives and cost-share programs, sticks such as forest practice regulations and water quality protections, and sermons related to outreach and extension activities. A key distinction between federal and state governments is that many states directly regulate private land use within their borders, whereas regulation of private lands by the federal government comes indirectly via policies like the Endangered Species Act and Migratory Bird Treaty Act. Further, some of these federal laws—like the Clean Water and Clean Air Acts—operate primarily through cooperative federalism, in which the states, tribes, and territories take the lead role in interpreting and implementing federal policy. States can regulate forest management directly through forest practice acts, and both states and local governments (counties, municipalities) can restrict or allow land subdivision, development, road construction, and other activities through zoning and related forms of land-use regulation. Other state policies indirectly influence forest management as well, for example those that license forestry professionals or those that include wood-based energy sources in statewide renewable energy mandates. Many states have developed voluntary programs that utilize economic incentives to achieve land conservation goals. Combined with their taxation authority (covered in Chapter 8), this means that state and local governments are often the principal source of policy influence on private forest management.

Running throughout these varied policies and programs is the persistent tension regarding tradeoffs between private landowner privileges, broad public benefits, and the proper role of governments in mediating these tradeoffs. Forest policy scholars Erin Kelly and Mindy Crandall identified four distinct discourses (broad paradigms communicated through language) of state roles in governing forest practices on private lands. One discourse emphasizes private landowners as stewards and promotes voluntary guidelines rather than regulations to achieve desired outcomes. A contrasting discourse positions government scientists as experts, favoring mandatory rules and regulations to serve public interests while allowing landowners to benefit from the use of their forestlands. A separate discourse sees non-governmental forestry professionals as key to promoting good stewardship; under this model, one of the most important roles for states is to ensure that only licensed foresters are providing guidance to landowners. A fourth discourse emphasizes participation in incentive-based conservation programs as key to achieving public well-being while also protecting the benefits that accrue to private landowners. These four discourses are not necessarily mutually exclusive; rather, they may be used by individual states in complementary or overlapping ways depending on the issue at hand.[11]

### *Cooperative Federalism and Best Management Practices*

As described in Chapter 4, best management practices (BMPs) created and enforced by states, tribes, and territories are the primary tools for addressing nonpoint pollution sources under the federal Clean Water Act. BMPs are operational measures designed to minimize water quality impacts from forest management and other ground-disturbing activities, addressing issues such as road construction and design, stream crossings, log skidding, use of pesticides, prescribed fire application, replanting after harvest, location of landings and log decks, and special considerations for streamside management zones adjacent to waterways.[12] BMPs typically vary based on factors such as slope, soil erodibility, and proximity to resources of particular importance. For example, many states define special required or recommended practices for operating in wetlands, and some include BMPs specific to trout-bearing streams. Important roles for states under the Clean Water Act are to define forestry BMPs; design policies and programs to ensure BMP compliance; and conduct monitoring of programs, compliance levels, and impacts. Forestry BMPs have been defined in all 50 states and are revised regularly.[13]

Four broad approaches have been adopted by states for encouraging or ensuring BMP compliance; these can be classified as voluntary, regulatory, quasi-regulatory, and local option systems.[14] Voluntary approaches rely on education and outreach, cost-share programs, tax incentives, and other policy carrots to encourage landowners and forest operators to adopt BMPs but do not require compliance. Regulatory approaches, on the other hand, require compliance with BMPs, employing policy sticks in the form of fines and even jail time for violations. Quasi-regulatory systems include elements of both voluntary and regulatory systems. For example, BMP compliance may not be required, but fines are still possible in the case of water quality violations, or, alternatively, BMP compliance may be mandatory for certain practices or in certain places and voluntary with respect to other activities or locations.[15] Finally, in local option systems, the state governments do not require BMP compliance, but local governments within those states may opt to do so. Table 6.1 lists the system used by each state.

Given the wide array of state approaches to BMP regulation and the variety of specific practices included in BMPs, two important questions are whether BMPs are in fact being implemented and whether existing BMPs work effectively to protect water quality. As part of understanding patterns of BMP implementation,

*Table 6.1* State adoption of four broad approaches to encouraging or ensuring BMP compliance

| *BMP Compliance Approach* | *States* |
|---|---|
| Voluntary | Arkansas, Colorado, Connecticut, Hawaii, Illinois, Indiana, Iowa, Kansas, Louisiana, Minnesota, Mississippi, Missouri, Nebraska, North Dakota, Oklahoma, Rhode Island, South Carolina, South Dakota, Tennessee, Texas, Wyoming |
| Regulatory | Alaska, California, Delaware, Idaho, Kentucky, Maryland, Massachusetts, Nevada, New Jersey, Oregon, Pennsylvania, Washington, West Virginia |
| Quasi-Regulatory | Alabama, Florida, Georgia, Michigan, Montana, New Hampshire, New Mexico, North Carolina, Ohio, Vermont, Wisconsin |
| Local Option | Arizona, Maine, New York, Utah, Virginia |

Data Source: National Association of State Foresters (2019).

state, federal, and non-governmental entities have all participated in monitoring efforts. However, given the diversity of monitoring approaches and of BMPs themselves, these state findings cannot easily be compared with one another. A 2013 survey found an overall BMP implementation rate of 91% across 34 states, with individual states ranging from 80% to 99% compliance[16], and a 2019 survey found similar results: from a statewide low of 73% compliance to a high of over 99% and an overall average of 92% from the 32 states that conduct compliance monitoring.[17] The question of BMP effectiveness has been somewhat more challenging to answer definitively. Research has found evidence that BMPs are generally effective in reducing water quality impacts from forestry operations, particularly impacts related to the introduction of sediment into waterways and the loss of streamside shading.[18] However, there is less evidence of BMP efficacy in the case of some nutrient inputs.[19] Many scholars studying efficacy have concluded that more research is needed before firm conclusions can be drawn regarding the adequacy of existing BMPs for a range of water quality impacts.[20]

### *Forest Practice Acts*

Beyond their role in setting BMPs for water quality protection, state governments may opt to craft more holistic policies that regulate nonfederal forest management within their borders. As early as the beginning of the twentieth century, various states had passed laws designed to ensure long-term timber supplies, for example by requiring minimum cutting diameters or the retention of seed trees.[21] Additional state-level forestry regulations were adopted in the 1940s and 1950s, and a more comprehensive suite of state forest practice acts began to appear in the 1970s. Statewide forest practice acts vary in terms of scope and specificity; they generally mandate BMP compliance while also incorporating broader concerns with environmental protection, wildlife habitat, aesthetics, reforestation, and the long-term sustainability of forest resources. For example, forest practice acts may prescribe maximum clearcut sizes, reforestation timelines, and requirements for road construction and maintenance. The process of writing and revising forest practice rules is typically led by state-level forestry boards that contain representation from various governmental and non-governmental entities with forestry expertise. Fourteen states currently have statewide comprehensive forest practice acts.[22] Most of these are the same states listed in Table 6.1 as taking a regulatory approach to BMP compliance, but also include some local option states like Maine and Utah as well as the quasi-regulatory state of New Mexico.[23]

Forest practice acts often regulate the application of pesticides and herbicides used in forestry, and even states that do not have comprehensive forest practice legislation (or that do not include pesticides and herbicides in their forest practice rules) nevertheless regulate pesticide use directly. At the federal level, the Environmental Protection Agency sets minimum standards for pesticide regulation; states are responsible for the certification of pesticide applicators and can impose pesticide rules that are stricter than federal standards. An issue of growing controversy in recent years is the aerial application of herbicides, which is a common practice in some parts of the country for controlling brush and competing vegetation following timber harvest. The practice has come under critical scrutiny by environmentalists and by rural residents whose homes and municipal watersheds lie within spray zones. The Pacific Northwest has frequently been at the center of this controversy: public outcry following herbicide-linked fish kills, human health impacts, and elevated rates of miscarriages in the Northwest during the 1970s and 80s led to policy changes that curbed the use of herbicides like 2,4,5-T—infamous as an element of Agent Orange used by the US military as a weapon of war in Vietnam.[24] Yet aerial application of a new generation of herbicides has also sparked concern, and states have adopted divergent approaches to regulating issues such as requirements to provide prior notification and to avoid aerial spraying near schools and wells. In 2017, residents of Lincoln County on the Oregon coast narrowly approved a ballot measure outlawing aerial spraying of forestlands altogether, but a 2019 court ruling found that state pesticide laws preempted counties from setting their own rules. In 2021, the Governor of Maine vetoed a bill passed by the legislature banning aerial forest spraying.

### *Fire Control and Use*

Wildfire control has long been a central concern of state forestry agencies and was a key issue driving federal support of state capacity through laws like the Weeks Act and Clarke-McNary Act. Continuing its role in support of state fire protection, the federal government provides funding and surplus equipment to states and to rural fire departments each year to maintain an integrated network of wildfire responders. Federal funding comes through a variety of mechanisms, including Fire Management Assistance Grants administered by the Federal Emergency Management Agency, grants from US Forest Service State and Private Forestry, and cost-share agreements with federal agencies. For many states, however, these payments represent only a small fraction of overall state firefighting costs, and states are ultimately responsible for fire protection on most nonfederal lands within their borders.[25] California alone has spent over one billion dollars on fire suppression within a single year, and state agencies nationwide spent a total of nearly two billion dollars in 2020 on fire protection and other fire-related activities.[26] These costs are ultimately paid through various sources, including general state funds, land or harvest taxes, mineral royalties, insurance, and, occasionally, recovery of costs from liable parties.[27]

Many state and local governments actively regulate and promote forest restoration and fire reintroduction on nonfederal land. These activities are believed to ultimately reduce the loss of lives, property, and ecological benefits due to uncontrolled wildfire burning outside of historic patterns (see Chapter 5). States set training and qualification standards for prescribed burn bosses and crew members along with standards for the burn plans they use to guide their activities. State law also determines liability standards for any damages resulting from prescribed fires or the smoke they generate. So-called "right to burn" states have laws protecting landowners and burn crew members from legal liability related to prescribed fires; these laws generally hold that as long as those applying prescribed fire meet minimum qualifications, are operating within an approved burn plan, and are not guilty of negligence, they are immune from legal liability for any resulting damage. As of 2018, 26 states had prescribed fire laws that included a "simple negligence" standard for liability, and 7 states (Colorado, Florida, Georgia, Michigan, Montana, Nevada, and South Carolina) included an even higher "gross negligence" standard that would have to be met before a landowner or burn crew member can be held legally liable.[28] On the other hand, 12 states impose "strict liability," holding landowners and burn crew members liable for any resulting damages even if the burn was conducted in full compliance with the law and with general safety considerations.[29]

States have also taken proactive measures to restore more resilient forest conditions in places where fire has been excluded. State forestry agencies lead planning and implementation for prescribed fires across multiple ownership categories: state lands, county and municipal lands, and, in some cases, private lands. They may also work in partnership with federal and non-governmental partners on forest restoration and fire reintroduction on priority landscapes managed by those entities. As discussed in Chapter 5, the reintroduction of fire is seen as a key strategy for forest restoration across varied forest landscapes nationwide, but implementation of this strategy comes with risks. In 2012, a prescribed fire on state forestlands led by the Colorado State Forest Service escaped from containment and blew up into the Lower North Fork Fire, which ultimately burned 27 homes and led to the deaths of 3 people.[30] State leaders responded to this tragic event with a five-year ban on prescribed fire within the state of Colorado.

### *Voluntary Programs*

State forestry professionals carry out diverse duties that can include setting and monitoring BMPs, enforcing mandatory regulations, administering permit programs, directly managing forestlands, and managing fires. They also operate on the frontlines of voluntary program administration and implementation. Their roles include providing advice on issues such as forest planning, management, pest and pathogen control, fire, and related topics, and they are typically the primary point of contact for private forest owners seeking advice or access to programs such as tax incentives, conservation easements, cost-share support, and other voluntary tools. Given the importance of these policy carrots for promoting sustainable forest management—particularly on private lands—state forestry agencies and the programs they implement can greatly influence how forests are ultimately managed.

In some cases, state forestry agencies provide critical capacity and oversight for third-party forest certification on private, state, and other lands. For example, enrollees in Wisconsin's Managed Forest Law program are eligible to receive group certification under both the American Tree Farm System and the Forest Stewardship Council. By acting as the group certificate holder for state forests, county forests, and private forests enrolled in the Managed Forest Law program, the Wisconsin Department of Natural Resources directly oversees 87% of the certified acres in the state.[31] Participants in Indiana's Classified Forest and Wildlands program, described in Chapter 8, are also eligible to participate in group forest certification under the Forest Stewardship Council. In cases such as these, state forestry agencies help to address financial and technical barriers that might otherwise impede the certification of large numbers of small-acreage forest owners (see Chapter 9 for more information on third-party certification).

Several states administer voluntary programs to support forest restoration and fire risk reduction projects. For example, New Mexico's Forest and Watershed Restoration Act, established in 2019, provides annual funding to projects that promote ecological restoration and fire risk reduction in forests and in watersheds supplying municipal or irrigation uses. Proposed projects are selected for funding by a multi-stakeholder advisory board, that operates under prescribed rules regarding project prioritization. Funded projects are intended to support economic development and the utilization of restoration by-products in addition to improving ecological conditions. The Texas A&M Forest Service provides several grant programs to support prescribed fire application in targeted landscapes statewide. California's state fire agency, CAL FIRE, administers grant programs related to forest improvement, fire protection, landscape-scale forest health restoration, and urban and community forestry. The latter two programs are part of the statewide Global Warming Solutions Act of 2006, which emphasizes forests as part of a comprehensive approach to reducing the state's carbon emissions (Box 6.2).

State-led private landowner support programs typically draw upon a mix of federal and nonfederal funding sources. States administer grant programs funded by the State and Private Forestry branch of the US Forest Service, including the Urban and Community Forestry Program (Chapter 10) and Forest Legacy Program (Chapter 8), and states are key partners in the Forest Stewardship Program (Chapter 8). Since 2008, states have been required to develop and regularly update State Forest Action Plans to continue to receive federal financial support through State and Private Forestry. These comprehensive plans bring together information on forest conditions, trends, and threats within the state and identify priority landscapes for conservation along with targeted policies and programs for addressing needs in each priority area. States also administer other federal grant programs, such as those targeting forest insect and disease control, and many state conservation grants are funded from a mix of federal, private, and non-profit organization sources. In other cases, state taxes, fees, and other sources of income are used to fund projects.

## Other State and Local Policies Affecting Forests

### *Energy and Climate Policies*

Despite the important role played by the federal government in setting energy policy—and in regulating air pollution from energy production via the Clean

Air Act—states retain substantial authority to set the rules of the game regarding the energy matrix. State energy policies can directly influence forest governance through support for wood-based bioenergy generation, and both energy and climate policies can affect the incentives for various approaches to forest conservation and management. There exists a wide array of policy tools that states can adopt to promote energy from wood-based sources, and these can influence markets for forest products (including wood chips and pellets) and associated forest management practices. A 2008 nationwide review of state-level wood-based energy policies counted 370 distinct incentives, standards, mandates, and other tools in use.[32] Tax credits are especially popular tools at the state level; these include credits for installing renewable energy systems as well as credits for hauling woody biomass materials, purchasing biomass harvesting equipment, and other activities. Some states explicitly include wood-based energy in renewable portfolio standards (policies that mandate the minimum levels of energy production that must come from renewable sources). Other tools include grants and cost-share programs, financing programs, public procurement policies, and technical assistance.[33] In theory, these state-level programs complement other federal-level rules and incentives, although research suggests that federal and state biomass energy programs do not always work well together in practice.[34]

As demand for wood chips and pellets has increased, so has scientific and public attention to the possible long-term impacts of biomass harvesting practices. Biomass production often utilizes smaller trees, slash, and the tops, branches, and other material that would otherwise normally be left on site, and concerns regarding the impacts of biomass harvesting on soil productivity and wildlife have been raised. Several states responded by creating new biomass harvesting guidelines or by adapting existing guidelines. These are variably incorporated into forest practice acts, BMPs, or issued as stand-alone voluntary guidance. Missouri's biomass harvesting BMPs, for example, address issues such as retention of live trees and saplings, the timing of biomass removal, the quantity of slash and tops to be left on site, and special considerations for features like streams, wetlands, caves, sinkholes, and springs. Wisconsin biomass harvesting guidelines, published in revised form in 2014 after initial guidelines were crafted in 2008, emphasize the need to leave harvest residuals in place on certain soil types—specifically thin soils, sandy soils, and some wetland soils. Other elements of the Wisconsin guidelines focus on retaining existing coarse woody debris, leaving the litter layer intact, and ensuring the retention of fine woody debris. Both Missouri's and Wisconsin's guidelines were crafted in response to an anticipated increase in demand for wood-based energy.[35]

Forests figure prominently in many comprehensive state climate plans and policies. These strategies typically focus on curbing emissions from energy sources while also addressing the carbon emissions associated with other activities such as transportation, construction, and agriculture. For example, the State of Michigan has committed to attain carbon neutrality by 2050, and Michigan state forests are included in the strategy for achieving this goal. In 2021, the state and one of its largest energy producers announced a deal with a private carbon offset firm to generate carbon credits through changes to management on the Pigeon River

Country State Forest in the northern lower peninsula. Key changes include tree planting and refraining from harvesting larger trees that serve as important carbon sinks.[36] Massachusetts climate policy also calls for carbon neutrality by 2050 and allows sequestration by forests and other land uses to account for up to 15% of progress toward this goal.

California's Global Warming Solutions Act is arguably the country's most ambitious and comprehensive state-level climate policy. The multi-pronged strategy is oriented around the goal of progressively reducing carbon emissions from all sources, and its carbon offset provisions have implications for forest governance across the US (Box 6.2). Multi-state climate change initiatives have also been crafted at regional scales: the two prominent examples to date are the Regional Greenhouse Gas Initiative (RGGI) and the Western Climate Initiative, both of which include at least some role for forests to serve as carbon offsets within their broader cap-and-trade programs.[37] For example, RGGI, which encompasses twelve northeastern and Mid-Atlantic states, includes reforestation, improved forest management, and avoided conversion projects within the US as qualifying carbon offsets. Permanent conservation easements (see Chapter 8), with the state listed as the easement holder, are required for participating avoided conservation projects. Like California's Global Warming Solutions Act, RGGI rules require carbon monitoring and verification for 100 years past the awarding of carbon credits.

## BOX 6.2 CALIFORNIA CLIMATE POLICY

The California Global Warming Solutions Act was passed as AB 32 in 2006. The objective of the law, as revised, is to reduce California's total carbon emissions to 1990 levels by 2020, 40% below 1990 levels by 2030, and 80% below 1990 levels by 2050. The 2020 target was met four years early in 2016. One important aspect of the law is a "cap and trade" program administered by the California Air Resources Board. Under the program, major polluters are given emissions allowances as part of an annually declining statewide cap. Businesses whose emissions exceed their allowances must purchase allowances from others or, alternatively, purchase carbon offsets (offsets can comprise no more than 8% of a company's total emissions budget). Forest projects anywhere in the US are eligible activities under the offset market—specifically, projects involving reforestation, improved forest management, and avoided conversion of forests. Eligible projects are those that provide "additionality" over a carbon baseline. In other words, carbon sequestration benefits must be greater than what would have occurred in the absence of the project. To participate, forest owners must conduct a baseline carbon inventory, commit to a management plan that provides carbon sequestration benefits, and provide for monitoring and verification that extends for 100 years from the issuance of any credits. Additionally, participating forest owners must maintain or promote native tree species and the retention of standing and downed dead woody material. Offset credits must be reviewed and certified as "real, additional, quantifiable, permanent, verifiable, and enforceable." Once offset credits have been verified and registered with the state, they can be bought and sold on the open market. As of May 2022, over 182 million credits—each representing one metric ton of sequestered carbon dioxide—had been issued under the forests program. The sale of carbon credits provides another option for financing forest conservation and sustainable management as well as an opportunity for tribal land repatriation (see Chapter 7). State revenues from the program are used to finance various projects that reduce overall emissions, including low-carbon transportation development, retrofitting of industrial facilities, and the implementation of forest restoration and fire risk reduction projects across ownerships.[38]

### *Land-Use Regulation and the Question of Takings*

In addition to regulating forestry-specific activities, state and county governments are responsible for setting the rules regarding rural land subdivision and development—rules that can exacerbate or ameliorate the fragmentation and forest conversion pressures detailed in Chapters 8 and 9. Forestland zoning, minimum lot sizes, and related measures utilize regulatory sticks rather than (or in addition to) incentive-based carrots like tax policy and easements. Rural land zoning was first implemented in Oneida County, Wisconsin, in 1933 to reduce public costs in the depths of the Great Depression and has been adopted in various forms by many state and county governments since that time.[39] Just as urban zoning designates distinct uses and their accompanying restrictions for different parts of a city (such as single-family residential, light industrial, commercial, etc.), forestland zoning designates what uses are and are not permitted in forestland regions outside of urbanized areas. This kind of rural zoning is generally conducted in the context of a comprehensive plan and is usually accompanied by limitations on subdivision and development as well as differential tax rates depending on designation. For example, there is a minimum lot size of 42 acres on most private forestlands within New York's Adirondack Park.[40] Counties in California can designate "timber production zones" (TPZs) in which forestry and compatible activities are the only permitted land uses; land in TPZs is taxed based upon the region and site quality, resulting in lower property taxes than comparable non-TPZ land. Since 1985, Florida has required local governments to develop legally binding comprehensive plans to reduce sprawl and conserve undeveloped natural and semi-natural areas,[41] and Vermont's Act 250, passed in 1970, empowers regional environmental commissions to review proposed development projects and allow or deny permission to move forward.

The vignette from Wallowa County at the beginning of this chapter illustrates one example of state and county influence on land use. Oregon has long been known for its statewide land-use planning law that generally restricts subdivision and residential development on land outside of designated urban boundaries. The principles undergirding Oregon's land-use framework are that urban development should be spatially contained and that residential and productive uses should be kept separate in order to conserve farmland, rangeland, and forestland. Under this system, rural landowners receive favorable tax treatment as well as assurances that their neighbors will also be rural landowners, thereby minimizing the kinds of conflicts over noise, lights, dust, heavy truck traffic, and related issues that tend to arise when residential and productive land uses intermix. Most forestlands in Oregon are zoned with default 80- or 160-acre minimum lot sizes, and most residential development on these lands is prohibited. However, the state's widely lauded system has not gone unchallenged. Measure 37 was a 2004 ballot initiative passed by the state's voters in what was widely seen as a rebuke of the statewide planning framework. The measure required state and local governments to compensate landowners for any regulation that reduces private property values or to waive enforcement of the regulation. A subsequent legislative referral, Measure 49, undid some of Measure 37's most aggressive provisions but still allowed many landowners to pursue residential development in places that would have been disallowed under the earlier state land-use rules.[42]

The Oregon story is a microcosm of the persistent tensions between private and public interests in forestland that play out nationwide on a daily basis. Forests generate irreplaceable public benefits by sequestering carbon from the atmosphere; stabilizing soil; providing water for drinking and irrigation; maintaining habitat for birds, wildlife, plants, fungi, and other components of biodiversity; providing places for recreation; producing timber and non-timber forest products; and performing many other so-called "ecosystem services." Yet the majority of the nation's forests are privately owned, and some forest owners may be able to generate greater private benefits through converting, subdividing, or developing forestlands than by maintaining them as forest. To what extent should state and county governments place restrictions on those activities to protect public interests, and to what extent should landowners expect monetary compensation when development activities are restricted? Proponents of greater regulation might argue that providing for general societal wellbeing without unreasonably burdening private options is the essence of public policymaking and that restricting some forms of forest subdivision and development accomplishes this duty. Advocates for landowner freedoms might point to a long history of private property protections in American legal history, going back as far as the influence of John Locke on the framers of the Constitution and the Fifth Amendment prohibition on taking private property for public use "without just compensation."

Courts examining property cases brought under the Fifth Amendment have long held that compensation is required for any physical taking of private property by government, but not necessarily for so-called "regulatory takings," which occur when regulations restrict landowner options. Land-use restrictions may be found to be compensable takings if they leave the owner without reasonable economic uses of the property, if they interfere with distinct, investment-backed expectations, or if they do not advance a legitimate public purpose. On the other hand, if the regulation advances a public benefit and the landowner retains some economically beneficial uses, a claim for compensation under the Fifth Amendment will generally be unsuccessful. State regulation of private forestry practices has been endorsed by state courts going as far back as 1907, and the US Supreme Court upheld a 1949 Washington State Supreme Court decision that denied a Fifth Amendment compensation claim by a landowner forced to comply with state forestry rules.[43] Zoning, in its various forms, has been upheld as a constitutionally legitimate practice by courts at both state and federal levels; a key decision came in 1926 when the US Supreme Court found constitutional one of the nation's first applications of urban zoning to a Cleveland suburb.[44] In short, those who argue that any restriction on private landowner freedoms constitutes a compensable taking under the US Constitution have found little support for their argument in court.

However, the US Constitution is not the only controlling policy when it comes to private property rights, and federal courts are not the only venue in which property owners can seek favorable judgments. With the passage of Measure 37 (and later Measure 49), Oregon joined at least six other states that had adopted laws requiring some form of compensation for land-use restrictions.[45] Even in states lacking compensation laws, county and state officials may choose not to utilize forest practice rules, zoning, minimum lot sizes, and other regulatory tools that have been found to be constitutional. In large part, this is because private property rights are more than simply a legal institution; they are a cultural institution as well. In many places, the cultural understandings of private property are tilted more strongly in favor of the landowner than are the strictly legal interpretations. State and county officials may personally oppose the idea of land-use regulation, they may believe it to conflict with laws or constitutional provisions at the state or national levels, or they may fear the political consequences of regulating private land uses. This still leaves states and counties with the voluntary tools described here and in Chapter 8, such as title purchases, conservation easements, and tax incentives. However, patterns of forest fragmentation in much of the country continue to outpace the reach and influence of these voluntary tools. Forest economist Lloyd Irland observed that "policies such as use-value taxation or other aids to private forestry were never designed to cope with today's land prices and real estate demands. They are a six-inch levee facing a 20-foot storm surge."[46]

## Conclusion

As partners under cooperative federalism, as regulators of private land use, as leaders in fire, energy, and climate governance, and as owners and managers of their own forestlands, states and counties set the trajectory for diverse forestlands within their borders. Through their adoption of varied approaches to issues such as prescribed fire liability, carbon markets, renewable energy policies, forest practices, and land-use regulation, they fulfill their role as "laboratories of democracy," demonstrating the tradeoffs and outcomes resulting from various combinations of policy sticks, carrots, and sermons. Although social conflict on state and county forests has generally not risen to the level seen on federal forestlands, cases like Oregon's Elliott State Forest (Box 6.1) and other scattered examples around the country show that even managers of lands that were once cut over, burned, abandoned, and thought to be nearly worthless are faced with complex and competing societal expectations for forests. Conflict has also been common when states and counties have drawn upon their regulatory powers to stem the loss of forestland or to require carbon emitters to reduce their contributions to global climate change. States and counties are increasingly finding themselves working in partnership with private landowners, the federal government, NGOs, and other entities to address issues that refuse to recognize ownership boundaries—a topic covered in more detail in Chapters 4 and 5.

## Questions for Discussion

1. Why might state and county forest agencies have been insulated for so many years from the kinds of social conflicts that traditionally played out on federal forestlands? Why might this be starting to change?
2. Regional patterns are evident in many policies—such as the prevalence of "right to burn" policies in the South, adoption of mandatory forest practice laws in West Coast states, and state leadership on climate change policy in the West and Northeast. What factors could explain these regional variations?
3. Private land-use regulation is one of the thorniest public policy issues faced by state and county governments. What do you believe is the correct approach for states and counties in the face of continued loss of forests to fragmentation and residential conversion? What are the tradeoffs associated with your preferred approach?
4. How well has cooperative federalism worked in the case of nonpoint source pollution regulation under the Clean Water Act? What would it look like for this model to be institutionalized more widely in other federal laws like the Endangered Species Act?
5. California established the nation's first mandatory carbon market and has allowed forest projects to count as carbon credits. What are some of the implications of this program for forest conservation?

## Notes

1 Abrams, 2016.
2 Davis, 2013.
3 Davis, 2013; Koontz, 2002, 2007.
4 Fleischman et al., 2022.
5 Davis, 2013, p. 220.
6 Farr & Brock, 2006.
7 https://tpl.quickbase.com/db/bbqna2qct?a=dbpage&pageID=8.

8 Ellefson et al., 2007.
9 Ellefson et al., 2007.
10 National Association of State Foresters, 2021.
11 Kelly & Crandall, 2022.
12 Ice, 2004.
13 Cristan et al., 2018.
14 Note that there are different ways of classifying state BMP models and not all analysts agree on how to classify individual states; see Cristan et al. (2018); Kelly & Crandall (2022); National Association of State Foresters (2019).
15 Binkley & Brown, 1993b; Cristan et al., 2018.
16 Cristan et al., 2018.
17 National Association of State Foresters, 2019.
18 Binkley & Brown, 1993a; Cristan et al., 2016; Schuler & Briggs, 2000.
19 Edwards & Williard, 2010; Xu & Xu, 2018.
20 Edwards et al., 2016; Edwards & Williard, 2010; Xu & Xu, 2018.
21 Cubbage & Ellefson, 1980.
22 Cristan et al., 2018.
23 New Mexico forestry BMPs include a mix of voluntary and mandatory standards.
24 Lewis & Miller, 2018.
25 Cook & Becker, 2017; Gorte, 2011; Pennick McIver et al., 2021.
26 National Association of State Foresters, 2021; Pennick McIver et al., 2021.
27 Cook & Becker, 2017; Pennick McIver et al., 2021.
28 Melvin, 2018.
29 Melvin, 2018; Wonkka et al., 2015.
30 Investigations after the event found that as the fire threatened residences, a reverse 911 system failed to function as intended.
31 Abrams et al., 2018.
32 Becker et al., 2011.
33 Becker et al., 2011.
34 Abrams et al., 2017.
35 Evans et al., 2013.
36 https://www.michigan.gov/dnr/0,4570,7-350-86469-566565--,00.html.
37 The Midwestern Greenhouse Gas Reduction Accord was also established, but a lack of political support prevented it from establishing a carbon market and it remains inactive as of the time of this writing.
38 Caldwell et al., 2014; Manning & Reed, 2019.
39 Rowlands, 1933.
40 Daniels, 2010.
41 Linkous & Skuzinski, 2018.
42 Walker & Hurley, 2011.
43 Cubbage & Ellefson, 1980; Freyfogle, 2003.
44 Note that the plaintiffs sued under the "Due Process" and "Equal Protection" clauses of the Fourteenth Amendment in this case.
45 Jacobs, 1999.
46 Irland, 2005, p. 20.

## References

Abrams, J. (2016). Rural residential development and its discontents: A political ecology of sprawl containment in Wallowa County, Oregon, USA. In L. E. Taylor & P. T. Hurley (Eds.), *A comparative political ecology of exurbia* (pp. 83–103). Springer.

Abrams, J., Becker, D., Kudrna, J., & Moseley, C. (2017). Does policy matter? The role of policy systems in forest bioenergy development in the United States. *Forest Policy and Economics*, *75*, 41–48.

Abrams, J., Nielsen, E., Diaz, D., Selfa, T., Adams, E., Dunn, J. L., & Moseley, C. (2018). How do states benefit from nonstate governance? Evidence from forest sustainability certification. *Global Environmental Politics*, *18*(3), 66–85.

Becker, D. R., Moseley, C., & Lee, C. (2011). A supply chain analysis framework for assessing state-level forest biomass utilization policies in the United States. *Biomass and Bioenergy*, *35*(4), 1429–1439.

Binkley, D., & Brown, T. C. (1993a). Forest practices as nonpoint sources of pollution in North America. *Journal of the American Water Resources Association*, *29*(5), 729–740.

Binkley, D., & Brown, T. C. (1993b). *Management impacts on water quality of forests and rangelands* (GTR-RM-239). USDA Forest Service, Rocky Mountain Forest and Range Experiment Station.

Caldwell, B. T., Holland, K. A., Barbane, Z., & Jonas, M. (2014). AB 32: A compliance carbon market for US forests, an opportunity for foresters. *Journal of Forestry*, *112*(1), 60–61.

Cook, P. S., & Becker, D. R. (2017). *State funding for wildfire suppression in the western US* (PAG Report #37; p. 74). Policy Analysis Group, University of Idaho.

Cristan, R., Aust, W. M., Bolding, M. C., Barrett, S. M., & Munsell, J. F. (2018). National status of state developed and implemented forestry best management practices for protecting water quality in the United States. *Forest Ecology and Management*, *418*, 73–84.

Cristan, R., Aust, W. M., Bolding, M. C., Barrett, S. M., Munsell, J. F., & Schilling, E. (2016). Effectiveness of forestry best management practices in the United States: Literature review. *Forest Ecology and Management*, *360*, 133–151.

Cubbage, F. W., & Ellefson, P. V. (1980). State forest practice laws: A major policy force unique to the natural resources community. *Natural Resources Lawyer*, 421–468.

Daniels, T. L. (2010). Integrating forest carbon sequestration into a cap-and-trade program to reduce net CO2 emissions. *Journal of the American Planning Association*, *76*(4), 463–475.

Davis, S. M. (2013). The forests nobody wanted: The politics of land management in the county forests of the Upper Midwest. *Journal of Land Use and Environmental Law*, *28*(2), 197–225.

Edwards, P. J., & Williard, K. W. J. (2010). Efficiencies of forestry best management practices for reducing sediment and nutrient losses in the eastern United States. *Journal of Forestry*, *108*(5), 245–249.

Edwards, P. J., Wood, F., & Quinlivan, R. L. (2016). *Effectiveness of best management practices that have application to forest roads: A literature synthesis* (General Technical Report GTR-NRS-163). US Department of Agriculture, Forest Service, Northern Research Station.

Ellefson, P. V., Kilgore, M. A., & Granskog, J. E. (2007). Government regulation of forestry practices on private forest land in the United States: An assessment of state government responsibilities and program performance. *Forest Policy and Economics*, *9*(6), 620–632.

Evans, A. M., Perschel, R. T., & Kittler, B. A. (2013). Overview of forest biomass harvesting guidelines. *Journal of Sustainable Forestry*, *32*(1–2), 89–107.

Farr, J. A., & Brock, O. G. (2006). Florida's landmark programs for conservation and recreation land acquisition. *Sustain*, *14*(2), 35–44.

Fleischman, F., Schmitz, M., & Poljacik, K. (2022). Laws, markets, and local politics drive outcomes of Minnesota's county managed forests. *Journal of Forestry*, *120*(1), 51–63.

Freyfogle, E. T. (2003). *The land we share: Private property and the common good*. Island Press.

Gorte, R. W. (2011). *Federal funding for wildfire control and management* (No. RL33990). Congressional Research Service.

Ice, G. (2004). History of innovative best management practice development and its role in addressing water quality limited waterbodies. *Journal of Environmental Engineering*, *130*(6), 684–689.

Irland, L. C. (2005). US forest ownership: Historic and global perspective. *Maine Policy Review*, *14*(1), 16–22.

Jacobs, H. M. (1999). *State property rights laws: The impacts of those laws on my land*. Lincoln Institute of Land Policy.

Kelly, E. C., & Crandall, M. S. (2022). State-level forestry policies across the US: Discourses reflecting the tension between private property rights and public trust resources. *Forest Policy and Economics*, *141*, 102757.

Koontz, T. M. (2002). *Federalism in the forest: National versus state natural resource policy*. Georgetown University Press.

Koontz, T. M. (2007). Federal and state public forest administration in the new millennium: Revisiting Herbert Kaufman's *The forest ranger*. *Public Administration Review*, *67*(1), 152–164.

Lewis, J. G., & Miller, C. (2018). Vast, incredible damage: Herbicides and the US forest service. *Forest History Today*, Spring/Fall, 3–12.

Linkous, E., & Skuzinski, T. (2018). Land use decision-making in the wake of state property rights legislation: Examining the institutional response to Florida's Harris Act. *Land Use Policy*, *77*, 603–612.

Manning, B. R. M., & Reed, K. (2019). Returning the Yurok forest to the Yurok Tribe: California's first tribal carbon credit project. *Stanford Environmental Law Journal*, *39*, 71–124.

Melvin, M. A. (2018). *2018 national prescribed fire use survey report* (Technical report 03-18). National Association of State Foresters and Coalition of Prescribed Fire Councils.

National Association of State Foresters (2019). *Protecting the nation's water: State forestry agencies and best management practices*. National Association of State Foresters.

National Association of State Foresters (2021). *State foresters by the numbers: Data and analysis from the 2020 NASF state forestry statistics survey*. National Association of State Foresters.

Pennick McIver, C., Cook, P. S., & Becker, D. R. (2021). The fiscal burden of wildfires: State expenditures and funding mechanisms for wildfire suppression in the western US and implications for federal policy. *State and Local Government Review*, *53*(4), 337–351.

Rowlands, W. A. (1933). County zoning for agriculture, forestry, and recreation in Wisconsin. *The Journal of Land & Public Utility Economics*, *9*(3), 272–282.

Schuler, J. L., & Briggs, R. D. (2000). Assessing application and effectiveness of forestry best management practices in New York. *Northern Journal of Applied Forestry*, *17*(4), 125–134.

Walker, P. A., & Hurley, P. T. (2011). *Planning paradise: Politics and visioning of land use in Oregon*. University of Arizona Press.

Wonkka, C. L., Rogers, W. E., & Kreuter, U. P. (2015). Legal barriers to effective ecosystem management: Exploring linkages between liability, regulations, and prescribed fire. *Ecological Applications*, *25*(8), 2382–2393.

Xu, Z., & Xu, Y. J. (2018). Assessing effectiveness of long-term forestry best management practices on stream water quality at a basin scale: A case study in southern USA. *Environmental Monitoring and Assessment*, *190*, 108.

# 7 Tribal Forests and Forestry

The treaty negotiations that took place in 1855 in the Walla Walla Valley of the Washington Territory brought together thousands of Plateau tribal members and representatives of the Euro-American colonial powers. Washington Territorial Governor Isaac Stevens and Indian Affairs Supervisor for Oregon Territory Joel Palmer represented the interests of the United States government in negotiations with leaders of the Nez Perce, Walla Walla, Cayuse, Umatilla, Yakama, and other tribes. The negotiations resulted in the establishment of reservations over tens of millions of acres in the Pacific Northwest, including the Umatilla Reservation (designated for the Walla Walla, Cayuse, and Umatilla Tribes) and a reservation of over seven million acres for the Nez Perce people that included nearly all of present-day Wallowa County, Oregon. Some 58 distinct chiefs assented to the Nez Perce Treaty, including Old Chief Joseph of the Wallowa Band. Although the Nez Perce leaders were reluctant to cede territory, they feared that refusing to negotiate would mean losing everything. In addition to establishing the boundaries of the lands reserved to tribal nations, the treaties also contained language acknowledging the importance of off-reservation lands and waters to Native livelihoods and guaranteeing continued rights of access. The US government made a commitment to prevent Whites from settling within the reservations and to provide education, health care, and other infrastructure to support the well-being and economic development of the tribal nations.

Despite assurances contained within the treaties and in the words of Isaac Stevens and Joel Palmer, the US government failed to uphold its end of the deal once gold was discovered on Nez Perce lands. White prospectors flooded into the region, establishing settlements in the heart of the reservation; far from discouraging these actions, the Washington territorial government assisted in setting up governmental infrastructure, and both territorial and national armies defended the prospectors' claims to property. A second Nez Perce Treaty in 1863 diminished the reservation to a tenth of its former size, wiping out lands such as those in the Wallowa region. Although the chiefs of some Nez Perce bands assented to the newer treaty, others such as Old Chief Joseph did not. This created a sharp division between the "treaty" and "non-treaty" bands, the latter insisting that they never gave up their territories even as they were pursued and placed in confinement by the US military. Young Chief Joseph, who led

DOI: 10.4324/9781003043669-7

the non-treaty bands' ultimately unsuccessful running battle to the Canadian border, was held in Fort Leavenworth and then the Oklahoma Territory following his capture before being sent to the Colville Reservation in northeastern Washington. He visited the Wallowa country in the years before his death to try to secure legal right to a homeland for his people, but his efforts were rebuffed.[1] Over a century later, the Nez Perce Tribe has a growing presence in Wallowa County through its fisheries program, the annual Tamkaliks celebration, ownership and management of the 16,000-acre Héte'wits Wétes (Precious Lands) wildlife area, and a small population of enrolled tribal members living full-time within the county.

The history of encounters between colonial powers and Indigenous nations in the United States is complex and often tragic. Yet recent decades have also presented opportunities for innovation, renewal, and greater sovereignty for tribal communities. In this chapter we sketch the outlines of some important tribal dimensions of forest policy and governance, including recent tribal history, tribal recognition, tribal land and resource tenure, and an examination of Indigenous traditional ecological knowledge.

## Policy History

No understanding of contemporary tribal forest issues would be complete without at least a brief review of the history of relationships between Native American nations and the governments led by European colonists and their descendants. Native nations represented a formidable obstacle to early colonial settlement and expansion even as they imparted vital knowledge that allowed Euro-American populations to survive on the North American continent.[2] From a legal standpoint, colonial powers expressed contradictory understandings of Native land title. On the one hand, a set of legal theories that came to be known as the "Doctrine of Discovery" framed land, resources, and Native people as subject to the dominion of whichever European nation laid claim to their territories. With a grounding in European natural law concepts (including John Locke's distinction between passive gathering on common lands and productive labor on private lands), these theories provided a legal justification for the subjugation of Native peoples and the usurpation of their resources.[3] An illustration of this paradigm comes from the Washington State Supreme Court, which weighed in on the question of "aboriginal title" to the land in 1916 and concluded in no uncertain terms that racial lineage was inseparable from property rights:

> At no time did our ancestors in getting title to this continent ever regard the aborigines as other than mere occupants, and incompetent occupants, of the soil. Any title that could be had from them was always disdained. From France, from Spain, from Mexico, and from England we have ever proclaimed our title by purchase, by conquest, and by cession, in all of which great transactions the migratory occupant was ignored. Only that title was esteemed which came from white men, and the rights of these have always

> been ascribed by the highest authority to lawful discovery of lands, occupied, to be sure, but not owned by any one before.[4]

The Doctrine of Discovery blended seamlessly with Euro-American beliefs in a racial hierarchy, one that assumed that Native Americans were destined to disappear under the influence of the "superior" newcomers. The physical and cultural genocide that followed came with both legal and philosophical grounding.

Somewhat paradoxically, however, Native nations were also treated as sovereign governments within the terms of European legal systems, leading colonial powers to forge agreements with them (for example, regarding territory or trade) via government-to-government treaties. These treaties came about through various processes, often as a means of resolving the many wars provoked by the incursions of colonial populations into Native territories.[5] The relative power of Euro-American governments and of individual Native nations changed over time, especially following the American Revolution and the consolidation of former colonies under the new Republic. This power imbalance led to a string of broken treaties driven by Euro-American populations' desires for land, water, minerals, forests, and other valuable resources.

Much of the current framework for tribal rights and authorities within the US legal system can be traced back to a series of Supreme Court rulings in the early nineteenth century that reflects this ambiguity between tribal sovereignty and the Doctrine of Discovery. Known as the Marshall Trilogy (named for the Chief Justice who wrote the majority opinions), these decisions positioned tribal nations as possessing a limited sovereignty without full title ownership to lands and as dependent upon the federal government. In *Johnson v. M'Intosh* (1823), the court ruled that any lands retained by tribes during Euro-American colonization fell into a class of property that includes possession and use without full alienable title; ultimate ownership, under this ruling, would remain with the federal government. *Cherokee Nation v. Georgia* (1831) established tribes as "domestic, dependent nations," finding that their relationship to the federal government was akin to that of a ward to its guardian. *Worcester v. Georgia* (1832) held that tribes were sovereign entities with respect to state governments and therefore not subject to state laws except when specifically authorized by treaty or congressional direction.[6]

### *Eras in Federal–Tribal Relationships*

Historians often refer to distinct eras in federal policy regarding Indigenous peoples, and these also reflect the ambiguity and self-contradictory impulses in federal "Indian law." The Trade and Intercourse Era, lasting from around the ratification of the Constitution until 1830, established the federal government as the sole entity authorized to negotiate with tribes and attempted to regulate contentious issues such as trade with Native nations and the occupation of Native lands by settler populations. This gave way to the Removal Era associated most strongly with President Andrew Jackson and his military campaigns to push Native nations off of territories previously guaranteed to them in treaties. Most Native Americans living east of the Mississippi River were forcibly removed to Indian Territory during this time. During the Reservation Era, beginning around 1850, the federal government's policy was to place Native populations on small, usually isolated, areas of reservation land where their presence would pose a minimal impediment to the occupation and use of lands by settler colonial populations.

The Allotment Era followed, triggered by laws like the 1887 Dawes Severalty Act (also known as the General Allotment Act or simply the Dawes Act) and its successor, the 1898 Curtis Act. The Dawes Act was designed to break up reservation lands held in common, allocating private title of between 40 and 160 acres of reservation land to each tribal family. The property could be sold or transferred after a period of time had passed, and reservation lands remaining after each family had received their allotment could be (but were not always) removed from tribal control and opened up for sale to non-Native buyers (Figure 7.1). Massachusetts Senator Henry Dawes, architect of the Act and self-proclaimed "friend of the Indian," explained his philosophy in terms that sound as if they were ripped from the pages of Locke's famous writings on property:

> The head chief [of the Cherokees in Indian Territory] told us that there was not a family in the whole nation that had not a home of its own. There is not a pauper in that nation, and the nation does not owe a dollar. It built its own capitol…and built its schools and hospitals. Yet the defect of the system was apparent. They have got as far as they can go, because they hold their land in common…there is no enterprise to make your home any better than that of your neighbors. There is no selfishness, which is at the bottom of civilization. Till these people will consent to give up their lands, and divide them among their citizens so that each can own the land he cultivates, they will not make much progress.[7]

Although the stated intent of the Dawes Act, to help assimilate Native Americans into the dominant American culture and allow them to reap the benefits of private property ownership, was not met, the unstated aim of transferring Native lands to Euro-American ownership was wildly successful. Approximately 90 million acres of land—around two-thirds of the Native estate at that time—fell out of Native control before the policy was discontinued in 1934.[8]

The New Deal-era Indian Reorganization period followed, under which tribes were encouraged to establish tribal governments and to reestablish autonomy that had been lost under the Allotment Era. This effort ended abruptly in 1940 with the start of the Termination Era, in which the federal government began the process of terminating both the sovereign status of tribes and their remaining reservation lands. A total of 109 tribes lost federal recognition and over one million acres of tribal lands were extinguished. Most of these tribal terminations were reversed under the Self-Determination Era, characterized by federal government support to tribes to help them build their capacity for self-government.[9] The Self-Determination Era was founded upon the 1975 Indian Self-Determination and

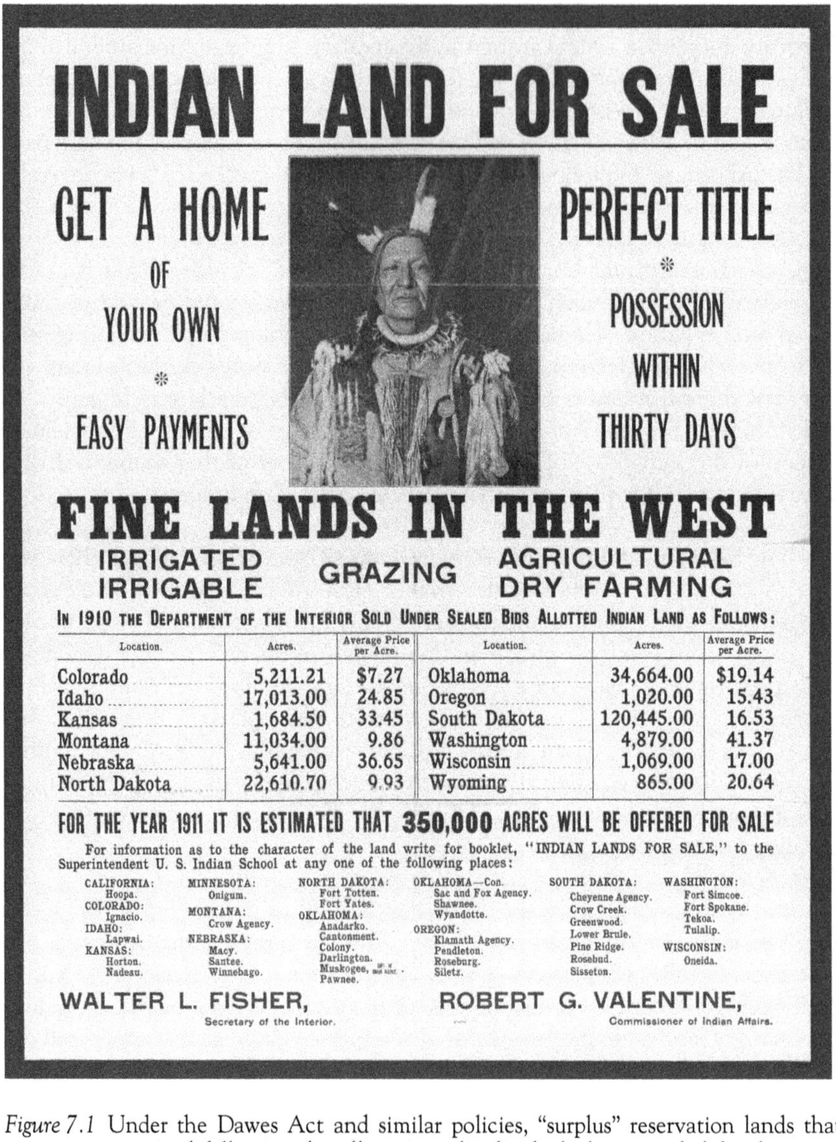


*Figure 7.1* Under the Dawes Act and similar policies, "surplus" reservation lands that remained following the allocation of individual plots to tribal families were often sold to non-Native settlers. Image: Library of Congress.

Education Assistance Act as well as a series of subsequent laws that required federal agencies to engage in formal consultation with tribes regarding resource management or other activities potentially affecting tribal interests.[10] Some observers argue that the Self-Determination Era was successful enough that it has given rise to a new era since the end of the 1980s, one of Tribal Self-Governance.

In this era, capacities are sufficiently developed to allow for tribes to fully execute some programs that were formally the purview of the federal government and to engage in co-management of other programs and resources.[11]

Collectively, these changing templates for federal–tribal relations have left their mark on patterns of tribal land ownership, tribal forest resources, and the challenges and opportunities faced by tribal entities in governing forests. Although the federal government and its agents attempted to erase Native languages, cultures, and land use practices throughout much of US history, Native peoples have demonstrated an amazing resilience—including sustaining traditional knowledges and land use practices despite the best efforts of outside agents to extinguish them. For example, even as federal foresters waged a joint physical and ideological war against fire throughout much of the twentieth century, tribal members in many places continued their traditional fire practices.[12] A forest surveyor visiting the Hoopa Valley Reservation in northern California in 1912 remarked that the forest was "over-run by fire" set by Indigenous inhabitants.[13] The Yakama and other Interior Northwest tribes were regularly admonished by national forest rangers for applying fire to help promote and maintain huckleberry fields on the national forestlands.[14] Several other tribes were applying prescribed fire on their reservations by the 1940s and 1950s,[15] in many cases supported by a small group of Bureau of Indian Affairs (BIA) foresters like Harold Weaver, whose research provided early evidence of the negative consequences of fire suppression in frequent-fire forests and of the potential benefits of fire reintroduction.[16]

## The Complexity of Tribal Recognition

Federal recognition is an important variable influencing tribal rights and opportunities, one of the legacies of the intergovernmental relations between Native nations and the United States (Box 7.1). Federally recognized tribes are those with an established government-to-government relationship with the United States. This formal relationship secures certain rights within the federal framework and makes the tribe eligible for services provided by the BIA. Federally recognized tribes are not subject to many state or local laws and may have the power to police their own territories.[17] As of 2022, the BIA counted 574 federally recognized American Indian and Alaska Native tribes and villages nationwide. Federal recognition does not necessarily imply that a treaty was signed and ratified; various federally recognized tribes have never had a ratified treaty with the federal government but are recognized through other means.

Further, not all federally recognized tribes possess reservation lands. For example, federal recognition of the Klamath, Modoc, and Yahooskin Tribes of the Klamath Reservation was rescinded during the Termination Era. The Klamath Reservation Tribes did not lose recognition and lands because of any failures on their part but rather because of their successes. Members of the Klamath Reservation were seen by government agents as models of prosperity and independence—in no small part due to the wealth they generated through their timber management—and were selected for termination under the belief that they would assimilate into the broader American society.[18] Some enrolled tribal members themselves supported termination while others sharply opposed it; such internal divisions made the tribe an easy target for termination campaigners in Congress.[19] A portion of what had once been a nearly 700,000-acre reservation south and east of Oregon's Crater Lake was sold to timber companies; most of the remainder went into federal ownership as the Fremont and Winema National Forests and the Klamath National Wildlife Refuge. Federal recognition of the Klamath Tribes was re-established in 1986, but this act did not restore the tribal land base.

Some Native American tribes remain unrecognized by the federal government. Dozens of Indigenous communities have petitioned the federal government for formal recognition since the establishment of a standardized process in 1978, but few have been successful.[20] In some cases, a tribe may be recognized by its respective state government but not by the federal government; examples include the Lumbee Tribe in North Carolina and the Nanticoke Lenni-Lanape, Ramapough Lenape, and Powhatan Renape Nations in New Jersey. The Winnemem Wintu Tribe, recognized by the State of California but not by the federal government, has seen its tribal lands flooded by federal dam projects and taken for federal fish hatchery expansion.[21] The tribe's lack of federal recognition limits its access to sovereign powers, excludes it from consideration for many federal programs aimed at supporting Native communities, and diminishes opportunities to be consulted on planned federal projects.

The concept of sovereignty is a central theme running throughout the history of tribal relationships with the US Government. William Johnson, the long-serving Chief Judge of the Tribal Court for the Confederated Tribes of the Umatilla Indian Reservation, defined sovereignty as "the ability of a government to exercise jurisdiction over a land base by making laws free of interference by other governments."[22] Lenape scholar Joanne Barker described sovereignty as "a multiplicity of legal and social rights to political, economic, and cultural self-determination."[23] Because colonial populations and their governments treated Native nations as peoples to be conquered, and because of the federal government's longstanding "trust" relationship to tribal populations regarding their resources (Box 7.2), tribal sovereignty has been a challenging ideal to achieve in practice. However, federal case law following the Marshall Trilogy has come to characterize the legal interactions between the US government and Native tribes as a form of government-to-government relationship.[24] Federal courts have established the principle that tribal governmental powers are limited by federal laws but not state laws (with some important exceptions). Further, states may have obligations to conserve fish, wildlife, and other resources important to tribal members and to allow tribal access to those resources. Importantly, however, sovereignty entails more than legal recognition; it also implies development of the human and technological capacities to access and manage resources in ways that are culturally appropriate. Tribal sovereignty is an ongoing process, one that has been advanced only slowly and with many setbacks within a history of resource control by non-Native populations.

**BOX 7.1 ALASKA NATIVES AND NATIVE HAWAIIANS**

Governmental relations with Indigenous populations in the states of Alaska and Hawaii differ in important ways from those in the contiguous 48 states. The rights of Alaska Natives were largely defined within the 1971 Alaska Native Claims Settlement Act (ANCSA), which established Native corporations at the village, regional, and urban levels as an alternative to the reservation system applied to western territories and states (one reservation was retained in Southeast Alaska, the Metlakatla Indian Community, Annette Islands Reserve). ANCSA extinguished Native title remaining from the US purchase of Alaska from Russia and allowed the Native corporations to select 44 million acres of public domain land for their own use.[25] Land owned by Native corporations is treated not as trust land but rather as fee-simple property. Further, Native corporations do not maintain the same principles of tribal sovereignty or reserved rights that many tribes in the lower 48 possess, although the Marine Mammal Protection Act and amendments to the Migratory Bird Treaty Act recognize specific harvest and co-management rights for Alaska Natives. Further, Alaska Native villages and communities—distinct from Native corporations—have established sovereign relationships with the federal government.[26]

Native Hawaiians have been subject to US control since a military-backed overthrow of the Hawaiian monarchy in 1893. No government-to-government compact exists between the federal government and Native Hawaiians, and a Department of Interior process to establish a Native government in 2016 was blocked by Supreme Court Justice Anthony Kennedy. However, Native Hawaiians have been included under the scope of over 160 pieces of federal legislation pertaining to Native Americans. Further, Native Hawaiians are identified as the beneficiaries of 200,000 acres of state lands (representing a tiny fraction of the total area of the state) designated for productive or residential use.[27] Tensions between Hawaiian land tenure regimes established by King Kamehameha III in the 1850s and Anglo-American property relations imposed via colonization continue to reverberate, and many key issues regarding the tenure rights of Native Hawaiians remain unresolved.[28]

## The Complexity of Tribal Land Ownership and Access

Not only are the relationships between Native American communities and various governmental entities complex, so are patterns of tribal land ownership. Most tribal reservation lands are legally classified as federal properties that are held in trust to members of particular federally recognized tribes. The term "trust" refers to an arrangement in which one entity (in this case the federal government) manages a resource on behalf of, and in the interests of, another entity (in this case a recognized tribe). The BIA is the federal entity charged with fulfilling this trust relationship (Box 7.2). Reservation lands have been established by various means, including treaties, legislation, executive orders, and via powers vested in the BIA.[29] There currently exist 56.2 million acres of reservation lands nationwide as well as various other categories of ownership and use rights pertaining to tribes and their members, as explained below. An estimated 18.6 million acres of forests and woodlands are under direct tribal control.[30]

Importantly, not all lands located within reservation boundaries are reservation lands. Other categories include "allotted trust land held by individual Indians, fee land held by individual Indians, fee land held by non-Indians, federal public land, and state and county land."[31] Substantial proportions of the territory within reservation boundaries may be owned by non-tribal individuals and families—one enduring legacy of the Dawes Act. For example, the Nez Perce Tribe and tribal members own only 13% of the lands within the boundaries of the Nez Perce Reservation in Idaho.[32] Tribal reservation land is just one of several categories of land controlled by tribes and tribal members. In addition to trust lands, tribes and tribal members may come to own non-trust lands through purchase, exchange, or other mechanisms. The fragmentation and inconsistency of tenure within reservation boundaries can complicate efforts to achieve tribal sovereignty.

Many tribal nations maintain legal rights of access and use on lands that are not owned by the tribe or its members.[33] These are often referred to as "retained rights" on "ceded lands." Upon signing treaties, various tribes retained the rights to access certain valued resources on the lands they ceded to the US government. For example, in addition to describing the boundaries of the Nez Perce reservation, the 1855 Walla Walla Treaty included the following language:

> The exclusive right of taking fish in all the streams where running through or bordering said reservation is further secured to said Indians: as also the right of taking fish at all usual and accustomed places in common with citizens of the territory, and of erecting temporary buildings for curing, together with the privilege of hunting, gathering roots and berries, and pasturing their horses and cattle upon open and unclaimed land.
>
> (Article 3)

These same retained rights were affirmed in the later 1863 Treaty. This language means that Nez Perce rights of access to fish, wildlife, plant, and grazing resources continue to be legally recognized within their traditional homelands, even when those lands are found off-reservation. Among other implications, this provides the Nez Perce with special status on the national forests within their traditional territory, allowing access to hunting and fishing sites and providing opportunities for the tribe to lead restoration projects on public lands that benefit salmon and other anadromous fish species. The authority provided by retained rights also empowers the Nez Perce Tribe to co-manage northwestern anadromous fisheries along with other Plateau tribes and non-tribal resource management agencies and serves as the tribe's basis for managing wolf reintroduction in the state of Idaho.[34]

## Management of Reservation Forests

The BIA and its predecessor agencies historically used their status as trustee to manage tribal resources on behalf of—and sometimes over the objections of—tribal members themselves (Box 7.2). The Office of Indian Affairs began overseeing the sale of timber on reservation and allotment land as early as the 1870s, and these operations often benefited White timber operators and loggers while returning little value to tribal members.[35] The Forestry Branch of the BIA was established in 1910, and for much of the twentieth century it largely instituted the same kinds of sustained yield and fire suppression practices that were being applied to the national forests.[36] However, there are important exceptions to this general pattern. For example, on the Fort Apache Indian Reservation in eastern Arizona, the White Mountain Apache Tribe has played an active role in forest management since the late 1940s, in many cases working in close partnership with the BIA. At a time when the US Forest Service was still abiding by the 10 a.m. rule, the White Mountain Apache Tribe and BIA foresters began an aggressive prescribed fire program on tribal lands, treating around 300,000 acres by 1966.[37] The tribe established its own mill and forest products enterprise during the 1960s, and the White Mountain Apache Timber Company remains in operation as of the date of this writing. By the 1990s, the tribe's natural resource management infrastructure was sufficiently developed to allow it to co-manage Endangered Species Act-listed species with the federal government.[38] Although its timber resources were devastated by the 2002 Rodeo-Chediski Fire, the tribe continues to actively manage forest resources for a mix of economic, cultural, and recreational uses to benefit tribal members.

The Menominee Tribe, based in the state of Wisconsin, has a history of sustainable forest management dating to the establishment of its reservation in 1854. The tribe's forestry philosophy is grounded in foundational concepts within the Menominee worldview.[39] The Menominee successfully resisted the imposition of the Dawes Act and benefited from laws passed in 1890 and 1908 that helped to limit the exploitation of Menominee timber resources by non-tribal interests. However, BIA forestry agents continued to harvest in ways that were not desired or supported by tribal members, leading to protracted court battles that were not resolved until midcentury. The tribe's relative success in forest-based economic development led it to be chosen for termination in the 1950s; the termination process was completed in 1961 but reversed twelve years later. Unlike the Klamath, the Menominee did not lose its land base as part of the termination process and was able to rebuild a forestry program under its own leadership. The Menominee Tribe has received global recognition for its holistic vision of sustainable forest management practices that include rotation periods of up to 200 years (much longer than the norm in this forest type on non-tribal lands), the use of selective harvesting rather than clearcutting, and a long-term forest inventory program that ensures that harvesting decisions are informed by site-specific data on forest conditions.[40]

The White Mountain Apache and Menominee are among the tribes with the greatest volume of on-reservation forest resources, yet their experiences mirror those of many other tribes as they fought to achieve sovereignty and to realize culturally appropriate forest management on tribal lands. Government foresters traditionally pushed to implement models built upon maximizing the sustained yield of timber, often over the objections of tribal members. Both the White Mountain Apache and Menominee Tribes, like many others, ultimately took legal action against the federal government for overriding tribal forestry approaches and failing to manage for the broad suite of values held by tribal members.[41] Both of these tribes have also been active members and leaders of the Intertribal Timber Council, an organization founded in 1976 that provides capacity-building for tribal forestry departments, advocacy for tribal forest management interests, and a point of cooperative engagement between tribes and agencies like the BIA and US Forest Service. Among many other accomplishments, the Intertribal Timber Council successfully pushed for passage of the National Indian Forest Resources Management Act of 1990, which increased tribal forest management capacity and led to a series of reports on the state of tribal forest resources.[42]

As tribes have increased their sovereignty through legal struggles and the development of natural resource management capacity, many have chosen to adopt forest management models that integrate the economic benefits of timber production with cultural values, fish and wildlife conservation, biodiversity, and other non-economic values. The Confederated Salish and Kootenai Tribes of the Flathead Reservation, for example, rejected a largely timber-oriented forest plan created by the BIA in the late 1990s and replaced it with a plan built around the pre-colonial fire regimes found in different forest types. The new plan included commercial timber harvesting, but harvests were planned to coincide with restoration objectives that included the accommodation of fire.[43] The Confederated Salish and Kootenai Tribes also famously established the first tribal wilderness area, the Mission Mountains Wilderness, in 1982. Language in the tribal ordinance establishing the wilderness borrows heavily from the 1964 Wilderness Act, with the primary difference being the role of wilderness in maintaining tribal cultural values. The ordinance states that wilderness "has served the Indian people of these Tribes as a place to hunt, as a place to gather medicinal herbs and roots, as a vision seeking ground, as a sanctuary, and in countless other ways for thousands of years."

## BOX 7.2 THE BUREAU OF INDIAN AFFAIRS

The Bureau of Indian Affairs (BIA), located in the Interior Department, is the federal agency directed to manage most aspects of federal–tribal relations, including implementing the federal trust responsibility regarding tribal lands and resources. Given the history of federal–tribal relations, it is not surprising that the BIA did not always act in furtherance of tribal sovereignty and self-determination. Legal scholar Charles Wilkinson describes the mid-twentieth century BIA as exercising "a nearly unfathomable degree of authority" over Native affairs, including controlling tribal budgets, manipulating tribal governments, and controlling the management and disposition of tribal forests in ways that did not serve Native people's interests.[44] For much of the twentieth century, the agency adopted a narrow view of forests as being primarily valuable for their commercial timber potential despite Native communities' complex and intimate value systems regarding forests;[45] this led to the imposition of forest management practices on reservations that were often sharply opposed by tribal members themselves. As well, the BIA came to adopt the fire exclusion paradigm of the US Forest Service, going so far as to criminalize the use of fire by Native Americans on their own lands.[46] Although framed as a means of benefiting tribal members, forestry activities on reservation lands were often contracted to White-owned firms, which meant that tribal members missed out on both income-generating opportunities and opportunities to build an Indigenous workforce.[47] Only after protracted legal battles and organizational efforts were many tribes able to develop their own tribal forest enterprises and build their own forestry capacity (Figure 7.2). This allowed tribes to implement forest management visions consistent with their own worldviews and relationships to the land while also providing economic and skill development opportunities to tribal members. In 2000, then-Assistant Secretary of the Interior and enrolled Pawnee member Kevin Gover apologized to the country's Native peoples for the historical conduct of the BIA, stating that "by accepting this legacy, we accept also the moral responsibility of putting things right."[48] Although the BIA was traditionally staffed largely by nontribal administrators and resource managers, it has increasingly hired enrolled tribal members; Indigenous scholar Donald Fixico noted that, as of 2010, "95 percent of the Bureau of Indian Affairs' 12,000 employees were American Indian or Alaska Native."[49] In recent decades, the BIA has taken a more cooperative and culturally responsive approach to resource management consistent with the principles of Native self-determination and self-governance, but questions have continued to linger over the agency's past management of its trust responsibility. The class-action suit *Cobell v. Babbitt* was brought in 1996, alleging federal mismanagement of tribal assets and resources stretching back to the nineteenth century. The suit was settled in 2009 for $3.4 billion, over half of which was dedicated to purchasing former reservation lands and reestablishing them under tribal ownership.

*Figure 7.2* The Walatowa Timber Industries mill run by the Jemez Pueblo in New Mexico processes timber, poles, firewood, wood pellets, and traditional beams for construction, using timber resources from Pueblo lands and from nearby national forests. Image: US Forest Service.

## Tribal Land Recovery

Most Indigenous nations lost control over the vast majority of their traditional homelands as part of the process of Euro-American colonization, and the recovery of traditional lands has long been a high priority for tribes. In recent years, some non-tribal entities, including many land trusts, have added tribal land repatriation to their missions and conservation activities. One mode by which some tribes have managed to repatriate traditional lands is through compensation from past federal public works projects that diminished tribal resources. Various Northwestern tribes, for example, received payments for the loss of salmon fishing opportunities when Celilo Falls on the Columbia River was dammed; each tribe included in the settlement determined how to allocate its respective share of the compensation. The Confederated Tribes of Warm Springs, whose reservation lies on the eastern slope of Oregon's Cascade Mountains, used most of the funding to purchase land within reservation boundaries along the Warm Springs River and at important fishing sites off-reservation on the Deschutes and John Day Rivers. The tribe also invested in mill infrastructure to increase tribal employment and process wood from the reservation's abundant forests.[50] Other Plateau Tribes have also repatriated land within their historic territories under compensatory deals with the federal government, most of them stemming from land and habitat losses from federal hydropower projects. This is the origin of the Héte'wits Wétes (Precious Lands) property in Wallowa County described at the beginning of this chapter.

Fee-simple purchases are another means of repatriating important lands. Although the economic struggles of Native American populations are well documented, many tribes have taken advantage of casinos and other gaming operations to fund land acquisitions (as well as to fund various social services). One example is the Oneida Nation, part of the Haudenosaunee (Iroquois) Confederacy, which used casino revenues to purchase tens of thousands of acres within the boundaries of its Wisconsin reservation that were lost during the Allotment Era.[51] Other repatriation processes are more complex in terms of their organization and financing. Northern California's Yurok Tribe took ownership of 57,000 acres of forestland in 2019 through a deal involving the tribe, the industrial timberland owner Green Diamond Resource Company, and the conservation NGO Western Rivers Conservancy. Financing for the purchase came through a complex amalgamation of tribal, state, federal, charitable, and nonprofit sources along with the sale of carbon offsets under the California Global Warming Solutions Act (see Chapter 6). The property includes Blue Creek, a salmon stronghold of "paramount cultural significance" to the tribe.[52] The land will be managed as a tribal community forest in which culturally appropriate land management practices are applied to favor species important for traditional foods and basket-making materials while protecting instream habitat for salmon. Tribes have also attempted to recover land through the use of the court system with mixed success. Onondaga scholar and writer Robin Wall Kimmerer, for example, describes the ultimately unsuccessful "land rights action" filed by the Onondaga Nation to secure legal recognition of traditional lands and waters in upstate New York.[53] The suit was driven by concerns about the slow pace of clean-up and restoration efforts on heavily polluted lands and waters around Lake Onondaga.

## Indigenous Traditional Ecological Knowledge

The concept of Indigenous traditional ecological knowledge, sometimes abbreviated as ITEK,[54] captures the idea that highly detailed knowledge of ecological patterns, processes, species, and relationships resides within the cultural, oral, and linguistic traditions of peoples with longstanding relationships to the land. Whereas western ecological knowledge emphasizes insights gained through the structured observation and analysis associated with the scientific method, ITEK is based upon knowledge generated through Indigenous peoples' long-term inhabitation and use of specific places. Examples include the knowledge of when, how, and how often to set controlled fires to achieve particular outcomes (such as promoting favored plant species[55]), how to harvest plants and animals to ensure their continued survival, and how to understand and interpret signals of environmental change.[56] ITEK often pertains to entire systems, whereas scientific knowledge has often focused on individual species and ecological processes thought to be of greatest commercial value.[57]

There has been growing interest in recent years in the integration of Indigenous and western knowledge systems for informing decision-making for forests. It is important to point out, however, that ITEK and western knowledge are not strictly comparable; they are qualitatively different in important ways. Unlike western scientific knowledge, ITEK is inextricably integrated into a larger set of moral considerations that structure people's obligations as users and stewards of the environment—considerations that may entail treating nonhuman species and even nonliving things as members of a moral community. ITEK is inseparable from language, pertains to specific ecosystems, and is part and parcel of other forms of knowledge and belief that might be classified as something other than strictly "ecological." This is not to say that western science and ITEK cannot be complementary or used in tandem; rather, the category of "science" in the western tradition implies a separateness and isolation from other moral, cultural, and spiritual considerations that would be unfamiliar—possibly even unthinkable—to practitioners of ITEK.

ITEK is important to tribal forest governance both on and off lands directly under tribal control. It is a key element of tribal sovereignty—the ability to make decisions regarding forests based in longstanding moral conceptions and relations of stewardship.[58] The application of Indigenous knowledge to forests represents the continuation of cultural transmission reaching back to "time immemorial." From a more narrowly utilitarian perspective, it also holds promise for recovering species of concern, improving the production of valued resources (such as species important for use as food or in basketmaking), and addressing issues such as the threat to human communities and infrastructure of high-intensity wildfire. As described below, recent years have seen some initial steps toward applying Indigenous knowledge and management practices to federal forestlands. ITEK is also a key element in the sustainable harvest and utilization of non-timber forest products, described in Chapter 11.

## Tribal Influence on Federal Lands

Nationwide, all federal properties (such as national forests, national parks, and national wildlife refuges) exist on lands that were once part of the homeland of one or more Native nations.[59] The founding of many of America's national parks—including "crown jewels" such as Yellowstone, Glacier, Grand Canyon, Yosemite, and Everglades—entailed displacing the Native communities that had been living in and making use of these landscapes that were later classified as "pristine."[60] In many cases these federal properties were established as part of, or shortly following, acts in which the federal government directly contravened its own commitments under treaty language (see Chapter 2). For example, the Black Hills of South Dakota and Wyoming were included as part of the Great Sioux Reservation under the 1851 and 1868 Treaties of Ft. Laramie. In violation of treaty language guaranteeing "absolute and undisturbed use" of these lands by the Sioux, the Black Hills were withdrawn from the reservation in 1877 following the discovery of gold and the subsequent influx of White prospectors to the region. The Black Hills Forest Reserve (now Black Hills National Forest) was established two decades later and is considered to be the site of the first federal timber sale. Federal courts have recognized the violations of the Treaties of Ft. Laramie and offered $100 million as compensation, but the Sioux Nation has refused to accept the settlement and continues to advocate for the return of its homeland. Sociologist Kari Norgaard notes that 98% of the lands in the traditional territory of Northern California's Karuk Tribe are now national forestlands.[61] She explains that because no treaty between the tribe and the federal government was ever ratified, many Karuk members see the US Forest Service as occupying tribal lands that were never legally ceded to the government.

In spite of this legacy, tribal communities and governments have worked to expand their ability to access and steward important lands and resources on federal lands. Federal agencies have been required to formally consult with tribes regarding issues or resources of tribal concern since the start of the Self-Determination Era. Formal consultation requirements were strengthened with 1992 revisions to the National Historic Preservation Act and by executive orders issued by Presidents Clinton, Obama, and Biden. Yet consultation in and of itself is limited in its ability to advance tribal interests, and federal agencies retain discretion regarding the degree to which they take tribal concerns seriously.[62] Amendments to the Indian Self-Determination Act in 1994 provided opportunities for designated "self-governance tribes" to manage federal programs pertaining to tribal services or interests in the Interior Department and Department of Health and Human Services.[63] This so-called "638" authority has been used, for example, to allow the Grand Portage Band of Chippewa Indians to take over some maintenance and management functions at Grand Portage National Monument in Minnesota, the Lower Elwha Tribal Community to engage in watershed restoration within Olympic National Park in Washington, and the Yurok Tribe to lead restoration, monitoring, and archaeological projects in Redwood National Park in California.[64]

Other programs provide opportunities for tribes to co-manage lands outside of the Department of Interior. For example, the Reserved Treaty Rights Lands program uses BIA funds to support tribal resource management on traditional lands within the National Forest System. The Tribal Forest Protection Act of 2004 offers the opportunity for tribes to manage national forestlands that directly abut tribal lands, though to date the policy has not been used to its full potential.[65] The Good Neighbor Authority, introduced in Chapter 5, was expanded in 2018 to authorize self-governance tribes to implement management activities on national forests and Bureau of Land Management lands. The first tribe to do so was the Tulalip Tribe in the State of Washington, which managed a beaver reintroduction program on the Mt. Baker-Snoqualmie National Forest in support of salmon recovery.[66] Several tribes have recently studied the opportunities associated with the "anchor forest" concept, in which the capacity and infrastructure on tribal lands would be used as part of large-landscape management and restoration initiatives on national forests and other non-tribal forestlands.[67]

Several national forests in Oregon, Washington, and California have entered into agreements with tribes to co-manage selected areas of national forestland of special ecocultural importance, emphasizing the reintroduction of fire and the promotion of species such as elk, huckleberry, and camas.[68] In northern California, US Forest Service staff on the Six Rivers National Forest have taken steps toward co-management with the Karuk Tribe and the inclusion of ITEK in project planning and implementation. An initial project built on limited co-management was initiated in the 1990s but later abandoned when incoming Forest Service staff no longer supported the project.[69] Since that time, the agency has increasingly worked in collaboration with local tribal and non-tribal entities via the Western Klamath Restoration Partnership. The Somes Bar project, signed in 2018, is built explicitly upon the concept of incorporating Karuk fire knowledge into forest restoration. This project demonstrates the potential for tribal co-management on federal lands, yet stands out as one of the few projects of its kind to date. As legal scholars Monte Mills and Martin Nie conclude, "federal agencies remain largely in the driver's seat when it comes to authorizing broader tribal control over federal activities on federal lands."[70]

Numerous tribal sacred sites are located on lands owned and controlled by federal land management agencies. For example, an area within the Lewis and Clark National Forest situated between Glacier National Park and the Blackfeet Indian Reservation is known as Badger-Two Medicine and is among the most important sacred sites to members of the Blackfeet Nation. It has been recognized as a traditional cultural district, which gives the Blackfeet the right to be consulted on management but does not require the Forest Service to manage in line with the tribe's wishes. The area had been extensively leased for oil and gas exploration, with the last leases extinguished as of 2020. A decision by the Coconino National Forest to allow the use of reclaimed municipal water for artificial snowmaking in the Arizona Snowbowl ski area was fiercely contested by southwestern tribes because of its impacts on the San Francisco Peaks, an area considered to be sacred to multiple tribal nations. A coalition of tribes failed in its initial legal challenge to the practice, and artificial snowmaking began in 2012; since that time, the Hopi Tribe has successfully brought a nuisance claim against the city of Flagstaff for damaging cultural resources with its wastewater sales to the ski resort. Further south in Arizona, Chi'chil Biłdagoteel (Oak Flat) on the Tonto National Forest is considered to be one of the most spiritually and culturally important sites to the San Carlos Apache Tribe. A 2014 defense authorization bill included language authorizing the transfer of the site to transnational mining interests; since that time, Native activists and allies have been attempting to stop a proposed copper mine, which would destroy the sacred site, from going forward. As of the time of this writing, the conflict continues to play out in the courts.

## Conclusion

Tribal forest governance in the US is shaped by thousands of years of forest inhabitation and use refracted through a more recent history of colonialism, attempted physical and cultural extermination, and an ongoing process of self-determination. The continuing effort to reclaim sovereignty across diverse and complex patterns of forest ownership and access is a central theme in tribal forest governance. The concept of sovereignty takes multiple forms, including development of tribal forest management capacity, recognition of rights within the US legal system, and the ability to apply knowledge and management approaches that depart from many of the core assumptions undergirding non-Native forest management. In many cases, tribal communities have had to struggle to reintroduce culturally appropriate management following the imposition of management models imported from nineteenth century Europe that emphasized fire exclusion and the simplification of complex ecosystems.

The struggle for sovereignty is especially reflected in ongoing tensions regarding the authority to make decisions for forests that are important to tribal members but are not under direct tribal control. On some lands, this has been at least partially realized via contracting or co-management arrangements on lands managed by other entities such as the National Park Service and US Forest Service. To date, these projects have been limited in number and scope. At the same time, a growing acknowledgment that America's forests lack resiliency to climate-induced stressors has sparked greater interest among non-Native populations in the application of Indigenous fire and forest management practices as models for sustainability.

## Questions for Discussion

1. What does it mean for a tribe to have federal recognition? Does it mean that the tribe has signed a treaty with the US government? Does it mean that the tribe possesses reservation lands?
2. What is the significance of reserved rights on ceded lands?
3. Why did several tribes ultimately take legal action against the BIA over the agency's management of tribal forestlands? Why was the agency not more responsive to tribal wishes for the management of reservation forests?
4. What factors ultimately changed the relationship between the BIA and tribal nations?
5. What does tribal sovereignty mean? How might sovereignty be expressed with regards to the management of tribal forests both on- and off-reservation?
6. What opportunities exist to integrate Indigenous traditional ecological knowledge into the management of federally owned forestlands? What are some obstacles that might complicate this integration?

## Notes

1 See Josephy (1997) for details on the Nez Perce treaties and Nez Perce War.
2 Hsueh, 2006.
3 Barker, 2005; Lind, 2020.
4 482 *State vs. Towessnute* Opinion Per Bausman, J. 89 Wash.

5 Blumm, 2003.
6 Blumm, 2003.
7 Quoted in Malcomson (2000), p. 15.
8 Wilkinson & Biggs, 1977.
9 Importantly, however, in most cases lands lost to terminated tribes have not been recovered.
10 Huntsinger & Diekmann, 2010; Mills & Nie, 2021.
11 Washburn, 2017; Wilkins, 2016.
12 Pyne, 1982, p. 81.
13 Huntsinger & McCaffrey, 1995, p. 163.
14 Catton, 2016, p. 77.
15 Kallander, 1969; Truesdell, 1969.
16 Weaver, 1947, 1959, 1974.
17 General Accounting Office, 2001.
18 Kelly et al., 2013.
19 Catton, 2016, pp. 91–92.
20 General Accounting Office, 2001.
21 Dallman et al., 2013; Garrett, 2010.
22 Luce & Johnson, 2006, p. 171.
23 Barker, 2005, p. 1.
24 Wilkinson, 1987.
25 Catton, 2016, pp. 233–234.
26 Catton, 2016, pp. 244–245.
27 Kauanui, 2005; Trask, 1991.
28 Lâm, 1989.
29 Wilkinson, 1987, p. 8.
30 Indian Forest Management Assessment Team, 2013, p. 18.
31 Wilkinson, 1987, p. 9.
32 https://critfc.org/member-tribes-overview/nez-perce-tribe/.
33 Wilkinson, 1987, p. 15.
34 Wilkinson, 1998.
35 Catton, 2016; Steen-Adams et al., 2010.
36 Steen-Adams et al., 2017.
37 Kallander, 1969.
38 Tomblin, 2016.
39 Trosper, 2007.
40 Davis, 2000; Trosper, 2007.
41 Yazzie-Durglo, 1998.
42 Trosper, 2007.
43 Trosper et al., 2012.
44 Wilkinson, 2005, p. 21.
45 Steen-Adams et al., 2017.
46 Huntsinger & Diekmann, 2010.
47 Steen-Adams et al., 2010.
48 Fixico, 2012, p. 160.
49 Fixico, 2012, p. 163.
50 Wilkinson, 2005.
51 Kiel, 2019.
52 Manning & Reed, 2019, p. 110.
53 Kimmerer, 2013.
54 Sometimes referred to as traditional ecological knowledge or TEK. ITEK distinguishes the knowledge of Indigenous practitioners from other holders of traditional knowledge.
55 Senos et al., 2006.
56 Kimmerer, 2013.

57 Long et al., 2020.
58 Coffey & Tsosie, 2001.
59 Washburn, 2022.
60 Keller & Turek, 1998.
61 Norgaard, 2019, p. 56.
62 Mills & Nie, 2021.
63 King, 2007; Washburn, 2022.
64 King, 2007.
65 Mills & Nie, 2021.
66 Washburn, 2022.
67 Jacobson et al., 2021.
68 Long & Lake, 2018.
69 Diver, 2016.
70 Mills & Nie, 2021, p. 111.

## References


Barker, J. (2005). For whom sovereignty matters. In *Sovereignty matters: Locations of contestation and possibility in indigenous struggles for self-determination* (pp. 1–31). University of Nebraska Press.

Blumm, M. C. (2003). Retracing the discovery doctrine: Aboriginal title, tribal sovereignty, and their significance or treaty-making and modern natural resources policy in Indian country. *Vermont Law Review*, *28*, 713–777.

Catton, T. (2016). *American Indians and national forests*. University of Arizona Press.

Coffey, W., & Tsosie, R. (2001). Rethinking the tribal sovereignty doctrine: Cultural sovereignty and the collective future of Indian nations. *Stanford Law & Policy Review*, *12*(2), 191–222.

Dallman, S., Ngo, M., Laris, P., & Thien, D. (2013). Political ecology of emotion and sacred space: The Winnemem Wintu struggles with California water policy. *Emotion, Space and Society*, 6, 33–43.

Davis, T. (2000). *Sustaining the forest, the people, and the spirit*. State University of New York Press.

Diver, S. (2016). Co-management as a catalyst: Pathways to post-colonial forestry in the Klamath Basin, California. *Human Ecology*, *44*(5), 533–546.

Fixico, D. L. (2012). *Bureau of Indian Affairs*. Greenwood Press.

Garrett, B. L. (2010). Drowned memories: The submerged places of the Winnemem Wintu. *Archaeologies*, 6(2), 346–371.

General Accounting Office (2001). *Improvements needed in tribal recognition process* (GAO-02–49). General Accounting Office.

Hsueh, V. (2006). Cultivating and challenging the common: Lockean property, indigenous traditionalisms, and the problem of exclusion. *Contemporary Political Theory*, 5(2), 193–214.

Huntsinger, L., & Diekmann, L. (2010). The virtual reservation: Land distribution, natural resource access, and equity on the Yurok Forest. *Natural Resources Journal*, 50(2), 341–369.

Huntsinger, L., & McCaffrey, S. (1995). A forest for the trees: Forest management and the Yurok environment, 1850 to 1994. *American Indian Culture and Research Journal*, *19*(4), 155–192.

Indian Forest Management Assessment Team (2013). *An assessment of Indian forests and forest management in the United States: Final report* (vol. I) Intertribal Timber Council.

Jacobson, M. A., Hajjar, R., Davis, E. J., & Hoagland, S. (2021). Learning from tribal leadership and the anchor forest concept for implementing cross-boundary forest management. *Journal of Forestry*, *119*(6), 605–617.

Josephy, A. M. (1997). *The Nez Perce Indians and the opening of the Northwest*. Houghton Mifflin.

Kallander, H. (1969). Controlled burning on the Fort Apache Indian Reservation, Arizona. *Proceedings of Tall Timbers Fire Ecology Conference*, 9, 241–250.

Kauanui, J. K. (2005). Precarious positions: Native Hawaiians and US federal recognition. *The Contemporary Pacific*, *17*(1), 1–27.

Keller, R. H., & Turek, M. F. (1998). *American Indians & national parks*. University of Arizona Press.

Kelly, E. C., Bliss, J. C., & Gosnell, H. (2013). The Mazama returns: The politics and possibilities of tribal land reacquisition. *Journal of Political Ecology*, *20*, 429–443.

Kiel, D. (2019). Nation v. municipality: Indigenous land recovery, settler resentment, and taxation on the Oneida reservation. *Native American and Indigenous Studies*, 6(2), 51–73.

Kimmerer, R. (2013). *Braiding sweetgrass: Indigenous wisdom, scientific knowledge and the teachings of plants*. Milkweed editions.

King, M. A. (2007). Co-management or contracting: Agreements between Native American tribes and the US National Park Service pursuant to the 1994 Tribal Self-Governance Act. *Harvard Environmental Law Review*, *31*, 475–530.

Lâm, M. C. (1989). The Kuleana Act revisited: The survival of traditional Hawaiian commoner rights in land. *Washington Law Review*, 64(2), 233–288.

Lind, D. (2020). Doctrines of discovery. *Washington University Jurisprudence Review*, *13*(1), 1–63.

Long, J. W., & Lake, F. K. (2018). Escaping social-ecological traps through tribal stewardship on national forest lands in the Pacific Northwest, United States of America. *Ecology and Society*, 23(2), 10.

Long, J. W., Lake, F. K., Goode, R. W., & Burnette, B. M. (2020). How traditional tribal perspectives influence ecosystem restoration. *Ecopsychology*, *12*(2), 71–82.

Luce, C. F., & Johnson, W. (2006). The beginning of modern tribal governance and enacting sovereignty. In J. Karson (Ed.), *Wiyaxayxt/Wiyaakaa'awn/As days go by: Our history, our land, our people–the Cayuse, Umatilla, and Walla Walla* (pp. 151–192). Tamastslikt Cultural Institute; Oregon Historical Society Press; University of Washington Press.

Malcomson, S. (2000). *One drop of blood: The American misadventure of race*. Farrar, Straus and Giroux.

Manning, B. R. M., & Reed, K. (2019). Returning the Yurok forest to the Yurok Tribe: California's first tribal carbon credit project. *Stanford Environmental Law Journal*, 39, 71–124.

Mills, M., & Nie, M. (2021). Bridges to a new era: A report on the past, present, and potential future of tribal co-management on federal public lands. *Public Land & Resources Law Review*, 44(2), 49–184.

Norgaard, K. M. (2019). *Salmon and acorns feed our people: Colonialism, nature, and social action*. Rutgers University Press.

Pyne, S. J. (1982). *Fire in America: A cultural history of wildland and rural fire*. Princeton University Press.

Senos, R., Lake, F. K., Turner, N., & Martinez, D. (2006). Traditional ecological knowledge and restoration practice. In D. Apostol & M. Sinclair (Eds.), *Restoring the*

*Pacific Northwest: The art and science of ecological restoration in Cascadia* (pp. 393–426). Island Press.

Steen-Adams, M. M., Charnley, S., & Adams, M. D. (2017). Historical perspective on the influence of wildfire policy, law, and informal institutions on management and forest resilience in a multiownership, frequent-fire, coupled human and natural system in Oregon, USA. *Ecology and Society, 22*(3), 23.

Steen-Adams, M. M., Langston, N. E., & Mladenoff, D. J. (2010). Logging the Great Lakes Indian reservations: The case of the Bad River Band of Ojibwe. *American Indian Culture and Research Journal, 34*(1), 41–66.

Tomblin, D. C. (2016). The White Mountain recreational enterprise: Bio-political foundations for White Mountain Apache natural resource control, 1945–1960. *Humanities, 5*(3), 58.

Trask, M. B. (1991). Historical and contemporary Hawaiian self-determination: A Native Hawaiian perspective. *Arizona Journal of International and Comparative Law,* 8(2), 77–95.

Trosper, R. L. (2007). Indigenous influence on forest management on the Menominee Indian Reservation. *Forest Ecology and Management, 249*(1–2), 134–139.

Trosper, R. L., Parrotta, J. A., Agnoletti, M., Bocharnikov, V., Feary, S. A., Gabay, M., Gamborg, C., Latorre, J. G., Johann, E., & Laletin, A. (2012). The unique character of traditional forest-related knowledge: Threats and challenges ahead. In J. A. Parrotta & R. L. Trosper (Eds.), *Traditional forest-related knowledge: Sustaining communities, ecosystems and biocultural diversity* (pp. 563–588). Springer.

Truesdell, P. S. (1969). Postulates of the prescribed burning program of the Bureau of Indian Affairs. In Proceedings Annual [9th] Tall Timbers Fire Ecology Conference, Tallahassee, FL (pp. 235–240).

Washburn, K. K. (2017). What the future holds: The changing landscape of federal Indian policy. *Harvard Law Review Forum, 130*(6), 200–232.

Washburn, K. K. (2022). Facilitating tribal co-management of federal public lands. *Wisconsin Law Review, 2022*(2), 263–328.

Weaver, H. (1947). Fire: Nature's thinning agent in ponderosa pine stands. *Journal of Forestry, 45*(6), 437–444.

Weaver, H. (1959). Ecological changes in the ponderosa pine forest of the Warm Springs Indian Reservation in Oregon. *Journal of Forestry, 57*(1), 15–20.

Weaver, H. (1974). Effects of fire on temperate forests: Western United States. In T. T. Kozlowski & C. E. Ahlgren (Eds.), *Fire and ecosystems* (pp. 279–319). Academic Press.

Wilkins, D. E. (2016). A history of federal Indian policy. In S. Lobo, S. Talbot, & T. L. Morris (Eds.), *Native American voices: A reader* (Third ed., pp. 104–112). Routledge.

Wilkinson, C. F. (1987). *American Indians, time, and the law: Native societies in a modern constitutional democracy.* Yale University Press.

Wilkinson, C. F. (1998). Indian tribal rights and the national forests: The case of the aboriginal lands of the Nez Perce Tribe. *Idaho Law Review, 34*(3), 435–464.

Wilkinson, C. F. (2005). *Blood struggle: The rise of modern Indian nations.* W. W. Norton & Company.

Wilkinson, C. F., & Biggs, E. R. (1977). The evolution of the termination policy. *American Indian Law Review, 5*(1), 139–184.

Yazzie-Durglo, V. (1998). The right to change tribal forest management. *Journal of Forestry,* 96(11), 33–35.

# 8 Family Forests

The Jackson-Goebel Tree Farm sits on a verdant hillslope surrounded by farmland in a part of Wallowa County known as Alder Slope. Bob Jackson and Leo Goebel, two long-time loggers, purchased in 1970 what was then a cutover piece of forestland with a vision to manage it differently than it had been managed in the past. Unlike on many of the federal and corporate forestlands on which they had worked for years, the focus on their 160 acres would be the forest ecosystem as a whole and not just the trees. For Bob and Leo, this meant paying close attention to the soil and the various creatures that live in and on it. Over the decades, the two spent countless hours moving and rearranging slash and fallen woody material by hand to encourage decomposition and prevent soil erosion. A common practice was creating "wildlife structures" by hand-piling slash and branches to attract voles and other small mammals. Rather than encouraging only the fastest growing and most economically valuable tree species, they managed for a forest of diverse species, sizes, and age classes. This is not to say, however, that they neglected timber harvesting altogether; to the contrary, their labors resulted in a highly productive forest that could be selectively harvested on an ongoing basis without putting other forest values at risk. Over the first three decades of the tree farm's existence, Bob and Leo sold 22 million board feet of timber even as the standing volume on the forest continued to increase.

The Jackson-Goebel tree farm, while unique in Wallowa County and the world, captures many of the key elements of forestlands owned by individuals and families. Its uses cannot be captured using simple terms like "production" or "preservation"; rather, it blends multiple objectives and uses in ways that are both complementary and constantly evolving. Its management defies reduction to economic rationality; for Bob, Leo, and their families, work on the tree farm was always a labor of love, where time spent planning, surveying, and managing is a fulfilling element of ownership rather than a cost to be minimized. The forest is also more than a strictly private asset—it has served as a public model for neighbors and the broader community. Bob and Leo won multiple state and regional awards for their stewardship, and their tree farm has been visited by thousands of students, forest owners, and others. The forest and its multifaceted values endure beyond its first generation of owners; both Bob and Leo have passed away since 2009, but the next generation of the Goebel family is carrying on their vision and

DOI: 10.4324/9781003043669-8

continuing to innovate and experiment. In this chapter we take a closer look at family-owned forestlands—those owned by millions of individuals and families, some of whom may have much in common with the Jackson and Goebel families and some of whom may be quite different. We start with an exploration of the category of family forest owners and then consider the varied policy and governance influences on the management and conservation of family forests.

## Understanding Family Forests

Privately owned forests held by individuals, families, trusts, and estates make up the single largest ownership category in the nation—representing more acreage than federal, state, tribal, or corporate ownerships alone. Family forests (sometimes called non-industrial private forests or NIPFs) are especially common in the eastern US, where in many cases they make up part of current farm properties or stand on lands that were previously cleared for farming. The proportion of forestland owned by families varies by state, with 89% of Kansas' 2.5 million acres of forestland under family ownership, while in Nevada only around 2% of forests are family owned.[1] These lands are highly diverse, ranging from peri-urban woodlots of just a few acres to expansive parcels containing thousands of acres of forestlands. They are put to a wide array of uses, from outdoor recreation to wildlife viewing to timber production to simply serving as peaceful places to "get away" and build new family memories. Indeed, the family dimension of these forestlands tends to be of paramount importance: they are often properties that are retained across multiple generations and serve as enduring elements of family identity.

The National Woodland Owner Survey, administered by the Family Forest Research Center, provides insights about the ownership patterns, activities, and characteristics of family forest owners in the US and their nearly 300 million acres of forestland.[2] The 2018 survey cycle found that around 5.9 million of the nation's 9.6 million family forest ownerships represent properties of less than ten acres. On the remaining 3.7 million ownerships of at least ten acres, 62% are within a mile of the owner's residence and 41% are part of a farm operation. For the vast majority (83%) of owners, the income generated by their forest holdings is effectively zero—and fewer than 1% of owners receive a majority of their income from their forest activities. According to this survey, 97% of respondents, representing 98% of family forest acres, are owners who self-identify as White, and only around 2% of forest owners self-identify as Hispanic or Latino.

The National Woodland Owner Survey also tells us something about the motivations of family forest owners. Although some properties are managed to maximize timber production, the overwhelming balance of evidence from the survey and decades of complementary research indicates that family forest owners tend to be driven by values and objectives that go well beyond timber production.[3] For example, "timber products" tends to rank near the bottom of family forest owners' reasons for owning woodland, with less than a quarter of respondents ranking this motivation as "Important" or "Very Important" in the 2018 survey cycle. At the top of the list are "beauty or scenery" (with 81% ranking it "Important" or "Very Important"), "wildlife habitat" (77%), and "nature protection" (72%). It is important to clarify, however, that many family forest owners do sell timber at some point; typically, it is only the smallest properties—and those specifically managed for preservation—that tend to be reserved from all commercial timber management over the long term. Family forest owners may choose to sell timber for many reasons: when in need of additional income (for example, to pay for a child's college tuition or for unexpected healthcare needs), in response to particularly high timber prices, or as part of a long-term plan to maximize income or to maintain forest density at desired levels[4] (Figure 8.1). Harvesting timber may also serve to maintain favorable tax treatment, a topic that is discussed in greater detail below.

In effect, it is just as inaccurate to portray family forest owners as generally averse to timber harvesting (although some owners may never harvest timber for income) as it is to portray them as scaled-down versions of industrial timber producers (although some owners may emphasize intensive, income-maximizing practices on their properties). The variety and complexity of family forest owners' management motivations is a defining feature of this group. Patterns of management of family-owned forestland are, consequently, varied and complex as well, although research has provided some broad findings helpful for understanding

*Figure 8.1* Part of this family forest in South Carolina is managed for maximum production of loblolly pine for sale; other parts of the forest are devoted to conservation, recreation, and wildlife habitat. Photograph by the author.

how decisions are typically made on family-owned forests.[5] One common pattern relates to the size of forest holdings: in general, larger properties tend to be more commonly oriented toward forest-based income generation than smaller properties—and the smallest properties are overwhelmingly dedicated to uses such as aesthetics, private enjoyment, and recreation.[6] Regardless of the management objectives or size of the property, family forest owners typically cherish their decision-making autonomy and look negatively upon policies or programs that limit their freedom to manage according to their wishes.[7] Across the great diversity of family forest owners, factors such as race, ethnicity, and gender can often be important not only as influences on management priorities but also for understanding past histories and future opportunities.[8]

There are few topics that have consumed forest social scientists and economists more than understanding family forest owners: what drives their decision-making, what values they hold for their properties, how they manage, how (or even if) they engage in long-term planning. Among other issues, forestry professionals have been concerned about the relatively small proportion of family forest owners that operate under a written plan (only 11% according to the 2018 National Woodland Owner Survey) or make use of forestry professionals in their timber management.[9] There is also widespread concern about the ongoing subdivision of forested properties into smaller ownership units, a process called parcelization. Complicating family forest governance are factors such as intergenerational transfer, the spread of suburbs and exurbs, and the ever-changing array of federal, state, and local policies that alternately incentivize and undermine the conservation of family forestlands.

## Federal and State Policies Affecting Family Forests

### *Taxes*

Taxes come in diverse forms, they are administered at multiple governmental levels, and as Benjamin Franklin famously quipped, they are one of the few certain features of life. Although policies like the Endangered Species and Clean Water Acts can greatly influence specific management activities or individual landowners, taxes may have a greater overall influence on how private forestlands are managed.[10] In part, this is because taxes at least potentially affect all landowners all the time and in part because various incentive programs have been developed to reduce or eliminate certain tax obligations—and these programs typically require or encourage forest owners to manage in particular ways. Forest tax policy confronts a complex public policy dilemma: how to protect the continued provisioning of public benefits (such as wildlife habitat, carbon sequestration, scenery, and the production of clean water) from private forests while also ensuring that forest owners contribute their fair share to tax coffers.

The taxes paid by private forest owners typically fall under four broad categories: property taxes (applied to the land itself), income taxes (applied to any income-generating activity), harvest taxes (also called severance taxes, applied to harvested timber), and gift and estate taxes (applied when land is gifted to or inherited by others). Property taxes, which are generally collected by local governments, are assessed based on the value of the forest property or, alternatively, according to formulas that take into account the productivity of the land and its land use zoning or designation. Income taxes are collected by the federal government and by most states and some localities; they are assessed when the landowner sells something of value and receives income in return. Although proceeds from sold timber may be taxed as income, many jurisdictions apply a separate harvest or severance tax based on the volume or value of timber sold. Gift and estate taxes are collected by federal and some state governments, generally based on a percentage of the transferred property's value above an exempted amount.

Of the four primary categories of taxes, property taxes have received the most scrutiny from economists and others concerned with forest conservation. Unlike the other categories of taxes, property taxes are typically assessed annually. Because trees take decades to grow and mature, and because the vast majority of family forest owners harvest timber only occasionally, this means that in most years family forest owners end up paying more in taxes than their forests generate from timber income (although it should be noted that there are other means by which forest owners can generate income, such as hunting leases, sales of non-timber forest products, and ecosystem service payments). Aggregated over multiple years, these taxes can make forest ownership an economically marginal activity—or even make it difficult for some owners to continue to hold onto their properties. For example, one analysis of the combined federal and state/local tax burden in southern states found that owners of less productive forestland may end up paying more in taxes than their land ultimately produces in timber value once discounting (the economic principle that future costs and benefits are valued less in present terms than immediate costs and benefits) is considered.[11] Although this was only the case under one possible scenario out of several analyzed, it demonstrates that taxes can greatly diminish the economic incentives to maintain forests in production. This can have the unintended consequence of pushing landowners to convert their forests to other land uses such as agriculture or residential development. Gift and estate taxes can also put pressure on the recipients of especially valuable properties to develop or subdivide lands.[12]

### BOX 8.1 EXAMPLES OF VOLUNTARY FEDERAL PROGRAMS FOR PRIVATE FOREST CONSERVATION

The *Forest Legacy Program* (FLP), administered by the US Forest Service, is designed to protect private forestlands of high conservation value through fee-simple purchase or purchase of conservation easements. FLP selects awardees annually through a competitive process. Among other requirements, an eligible property must be part of a landscape designated as a Forest Legacy Area by the state in which it is located.

The *Environmental Quality Incentives Program* (EQIP) is a cost-share and technical assistance program administered by the Natural Resources Conservation Service. Starting with a conservation plan, EQIP funds are used to support specific management practices that deliver conservation benefits.

The *Conservation Stewardship Program* (CSP), also administered by the Natural Resources Conservation Service, provides annual payments and technical assistance for pursuing conservation objectives as part of five-year contracts. Funded projects are required to address "resource concerns" specified by NRCS.

The *Regional Conservation Partnership Program* (RCPP) is a Natural Resources Conservation Service program designed to engage partner organizations in increasing conservation impacts on private lands through a mix of approaches that can include conservation easements, conservation-oriented management practices, or other changes.

The Natural Resources Conservation Service manages the *Healthy Forests Reserve Program*, designed to support threatened or endangered species recovery, biodiversity protection, and carbon sequestration on private lands. Landowners can choose from a ten-year restoration cost-share agreement or a 30-year or permanent conservation easement. A 30-year contract (rather than easement) option is available for lands owned by Native American Tribes.

Wetland reserve easements are available as part of the *Agricultural Conservation Easement Program*, administered by the Natural Resources Conservation Service. Federal support pays for both the purchase of a conservation easement and associated restoration costs on converted wetlands that can be successfully restored. The proportion of financial support depends on the term of the easement.

The *Conservation Reserve Program*, administered by the Farm Service Agency, has been in operation since 1985. It pays landowners on ten- to fifteen-year contracts to plant certain tree and plant species on environmentally sensitive lands (especially wetlands and highly erodible lands) and to withdraw those areas from agricultural production. Around two million acres of CRP lands have been planted to forest since the program's initiation, and the federal government provides incentives for thinning and other management practices on those forestlands.

The *Joint Chiefs' Landscape Restoration Partnership* (see Chapter 5) is designed to accomplish restoration at a landscape scale in places characterized by mixed national forest and private ownership. The program is administered by the US Forest Service on national forestland and by the Natural Resources Conservation Service on adjacent private lands. Projects selected through a competitive proposal process are eligible to receive up to three years of funding.

### *Incentive Programs and Tools for Private Forest Conservation*

Although taxes, endangered species laws, clean water laws, and other policies can complicate the financial and managerial options for forest landowners, policymakers have also developed numerous programs that provide positive incentives for stewardship and forest conservation. Current use taxation is an incentive-based or "carrot"-type tool that has been adopted in numerous states. In essence, it reduces the property tax burden on landowners by assessing the value of the property at a lower rate compared to its "highest and best" value on the open market. The concept of highest and best use (sometimes abbreviated to HBU) refers to the land use that will generate the greatest short-term income, which in many cases is residential real estate or other non-forest uses. In order to take advantage of current use taxation, landowners may need to meet acreage qualifications, commit to maintaining their property as undeveloped forest for a certain period of time, or engage in certain activities such as complying with Best Management Practices or completing a forest management plan. For example, Indiana's Classified Forest and Wildlands program is a voluntary effort under which owners of at least ten acres of forestland receive a state property tax deduction, technical assistance, and priority consideration for cost-share programs in exchange for following a professionally written management plan and complying with certain restrictions, including a restriction on building permanent structures. When it was first established in 1921, the program assessed the value of forestland at $1 per acre for tax purposes; assessment rates have since been updated but still represent a substantial discount compared to lands not enrolled in the program.

At the federal level, multiple options exist to reduce forest owners' income taxes. These include, among other provisions, the treatment of sold timber not as ordinary income but rather as capital gains (which is taxed at a lower rate) and tax deductions for items such as reforestation costs, equipment expenses, and forest losses due to natural events.[13] The federal government also administers various programs designed to incentivize forest conservation and stewardship through payments, cost-sharing, support for planning, and tax benefits. These programs are managed by the US Forest Service, Natural Resources Conservation Service, US Fish and Wildlife Service, and other agencies (Box 8.1). Many of these incentive programs are authorized and funded through the Farm Bill, which provides around $6 billion in annual conservation spending on private farms, forests, and rangelands.

Beyond the carrots and sticks associated with legal prohibitions, taxes, tax reductions, and financial cost-sharing, state and federal governments often attempt to nudge family forest owners in the direction of sustainable management through policy sermons that draw upon the power of advice and information. University-affiliated forestry extension programs and many state forestry programs (see Chapter 6) are designed to deliver technical assistance, support, and advice to forest owners through a network of foresters with expertise in topics such as forest health, silviculture, and planning. The US Forest Service also connects to family forest owners through its State and Private Forestry branch, continuing a role it has played since the earliest days of the agency under Gifford Pinchot. The agency's Forest Stewardship Program (FSP), a product of the 1978 Cooperative Forestry Assistance Act, is designed to work in partnership with states to provide outreach and technical assistance. A cornerstone of FSP is encouraging family forest owners to establish forest stewardship management plans for their properties. The plans themselves are written by consulting foresters, state agency foresters, or other forestry professionals under guidelines established by the state and with the financial support of the FSP. Although participating forest owners are not required to implement the activities recommended within the plan, they may be eligible for various financial and technical support opportunities to do so. Further, having an approved plan can make forest owners eligible to participate in other federal and state incentive programs.

Conservation easements are a tool of increasing importance in private land conservation and are included as options in many federal and state incentive programs. Unlike fee-simple acquisition, where the entirety of a property and associated rights are sold by one owner and purchased by another, conservation easements work by transferring only specific rights while leaving the core property ownership unchanged. These rights, such as the right to subdivide, extract mineral or petroleum resources, exclude the public, or develop residential or other structures, are purchased by a federal or state agency or by a non-governmental organization such as a land trust; the rights are then "extinguished" so that they can no longer be used by anyone. For example, Woody and Megan Wolfe's multi-generational farm in the Wallowa Valley contains the confluence of the Lostine and Wallowa Rivers, a place that has long been important to Plateau Tribes and that was once the burial site of Old Chief Joseph. The Wolfe family worked with the Wallowa Land Trust to secure two conservation easements totaling 463 acres surrounding the confluence. The Wolfe family still retains exclusive access rights to the land and can continue managing the mix of riparian forest, wetland, and farmland as they always have, but they have in essence "sold" their rights to convert the land to residential property. In return, the family received property tax benefits, as the market value of the land is lower than it would be if it were available for development. Because the transaction is considered a "qualified conservation easement" under federal law, it may also provide estate tax benefits when the land is inherited by the next generation.

As voluntary, market-based mechanisms, easements can be an attractive alternative to regulation for achieving conservation goals. A 2020 census by the Land Trust Alliance found that over 20 million acres of conservation easements were held by non-profit land trusts alone[14], and estimates of the total conservation easement acreage held by all governmental and non-governmental entities exceeds 40 million acres.[15] Land trusts have proliferated nationwide to help landowners conserve forests and other undeveloped lands. However, critics have pointed to questionable accounting practices associated with some easements, accusing a small proportion of landowners and intermediaries of abusing the process in order to generate the associated tax benefits (Box 8.2). Although most easements are established in good faith, courts have in certain instances found both conservation organizations and property owners to have engaged in "overvaluation of easements, failure to satisfy the deduction's conservation purposes test, and failure to satisfy the deduction's granted-in-perpetuity and protected-in-perpetuity requirements."[16] Estimates of the foregone federal tax collections associated with conservation easements range from the hundreds of millions to over one billion dollars annually,[17] with substantial tax benefits flowing to some of the nation's wealthiest landowners.[18] Easements are clearly an important conservation tool, but their continued social and political support will depend on ensuring that their tax advantages align with clear and compelling public benefits.

Other incentive programs designed to support forest conservation and stewardship—such as third-party certification and various payment for ecosystem service schemes (see Chapter 9)—may be difficult for family forest owners to access due to questions of scale. In essence, the "transaction costs" entailed in establishing and monitoring these kinds of voluntary programs on individual small properties often outweigh any resulting economic benefits. State governments and non-governmental organizations can help address this challenge by aggregating landowners under supervised programs. For example, The Nature Conservancy and the American Forest Foundation have piloted a forest carbon program in Pennsylvania that pays family forest owners to enroll in 10- or 20-year contracts to improve carbon storage through specified forest management and planting practices. State forestry agencies may be essential in providing the oversight and regulatory framework needed to enroll large numbers of family forest owners in third-party forest certification, as described in Chapter 6.

### BOX 8.2 CONSERVATION EASEMENT CRITIQUES AND CONTROVERSIES

Conservation easements are a firmly established part of the private land conservation toolkit in the US. Through their use of market mechanisms and the participation of willing sellers and buyers, conservation easements are framed as a "win-win" solution that accomplishes conservation outcomes without the use of heavy-handed government regulations or restrictions. Although the overall benefits of conservation easements are indisputable, serious concerns have been raised regarding some kinds of transactions. A particularly controversial model is one known as a syndicated conservation easement, which entails a group of investors purchasing a property via a limited liability company (LLC), securing a conservation easement, and obtaining the tax deduction benefits according to their stake in the LLC. Although this arrangement is neither illegal nor fraudulent in and of itself, accusations have been made regarding its use in specific cases to generate hefty tax deductions for investors without any clear conservation benefit. This can occur, for example, through inflated appraisals that overstate the value of the land absent a conservation easement; the gap between that value and the value of the land with development restrictions forms the basis of the resulting charitable donation and tax benefits.[19] The Internal Revenue Service has been aware of problems like these since the early 2000s and has taken steps to scrutinize syndicated easements and penalize those involved in fraudulent transactions. Many conservation advocates have defended the syndicated conservation easement model, even if instances of its application have been questionable.

Conservation easements have been criticized for other reasons as well, particularly when they are used to reduce the tax burdens of wealthy landowners who were unlikely to develop their properties even in the absence of an easement. Some critics have pointed out that an easement-focused approach to conservation works against the concept of "smart growth" due to its opportunistic, rather than strategic, nature.[20] A separate issue relates to the transfer of properties with conservation easements: the new property owners (whether family member inheritors or unrelated buyers) are often unfamiliar with the purposes or restrictions associated with the easement and may attempt to avoid compliance with its terms.[21] In general, the long-term monitoring and enforcement of easements creates a substantial burden that will fall on numerous land trusts, not all of which will necessarily have the resources to fulfill this duty over the long term. Finally, some scholars and practitioners have questioned the wisdom of the perpetual nature of most conservation easements, asking whether future adaptation (including adaptation to a changing climate) will be limited due to decisions made by willing sellers and buyers decades earlier.[22]

## Residential Development Trends and Consequences

The incentive programs previously described have multiple, complementary goals: to encourage family forest owners to proactively plan, to support active forest management, and to help slow the pace of residential development of forestlands. This last objective has been especially difficult to achieve as urbanized areas continue to expand and as homes are increasingly constructed in and near formerly undeveloped forests. Ecologist Volker Radeloff and colleagues reported that the wildland-urban interface (WUI)—the area where homes and other structures meet natural vegetation—is the fastest-growing land cover category in the nation, representing 9.5% of the contiguous US. Over 25 million people and 12.7 million houses were added to the nation's WUI area over the 20-year period of their study.[23]

The phenomenon of forest conversion to low-density residential uses is driven by multiple factors. On the demand side, many individuals and families like the idea of owning a primary or secondary home in forested areas away from both cities and suburbs. The demand for these homes comes from retirees, those who can telecommute to work, and those with the disposable income to afford a rural alternative to their primary residence. On the fringes of urban areas, new developments represent an expansion of the medium-density suburbs into low-density "exurbs."[24] Taken together, this means that many WUI forest owners come from urban or suburban backgrounds and have had relatively little prior experience owning or managing forests. The continued supply of WUI housing is facilitated by lax zoning standards in most states, the continued parcelization of forested properties, and strong economic demand for rural or peri-urban residential properties. Perhaps the strongest driver is the discrepancy between the modest long-term income opportunities associated with land uses such as forestry and the more lucrative short-term income from residential conversion. This phenomenon is not limited to family forestlands; in many parts of the country, new forested residential developments are being installed on lands that were formerly part of industrial timberland estates (this topic is covered in Chapter 9).

The US Forest Service's "Forests on the Edge" project took an in-depth look at the patterns of existing and projected future residential development on private forestlands. The research team's analysis showed that tens of millions of acres of private forestlands were on track to experience housing density increases in the coming decades.[25] Because most of the forestlands in the eastern US are privately owned, they were assessed as being the most susceptible to parcelization and residential development overall. Watersheds in Florida, Michigan, North Carolina, and southern Appalachia more broadly emerged as having the highest risk, but some western forests were rated as highly vulnerable as well. Watersheds along the Colorado Front Range and on the urban fringes of Washington, Oregon, and California were all projected to experience substantial increases in housing density and associated forest fragmentation.

Forest managers and conservationists have been alarmed at the ongoing parcelization and development of private forestland nationwide. A prime concern is that the expansion of low- to medium-density housing and other infrastructure into forestlands creates an increasingly challenging scenario for wildfire safety and management. Combined with a changing climate and past forest management practices, the expansion of the WUI places millions of people and their belongings in zones of elevated fire danger[26]—in some cases with limited options for escape in the case of an emergency. This complicates wildfire management and increases the cost and risk to firefighters of suppression activities. As people move into the forest fringe, they also introduce activities and infrastructure that increase the likelihood of new fire starts[27] and contribute additional hazards (such as gas tanks and storage sheds filled with fuel and chemicals) that make firefighting more dangerous. The use of prescribed fire also becomes more complicated in areas of expanding WUI—not only because of the increasing amount of built infrastructure, but also because of challenges regarding smoke management.

Beyond fire issues, the parcelization and development associated with an expanding WUI also tend to degrade water quality, reduce habitat for many sensitive wildlife species, and introduce invasive plants and animals.[28] Because timber management is less common on smaller properties, parcelization can complicate the continued supply of wood products[29] and close off informal access to non-timber forest products (a topic discussed in greater detail in Chapter 11). However, the details of each property are important; relevant factors include the land's previous uses as well as decisions made by WUI landowners. In some cases, new landowners can contribute to forest restoration and conservation on their own properties,[30] and they may serve as leaders in organizing their neighbors to be better stewards and to better prepare for wildfire, invasive species, and other hazards. The regulation of private land subdivision and development is formally the purview of state and local governments, but few places have adopted mandatory rules to improve the safety of residents and firefighters.[31] More common are initiatives such as FireWise USA, a voluntary program that encourages residents to create defensible space, reduce hazards in the immediate home environment, and improve ingress and egress. Many communities have also created Community Wildfire Protection Plans, discussed in Chapter 5.

## The Diversity of Forest Owners and Tenure Patterns

Family forest owners are a varied and diverse group, encompassing everything from small parcel owners in forested WUI regions to absentee owners of forests intensively managed for timber income to holistic forest stewards like the Jackson and Goebel families. Although the National Woodland Owner Survey reveals that family forest owners are largely male (76%) and overwhelmingly White (97%), there are important dimensions of gender and racial/ethnic diversity that deserve a closer look. Recent scholarship has taken a closer look at the unique histories and complexities of African American forest ownership and management.[32] Land ownership may hold special importance for African American families in light of a long history of political and economic marginalization. Further, land may serve as an important connection to the past. Anthropologist Sarah Hitchner and her colleagues, for example, explain that

> Family land left to [Black landowners] by their predecessors, who obtained the land during times of slavery, Reconstruction, or intense Jim Crow racism, and land that often literally contains the bones of their ancestors, has more than sentimental value; ties to the land are deeply embedded in their identity as individuals, families, and communities.[33]

African Americans and other nonwhite groups were legally prohibited from owning land throughout much of the antebellum South and faced barriers to land ownership ranging from discrimination to organized violence after the end of legal slavery.[34] Despite these challenges, many African Americans moved quickly after the close of the Civil War to purchase farmland—often the same lands that they or their forebears had worked under conditions of enslavement. Sixteen million acres of land, almost entirely consisting of southern farms, were owned by African American families in 1920. By the mid-1990s, only three million acres remained in Black hands.[35] The drivers of African American land loss are complex, but important factors include Black families' disproportionate lack of access to legal protection and government support programs as well as the vulnerabilities created by land inheritance patterns.

The term "heir property" (also called "heirs' property" or "heir's property") refers to land that is passed down intergenerationally without a legally probated will. This category represents approximately 35 to 40 percent of Black-owned land in the South[36] and is common among other marginalized groups as well.[37] Under this system, individuals in each successive generation receive a fractional but undivided share of the inherited property; after several generations, this process can result in a clouded title as dozens or even hundreds of individuals can legally claim partial ownership. In some cases, heirs may not even realize that they have legal claims to a piece of land, and some may never have seen or visited the property. Heir property presents several challenges to its owners, including making it difficult or impossible to use property as collateral for loans, to secure a mortgage, or even to make regular management decisions without the consent of all legal heirs.[38] Most important, heir property has traditionally been easy for predatory developers to acquire through forcing a sale of the entire property. Although eighteen states have now passed laws designed to prevent this from occurring, past abuses have contributed to substantial loss of Black-owned land in places like coastal Georgia and South Carolina.[39] In addition to creating susceptibility to predatory practices, the existence of multiple fractional owners leads to insecure tenure as any owner can force a sale, and it raises complicated issues related to taxes and the distribution of income from management activities. However, it is important to point out that some owners have chosen to retain heir property status despite its drawbacks, believing that its benefits to the extended family outweigh the complications it presents.[40]

Research has found that minority forest owners may be disproportionately unlikely to benefit from incentive programs designed to support working forest conservation—in part due to difficulties contributing required cost-share dollars as well as fear of land loss in the case of the inability to repay loans.[41] Yet African American landowners' lack of benefits from support programs is the result of forces that go well beyond these factors. Investigations going back to the 1960s uncovered evidence of discriminatory farm lending practices by the federal government, making Black landowners more susceptible to bankruptcy and foreclosure than their White counterparts.[42] A 1997 class-action lawsuit against the USDA, *Pigford v. Glickman*, resulted in the largest federal civil rights settlement to date. The settlement provided for economic compensation and limited avenues for land recovery, but ultimately accomplished little to reverse the decades-old trend of Black land loss.[43] Organizations such as the Federation of Southern Cooperatives/Land Assistance Fund, the Center for Heirs' Property Preservation, the Land Loss Prevention Project, and others have been founded across the South specifically to provide legal assistance, planning advice, management support, and other resources to existing African American landowners and help stem the further loss of Black-owned land.

The diversity of family forest owners is also reflected in the changing gender balance for ownership and decision-making. Women are the primary decisionmaker on 24% of family forest ownerships, representing 20% of family forest acreage, according to the 2018 National Woodland Owner Survey. This proportion has more than doubled since 2006. To date there has only been a limited amount of research done on women forest owners in the US, but some patterns are beginning to emerge. First, women are more likely than men to have inherited (rather than purchased) their forested properties, whether from parents or from a spouse. Women forest owners tend to have less interest in hunting than men and are less likely to manage their properties for wildlife—indeed, they are less likely to actively manage in general, although the reasons for this discrepancy are not entirely clear.[44] Women forest owners are also more likely than men to place value on the importance of family legacy in their management decision-making.[45] Outreach and extension activities have traditionally targeted male landowners, but new organizations and strategies have emerged to engage women. An example is the nationwide Women Owning Woodlands network, a project of the National Woodland Owners Association and US Forest Service State and Private Forestry, which provides trainings, resources, advice, and networking specifically oriented toward women forest owners.[46]

## Conclusion

As the largest single category of forest owners in the country, family forest owners are a vital component of the conservation landscape. Beyond their importance in providing wildlife habitat, wood products, open space, water supplies, carbon storage, and myriad other benefits, family forest owners are also often part of the fabric of their respective communities, and their properties represent sites of experimentation, stewardship, and cultural continuity. Family forest owners and the lands they control are defined by their diversity, manifested in terms of the broad variety of tenure patterns, ownership objectives, management motivations, decision-making processes, and sources of knowledge and inspiration. While family forest owners have long been a diverse and complex group, recent trends have created additional layers of complexity and raise questions regarding the design and implementation of policies and programs. The continued parcelization and development of family forestlands, particularly in the WUI, and the changing demographics of family forest owners complicate some of the assumptions upon which longstanding incentive programs are built. A growing need will be attending to the unique needs of heir property owners, women forest owners, and other owner categories that have traditionally been on the margins of conservation programs.

## Questions for Discussion

1. Why do family forest owners account for greater proportions of the forested landscape in the eastern and southern US than in western states?

2. Why is tax policy so important in encouraging or discouraging the conservation of long-term forest cover on family forest land? Do you think contemporary tax incentive programs are adequate to address the unintended consequences of taxes on forest conservation?
3. Where on the landscape would you expect to see the greatest proportions of smaller forest parcels, and why? What are some of the challenges associated with managing smaller forest parcels? Are there possible benefits from parcelization?
4. Foresters and conservationists have been concerned about the expansion of the wildland-urban interface. What are some of the drivers and possible consequences of this phenomenon? What policy or governance tools might help to address these concerns?
5. What is heir property? What complications does it present for property owners? What kinds of policies might help to prevent the further loss of land in heir property status?

## Notes

1 B. J. Butler et al., 2016.
2 B. J. Butler, Butler, et al., 2021.
3 Beach et al., 2005; Bengston et al., 2011.
4 Kittredge, 2004.
5 E.g., Beach et al., 2005; Bliss, 2008; Hodgdon et al., 2007; Sampson & DeCoster, 2000.
6 B. J. Butler, Caputo, et al., 2021; Silver et al., 2015.
7 Bliss, 2008.
8 Bailey et al., 2019; Bliss & Martin, 1988; Mook et al., 2022.
9 B. J. Butler & Leatherberry, 2004.
10 B. J. Butler, 2008; Giampaoli & Bliss, 2011; Greene et al., 2014.
11 Cushing & Newman, 2018.
12 Greene et al., 2014.
13 Greene et al., 2014; Kimbell et al., 2010.
14 https://findalandtrust.org/land-trusts/gaining-ground/united-states#land-protected.
15 McLaughlin & Cheever, 2015, p. 110.
16 McLaughlin, 2018, p. 225.
17 McLaughlin, 2018, p. 223.
18 Farrell, 2020.
19 Godin, 2022.
20 Roddewig, 2019.
21 Rissman & Butsic, 2011.
22 Owley et al., 2018; Roddewig, 2019.
23 Radeloff et al., 2018. Note that not all of the WUI included in this analysis is forestland.
24 Taylor & Hurley, 2016.
25 Stein et al., 2009.
26 Schoennagel et al., 2017.
27 Bar-Massada et al., 2014; Downing et al., 2022.
28 Abrams et al., 2012; Stein et al., 2009.
29 Sampson & DeCoster, 2000.
30 Walker et al., 2003.
31 Haines et al., 2008.

32 Bailey et al., 2019; Barlow & Bailey, 2017; Dyer & Bailey, 2008; Gaither, 2016; Gaither et al., 2019; Goyke & Dwivedi, 2021; Hitchner et al., 2017.
33 Hitchner et al., 2021, p. 2.
34 Copeland, 2013.
35 Daniel, 2013, p. 260.
36 Bailey et al., 2019.
37 Dyer & Bailey, 2008.
38 Bailey et al., 2019; Gaither & Zarnoch, 2017.
39 Chandler, 2005; Mitchell, 2001.
40 Dyer & Bailey, 2008.
41 Gan et al., 2005; Gordon et al., 2013.
42 Daniel, 2013; Mitchell, 2001.
43 Mitchell, 2001.
44 S. M. Butler et al., 2018.
45 Mook et al., 2022.
46 Huff, 2017; Redmore & Tynon, 2011.

## References


Abrams, J., Gosnell, H., Gill, N. J., & Klepeis, P. J. (2012). Re-creating the rural, reconstructing nature: An international literature review of the environmental implications of amenity migration. *Conservation and Society*, *10*(3), 270–284.

Bailey, C., Barlow, B., & Dyer, J. (2019). Practical constraints to timber management among African American owners of heir property. *Landscape and Urban Planning*, *188*, 180–187.

Barlow, B., & Bailey, C. (2017). The potential impact of heir property on timber management in the southeastern United States. *Professional Agricultural Workers Journal*, *5*(1), 5–13.

Bar-Massada, A., Radeloff, V. C., & Stewart, S. I. (2014). Biotic and abiotic effects of human settlements in the wildland–urban interface. *Bioscience*, *64*(5), 429–437.

Beach, R. H., Pattanayak, S. K., Yang, J.-C., Murray, B. C., & Abt, R. C. (2005). Econometric studies of non-industrial private forest management: A review and synthesis. *Forest Policy and Economics*, *7*(3), 261–281.

Bengston, D. N., Asah, S. T., & Butler, B. J. (2011). The diverse values and motivations of family forest owners in the United States: An analysis of an open-ended question in the National Woodland Owner Survey. *Small-Scale Forestry*, *10*(3), 339–355.

Bliss, J. C. (2008). Family forest owners. In E. M. Donoghue & V. E. Sturtevant (Eds.), *Forest community connections: Implications for research, management, and governance* (pp. 205–218). Resources for the Future.

Bliss, J. C., & Martin, A. J. (1988). Identity and private forest management. *Society & Natural Resources*, *1*(1), 365–376.

Butler, B. J. (2008). *Family forest owners of the United States, 2006* (GTR-NRS-27). USDA Forest Service Northern Research Station.

Butler, B. J., & Leatherberry, E. C. (2004). America's family forest owners. *Journal of Forestry*, *102*(7), 4–14.

Butler, B. J., Butler, S. M., Caputo, J., Dias, J., Robillard, A., & Sass, E. M. (2021). *Family forest ownerships of the United States, 2018: Results from the USDA Forest Service, national woodland owner survey* (GTR-NRS-199). USDA Forest Service Northern Research Station.

Butler, B. J., Hewes, J. H., Dickinson, B. J., Andrejczyk, K., Butler, S. M., & Markowski-Lindsay, M. (2016). Family forest ownerships of the United States, 2013: Findings from the USDA Forest Service's National Woodland Owner Survey. *Journal of Forestry, 114*(6), 638–647.

Butler, B. J., Caputo, J., Robillard, A. L., Sass, E. M., & Sutherland, C. (2021). One size does not fit all: Relationships between size of family forest holdings and owner attitudes and behaviors. *Journal of Forestry, 119*(1), 28–44.

Butler, S. M., Huff, E. S., Snyder, S. A., Butler, B. J., & Tyrrell, M. (2018). The role of gender in management behaviors on family forest lands in the United States. *Journal of Forestry, 116*(1), 32–40.

Chandler, A. B. (2005). The loss in my bones: Protecting African American heirs' property with the public use doctrine. *William & Mary Bill of Rights Journal, 14*, 387–414.

Copeland, R. W. (2013). In the beginning: Origins of African American real property ownership in the United States. *Journal of Black Studies, 44*(6), 646–664.

Cushing, T. L., & Newman, D. (2018). Analysis of relative tax burden on nonindustrial private forest landowners in the southeastern United States. *Journal of Forestry, 116*(3), 228–235.

Daniel, P. (2013). *Dispossession: Discrimination against African American farmers in the age of civil rights*. University of North Carolina Press.

Downing, W. M., Dunn, C. J., Thompson, M. P., Caggiano, M. D., & Short, K. C. (2022). Human ignitions on private lands drive USFS cross-boundary wildfire transmission and community impacts in the western US. *Scientific Reports, 12*(1), 1–14.

Dyer, J. F., & Bailey, C. (2008). A place to call home: Cultural understandings of heir property among rural African Americans. *Rural Sociology, 73*(3), 317–338.

Farrell, J. (2020). *Billionaire wilderness: The ultra-wealthy and the remaking of the American West*. Princeton University Press.

Gaither, C. J. (2016). *"Have not our weary feet come to the place for which our fathers sighed?": Heirs' property in the southern United States* (GTR-SRS-216). USDA Forest Service Southern Research Station.

Gaither, C. J., Carpenter, A., McCurty, T. L., & Toering, S. (Eds.). (2019). *Heirs' property and land fractionation: Fostering stable ownership to prevent land loss and abandonment*. USDA Forest Service Southern Research Station.

Gaither, C. J., & Zarnoch, S. J. (2017). Unearthing 'dead capital': Heirs' property prediction in two US southern counties. *Land Use Policy, 67*, 367–377.

Gan, J., Onianwa, O. O., Schelhas, J., Wheelock, G. C., & Dubois, M. R. (2005). Does race matter in landowners' participation in conservation incentive programs? *Society and Natural Resources, 18*(5), 431–445.

Giampaoli, P., & Bliss, J. C. (2011). Landowner perceptions of habitat protection policy and process in Oregon. *Western Journal of Applied Forestry, 26*(3), 110–118.

Godin, J. (2022). A Sand County tax shelter: Syndicated conservation easements and their toll on the American taxpayer. *Utah Law Review, 1*, 213–234.

Gordon, J. S., Barton, A., & Adams, K. (2013). An exploration of African American forest landowners in Mississippi. *Rural Sociology, 78*(4), 473–497.

Goyke, N., & Dwivedi, P. (2021). Ascertaining differences between farmer and non-farmer African American forest landowners in Georgia, United States. *Trees, Forests and People, 5*, 100118.

Greene, J. L., Butler, B. J., Catanzaro, P. F., Hewes, J. H., Kilgore, M. A., Kittredge, D. B., Ma, Z., & Tyrrell, M. L. (2014). Family forest owners and federal taxes. *Forest Policy and Economics, 38*, 219–226.

Haines, T. K., Renner, C. R., & Reams, M. A. (2008). A review of state and local regulation for wildfire mitigation. In T. P. Holmes, J. P. Prestemon, & K. L. Abt (Eds.), *The economics of forest disturbances* (pp. 273–293). Springer.

Hitchner, S., Schelhas, J., & Dwivedi, P. (2021). Safe havens: The intersection of family, religion, and community in black cultural landscapes of the southeastern United States. *Landscape and Urban Planning*, *214*, 104136.

Hitchner, S., Schelhas, J., & Gaither, C. J. (2017). "A privilege and a challenge": Valuation of heirs' property by African American landowners and implications for forest management in the Southeastern US. *Small-Scale Forestry*, *16*(3), 395–417.

Hodgdon, B., Cusack, C., & Tyrrell, M. (2007). *An annotated bibliography on family forest owners* (GISF Research Paper 002-R). Yale University.

Huff, E. S. (2017). A national perspective on women owning woodlands (WOW) networks. *Journal of Extension*, *55*(2), 2RIB6.

Kimbell, A., Hickman, C., & Brown, H. (2010). How do taxes affect America's private forest landowners? *Journal of Forestry*, *108*(2), 93–102.

Kittredge, D. B. (2004). Extension/outreach implications for America's family forest owners. *Journal of Forestry*, *102*(7), 15–18.

McLaughlin, N. (2018). Tax incentives for conservation easement donations. In R. Mann & T. Roberts (Eds.), *Tax law and the environment: A multidisciplinary and worldwide perspective* (pp. 219–253). Lexington Books.

McLaughlin, N., & Cheever, F. (2015). An introduction to conservation easements in the United States: A simple concept and a complicated mosaic of law. *Journal of Law, Property, and Society*, *83*, 107–186.

Mitchell, T. W. (2001). From reconstruction to deconstruction: Undermining Black landownership, political independence, and community through partition sales of tenancies in common. *Northwestern University Law Review*, *95*(2), 505–580.

Mook, A., Goyke, N., & Dwivedi, P. (2022). Conservation intentions and place attachment among male and female forest landowners. *Rural Sociology*. https://doi.org/10.1111/ruso.12434

Owley, J., Cheever, F., Rissman, A. R., Shaw, M. R., Thompson, B. H. Jr., & Weeks, W. W. (2018). Climate change challenges for land conservation: Rethinking conservation easements, strategies, and tools. *Denver Law Review*, *95*(3), 727–780.

Radeloff, V. C., Helmers, D. P., Kramer, H. A., Mockrin, M. H., Alexandre, P. M., Bar-Massada, A., Butsic, V., Hawbaker, T. J., Martinuzzi, S., & Syphard, A. D. (2018). Rapid growth of the US wildland-urban interface raises wildfire risk. *Proceedings of the National Academy of Sciences*, *115*(13), 3314–3319.

Redmore, L. E., & Tynon, J. F. (2011). Women owning woodlands: Understanding women's roles in forest ownership and management. *Journal of Forestry*, *109*(5), 255–259.

Rissman, A. R., & Butsic, V. (2011). Land trust defense and enforcement of conserved areas. *Conservation Letters*, *4*(1), 31–37.

Roddewig, R. J. (2019). Conservation easements and their critics: Is perpetuity truly forever...and should it be? *UIC John Marshall Law Review*, *52*(3), 677–714.

Sampson, N., & DeCoster, L. (2000). Forest fragmentation: Implications for sustainable private forests. *Journal of Forestry*, *98*(3), 4–8.

Schoennagel, T., Balch, J. K., Brenkert-Smith, H., Dennison, P. E., Harvey, B. J., Krawchuk, M. A., Mietkiewicz, N., Morgan, P., Moritz, M. A., Rasker, R., Turner, M. G., & Whitlock, C. (2017). Adapt to more wildfire in western North American forests as climate changes. *Proceedings of the National Academy of Sciences*, *114*(18), 4582–4590.

Silver, E. J., Leahy, J. E., Weiskittel, A. R., Noblet, C. L., & Kittredge, D. B. (2015). An evidence-based review of timber harvesting behavior among private woodland owners. *Journal of Forestry*, *113*(5), 490–499.

Stein, S. M., McRoberts, R. E., Mahal, L. G., Carr, M. A., Alig, R. J., Comas, S. J., Theobald, D. M., & Cundiff, A. (2009). *Private forests, public benefits: Increased housing density and other pressures on private forest contributions* (PNW-GTR-795). USDA Forest Service, Pacific Northwest Research Station.

Taylor, L. E., & Hurley, P. T. (Eds.). (2016). *A comparative political ecology of exurbia: Planning, environmental management, and landscape change*. Springer.

Walker, P. A., Marvin, S. J., & Fortmann, L. P. (2003). Landscape changes in Nevada County reflect social and ecological transitions. *California Agriculture*, *57*(4), 115–121.

# 9 Corporate Forests and Forestry

The Maxville Heritage Interpretive Center on Main Street in Joseph is the creation of Gwen Trice, daughter of one of the African American loggers who worked in Wallowa County's Maxville logging camp. The town of Maxville was established in 1923 by Kansas City-based Bowman-Hicks following its purchase of over 100,000 acres of forestland. Maxville is unique in being one of the only logging camps in Oregon history that included White and Black laborers working together and sharing a common, if segregated, living space. Forest workers were brought by the company to remote northeastern Oregon from the South and Southwest to harvest the region's abundant ponderosa pine, which was transported by rail to be processed at the Bowman-Hicks mill in the town of Wallowa. Maxville grew to around 400 residents by 1924 but began losing population during the Great Depression as Bowman-Hicks ceased its Wallowa County operations; the town was abandoned by the end of the 1940s.[1] The Maxville townsite was bought and sold as part of various corporate ownership transitions in the following decades. Starting in the early 2000s, Gwen began working with an ever-changing roster of corporate owners to retain access to the site; the Maxville Heritage Interpretive Center was eventually able to purchase the only remaining structure, a lodge that once housed the Bowman-Hicks headquarters. In 2022, Gwen announced the pending purchase of the Maxville townsite and the surrounding 240 acres. As part of plans for a rededication of Maxville, Gwen has collaborated with the Nez Perce Tribe and the Confederated Tribes of the Umatilla Indian Reservation to provide a deep and holistic interpretation of the site's long and storied history. The collaboration also reflects the close ties between Gwen's father and Plateau Tribes.

Corporate forest owners had operated in Wallowa County since the early twentieth century. Companies like the George Palmer Lumber Company and Nibley-Mimnaugh were financed by capital from the vicinity of Wallowa County, but they were soon replaced by larger nonlocal firms like the East Oregon Lumber Company and Bowman-Hicks. Corporate forest holdings in Wallowa County were consolidated later in the twentieth century, with the international forest products giant Boise-Cascade ultimately amassing around 150,000 acres; the smaller Montana-based R-Y Timber company acquired 15,000 acres. Both companies operated their own mills in the town of Joseph and supplied them with a

DOI: 10.4324/9781003043669-9

mix of timber from their respective land holdings, from family forestlands, and from the Wallowa-Whitman National Forest. Timber from national forestland was important enough that both mills folded in the wake of the timber program shutdown of the national forests in the 1990s. By the first decade of the twenty-first century, Boise-Cascade had restructured entirely, selling its forestlands in 2004 but retaining its mills and product lines. The new management vehicle for what had been Boise-Cascade forestlands in Wallowa County and beyond was a timber investment management organization called Forest Capital Partners. This firm, first assembled by investors in 2000, did not own the former Boise-Cascade lands directly but rather acquired and managed them on behalf of institutional investors. Forest Capital Partners continued to harvest timber as part of delivery contracts with Boise-Cascade mills while also identifying parcels to be sold as recreational properties under the company's real estate arm, Westslope Properties. Forest Capital Partners dissolved almost as quickly as it had appeared, selling its nearly two million acres of forest properties to other investor-based firms in 2012.

This chapter focuses on the policy and governance dimensions of forests owned and managed by a broadly defined forest industry. Many of the same policies relevant to family forests also apply to these forestlands, but the corporate organizational form itself and the sheer size of corporate forestland holdings introduce additional considerations. What modern forest corporations have in common is the ownership of productive forestlands on a large scale (typically in the hundreds of thousands to millions of acres) and an orientation toward maximizing economic returns, often—but not exclusively—through the production of wood fiber. The corporate model contrasts with that of family forest owners, who typically possess smaller properties and who report varied and diverse motivations for owning forestland. The governance of these extensive forestlands is driven by a combination of private corporate decision-making, public policies at local, state, and national levels, and engagement with non-governmental certification schemes that involve national- and even global-level organizations and actors.

## Industrial and Non-Industrial Forestlands

As Chapter 2 discussed, some of the earliest industrial forest firms in the US were established in the Upper Midwest in the late nineteenth century as companies began to integrate the cutting, shipping, and milling of eastern white pine. The industrial forest model went nationwide as those companies, along with many new firms funded by eastern capital, took control of western and southern timberlands. In the West, railroad land grants and other public land disposal policies were used—sometimes fraudulently—to piece together large holdings of prime old-growth Douglas-fir, ponderosa pine, western white pine, and other high-value species. Southern corporate lands began as properties for naval store production and, later, for liquidating extensive stands of longleaf and other pines. After being heavily cut over in the early twentieth century, they later became the center of a successful public and private experiment in intensive, short-rotation production for both timber and pulp uses.[2] Timber interests had controlled much of Maine's northern forestland dating back to statehood; the turn of the twentieth century saw a small number of companies take ownership of millions of acres used to supply paper mills that fed a growing national demand.[3] The South, the Pacific Northwest, and northern Maine would come to be the epicenters of corporate timberland ownership in the country.

For decades, private forests were considered to fall into one of two categories: industrial and non-industrial. Non-industrial private forestlands—what we now refer to as family forestlands (Chapter 8)—were defined as forests whose owners did not possess any significant processing infrastructure (such as a sawmill). Industrial forests, conversely, were defined as forests held by a company that also owned commercial-scale milling infrastructure. For much of the twentieth century, industrial forests were managed to maximize wood product-related income through intensive harvesting, planting of genetically preferred stock, fertilization, targeted thinning, and related practices. Harvests on company land, typically combined with timber purchased from public, tribal, and family forest owners, supplied company mills in a business model known as "vertical integration." By owning both mills and forests, vertically integrated firms benefited from a guaranteed timber supply, a reduction in volatility risk, and other economic advantages.[4] Since the 1980s, however, the definition of industrial forestlands as elements of vertically integrated firms has become less useful due to the emergence of new investor-driven approaches. As described in detail below, land and mills are typically kept separate under the model that has come to dominate the US forest industry since the late twentieth century. For this reason, it now makes more sense to talk about the category of corporate forestlands: those owned by a variety of corporate organizations, some of which operate their own mills and some of which do not. The extent of corporate forest ownership has waxed and waned over the years; it peaked at 71 million acres in 1987, but for much of the twentieth century it amounted to around 60 million acres.[5]

## A Crisis of Legitimacy and a Possible Solution

The global forest industry came under heightened scrutiny in the final decades of the twentieth century as part of growing international concern over the fate of forests. Environmental advocacy organizations waged a public campaign to reform forestry practices, emphasizing the protection of high conservation value forests and Indigenous peoples' rights as central themes. Environmental campaigners from mainstream organizations such as the Sierra Club, Greenpeace, and Natural Resources Defense Council—along with more radical groups like Earth First!—successfully drew public attention to the loss of high conservation value forests in places like the Amazon Basin, Indonesia, British Columbia, Pacific Northwest national forestlands, and, in some cases, private corporate lands in the US that still contained older forests. An example of the latter is the Headwaters tract in northern California, which included some of the last old-growth coast redwood trees remaining on private land. Pacific Lumber Company, the firm that had long owned Headwaters, was acquired in 1985 in a debt-financed hostile takeover by a corporate raider.[6] The new owner substantially increased logging to pay off the incurred debt, triggering a national protest campaign that included tree sits, road blockades, and participation by national celebrities such as Bonnie Raitt and Woody Harrelson. Among the activists involved in this campaign was Julia "Butterfly" Hill, who lived for two years on a tree platform high in a thousand-year-old redwood and became the public face of opposition to Pacific Lumber's activities.

In addition to direct-action protests in forests slated for harvest, environmental activists also initiated a "naming and shaming" campaign targeting the retailers whose shelves featured products sourced from places like the Amazon and northern California. No US company was targeted more intensively than Atlanta-based The Home Depot, which by the 1990s had emerged as a leading domestic retailer of wood products. According to an activist with the Rainforest Action Network,

> The company was confronted with shareholder resolutions, religious community organizing and zoning permit denials for new stores, while existing stores were hit with more than 600 demonstrations involving information flyering, guerilla theater, banner hangs, civil disobedience, old growth product stickering, wood product dumps, ethical shoplifting, dead rainforest tours and intercom takeovers.[7]

On a single day in October 1998, 75 Home Depot stores in 35 states became protest sites in a massive day of action by environmental activists. Following a nationwide campaign and a shareholder resolution, the company announced that it would end the sale of wood from "endangered forests" including old-growth redwood forests—by 2002.

These dramatic confrontations in the woods and in retail stores took place at the same time that international conservation organizers were working through more conventional channels to strengthen forest protections. The 1992 United Nations Conference on Environment and Development in Rio de Janeiro, Brazil (better known as the Rio "Earth Summit") held out the possibility of a binding international agreement to slow the ongoing degradation of high conservation value forests. Several factors ultimately stood in the way of such an agreement, however. Representatives of some developing countries argued that they would bear a disproportionate share of the costs in comparison with countries in the global North, many of which had long ago converted native forests to other land uses. More broadly, countries in both the global North and South with large forest products industries were opposed to any rule that could lead to higher monitoring and compliance costs. Finally, the Earth Summit came at a time when international trade barriers were being dismantled worldwide; a binding agreement on forest management would have run counter to the predominant paradigm encouraging global free trade.[8] The Earth Summit negotiations drew to a close after participants agreed to a non-binding set of forest conservation principles.

Although some forest industry representatives and free-trade advocates were heartened by this outcome, forest conservation advocates and Indigenous communities pushing for binding rules saw it as a defeat. A coalition of international environmental organizations including Greenpeace, the Worldwide Fund for Nature, and the Rainforest Alliance had been working on an alternative forest governance model since 1990, and they introduced it publicly in 1993 under the name Forest Stewardship Council (FSC). Rather than forcing changes through international law, this new model was envisioned to work within the forest products supply chain: products independently certified to have come from well-managed forests would be eligible to receive the seal of FSC approval, and consumer demand for labeled products was envisioned to drive improvements in forest management. As an organization, FSC began with a diverse coalition that included environmental advocacy organizations, rural and Indigenous community representatives, and some large retailers. The FSC General Assembly was designed to give equal weight to environmental, social, and economic concerns and to include equal representation from global North and global South actors.

FSC was built upon a third-party model—meaning that the FSC organization itself establishes standards, forest owners and operators attempt to meet those standards, and independent auditors accredited by FSC perform the on-the-ground work to verify compliance. Since its inception it has offered two types of certificates: a forest management certificate applies to forests that have been audited and verified, whereas a chain of custody certificate indicates that wood from certified forests can be reliably tracked through the milling and manufacturing process to an end product authorized to carry the FSC logo. FSC also developed group certification options under which multiple forest owners or operators can be certified with oversight by a private, public, or non-profit manager. By fostering a model that rewards progressive forestry practices through consumer labels (and the price premiums those labels were assumed to deliver), the FSC model was designed to drive improvements to forest management while steering clear of free trade conflicts and avoiding a reliance on government policies and enforcement.[9]

Looking back almost three decades after the Rio Earth Summit, it is clear that the emergence of this new model changed forest governance in the United States, although outcomes differ greatly from initial expectations in several important respects. First, FSC soon found that it was not the only certification scheme in town; forest industry actors moved quickly to establish competitor labeling schemes following the rollout of FSC. The American Forest and Paper Association unveiled the Sustainable Forestry Initiative (SFI) in 1994, just a year after the launch of FSC. Although SFI did not begin as a third-party certification and labeling scheme, it came to adopt this model by 2002. As an industry-originated system, SFI was designed to be less prescriptive than FSC; this approach led some environmental advocates to accuse SFI of "greenwashing" what they saw as unsustainable status-quo practices.[10] The availability of SFI certification, along with the family forest-oriented American Tree Farm System

(ATFS) certification, meant competition for FSC—and this competition intensified pressure on FSC to make compromises to retain its market influence.[11]

Second, the retail price premiums for certified wood products that were expected to support the entire certification system largely failed to materialize. Some certified wood products appear to be able to sustain modest retail price premiums, but consumer demand for certified products in the US has been modest at best and many product lines are still dominated by non-certified material. Certification persists despite this dilemma largely due to the pressure placed on buyers that are one step removed from individual end consumers. For example, The Home Depot reports that it provides "preferential treatment" to FSC-certified wood and requires that products be certified if they are sourced from some global areas of conservation concern.[12] NGOs have urged other retailers to carry certified product lines, and numerous government procurement policies mandate baseline levels of certified products in the paper and wood products purchased by public entities. At the same time, displaying certification labels has become institutionalized as a practice for some businesses to demonstrate their green credentials and burnish their overall corporate image.

Third, the spatial and ecological patterns of certification adoption are somewhat different than early certification proponents anticipated. At a global level, the highly biodiverse tropical forests that were at the center of global concern and activism in the 1990s have remained largely outside of the certification system. The vast majority of certified acres to date are found in North America, Europe, Russia, and former Soviet Republics, with tropical regions accounting for less than 15% of certified area. In the United States, a sizeable proportion of FSC-certified forest management units are owned not by private firms but by state governments. Roughly 1.5 million acres of FSC's national total are made up of Wisconsin state lands, 2.2 million acres are Pennsylvania state lands, 3.9 million acres are Michigan state lands, and 5 million acres are Minnesota state lands.

Many corporate forest owners—the original domestic target of sustainable forestry organizing—ultimately signed on to third-party certification, although SFI has proven to be more popular than FSC on these lands within the US. The SFI standard struggled with credibility for several years as it was perceived to lack transparency and to require few changes on the part of corporate forest managers; this lack of credibility translated into difficulties selling SFI-labeled products in some markets. However, research has shown that some SFI environmental and social standards have ratcheted up over the years since its inception,[13] and SFI products are regularly found on the shelves next to those certified by FSC. Family forest owners, who were generally not the targets of the kind of activism and organizing that ultimately led to the proliferation of certification programs, have been enrolled in smaller proportions, some under group certification led by state entities (see Chapter 6). As of 2022, 36 million acres of forestland in the US are certified by FSC, 67 million acres are certified by SFI, and 19 million by ATFS (with some forests enrolled in multiple schemes simultaneously).

An important question is whether forest certification resolved the crisis of legitimacy that plagued the forest industry for much of the 1980s and 1990s.

The answer is complicated. On the one hand, the protests, banner hangs, and other actions that were once commonplace in home improvement retail stores and their corporate headquarters have mostly disappeared. Retailers regularly offer certified products in their stores, and companies like IKEA (itself a founding member of FSC) highlight their certified supply chains as part of sustainability-oriented corporate branding. Corporate scandals do continue to erupt periodically; for example, in 2015 Lumber Liquidators pled guilty to Lacey Act violations for importing wood flooring illegally harvested in the Russian Far East in Siberian Tiger and Amur Leopard habitat and paid fines of over $13 million plus other penalties.[14] Direct-action protests of logging operations, while still occurring from time to time, are less common than they once were, and those that do occur tend to target state and federal forestlands rather than corporate lands such as Pacific Lumber's redwood tracts. Quieter efforts by shareholders concerned with "environmental, social, and governance" (often called ESG) issues have also become more common. In 2022, for example, the shareholders of Home Depot voted in favor of a resolution requiring the retailer to report on the sustainability of its wood sourcing policies; this move came following accusations that wood on store shelves could be traced back to unsustainably managed high conservation value forests overseas.[15]

On the other hand, it is unclear how much credit certification can take for these trends. US corporate timberlands contained relatively little in the way of old-growth forest by the dawn of the new millennium; for example, most of the old-growth redwood stands that remained following battles on Pacific Lumber property were purchased with public funds and are now owned and managed by the Bureau of Land Management. The federal land management agencies also substantially decreased old-growth harvests in the contiguous 48 states in the years following the establishment of the 1994 Northwest Forest Plan, and President Clinton's 2001 Roadless Area Conservation Rule effectively preserved most of the remaining roadless forests on federal lands.[16] Addressing the degradation of tropical forests has proven more difficult, but the entry of illegally harvested timber into US supply chains has slowed since passage of Lacey Act amendments in 2008, the law under which Lumber Liquidators was prosecuted in 2015. Additionally, the primary threat to tropical forests in many places is their conversion to oil palm, livestock ranches, and other land uses—issues that are difficult to address via forest management certification. Collectively, these changes may have done as much to resolve the crisis of legitimacy faced by the forest industry in the 1980s and 1990s as did the advent of forest certification itself.

## Forest Industry Restructuring

As forest products companies adjusted their strategies for growing and harvesting wood fiber throughout the twentieth century, the vertically integrated corporate model reigned supreme. That model allowed corporate forest owners to control the entire supply chain: the land on which trees were grown; labor in the woods for planting, preparation, and harvesting; transportation of logs to the mills, the mills themselves and the labor to run them, transportation and distribution of the finished products; and the in-house expertise needed to plan and optimize these various activities to satisfy corporate owners and shareholders.[17] Some of these elements—particularly logging and transportation—were frequently contracted out, and timber supplies were also regularly sourced from non-industry lands, but vertical integration remained the predominant model. The domestic forest industry invested in research and development intended to produce an increasing amount of wood volume from each acre of forestland and to develop new technologies for tree species and sizes that would otherwise be considered unmerchantable. Throughout the twentieth century, the forest industry funded its own research and development, and it partnered with both universities and the US Forest Service. These collective efforts helped to develop new products, processes, and technologies that transformed everything from the genetic stock of planted trees to the layout of logging road networks to the chemical processes used to make paper from tree pulp to the glues used to create large panels and boards from small pieces of wood.

The forest industry began shifting away from the vertically integrated corporate model in the final decades of the twentieth century. The drivers of this change had less to do with the heightened social scrutiny on forest management practices described above and more to do with federal tax policy, accounting practices, and what has been called the financialization of the timber industry. The concept of financialization refers to the rise of finance mechanisms and their associated logics as organizing principles for economic activity.[18] To provide a simple example, financialization is reflected in the growing importance of shareholder pressure on publicly traded firms: demands for short-term growth in stock prices can drive various changes to corporate practices so as to generate profits to keep shareholders happy. Timberlands became attractive assets for investors in the 1980s and 1990s because they were perceived as having untapped potential for growth and because they balanced out other more traditional investments; they were thus considered an important element of a diversified portfolio.[19]

Historian Michael Hillard describes the view of corporations under financialization as "bundles of assets to be blown up, reconfigured, put to death for tax advantages, or heavily financed if sufficiently promising."[20] In the case of vertically integrated timber companies, this meant taking a critical look at assets (such as timberlands and mills), costs (such as labor and research and development), and corporate strategies and adjusting to maximize short-run value creation. Timberlands were typically undervalued on corporate balance sheets due to prevailing accounting practices, creating a ripe environment for hostile takeovers such as the one that befell Pacific Lumber as well as other forest industry giants like Diamond International and Crown Zellerbach.[21] Changes to tax policies stemming from laws such as the 1986 Tax Reform Act helped set the stage for the financialization of the forest industry. Under these laws, vertically integrated companies found themselves at a disadvantage because they were taxed on their corporate income at the (relatively) high corporate tax rate, and the dividends they paid to investors were taxed again. The 1974 Employee Retirement Income Security Act encouraged pension managers to invest in real estate, creating a new institutional demand driver for timberland investments.[22]

As part of taking a critical look at their costs and growth strategies, investors and industry leaders saw that other corporate models could be more lucrative—especially those that avoided a "double taxation" scenario or that classified income streams in ways that lowered the overall tax burden. Two models in particular rose to the top: the timber investment management organization (TIMO) and the real estate investment trust (REIT). TIMOs manage timberland on behalf of clients—typically large investors that may include pension funds, insurance companies, foundations, universities, and wealthy individuals. TIMOs tend to have short investment horizons (typically ten to fifteen years), and, most importantly, their investments are only in the timberland itself and not in the processing infrastructure typical of vertically integrated firms. REITs, on the other hand, own landed property (such as timberland) outright and may own a limited amount of processing infrastructure as well. By law they must invest at least 75 percent of their assets in real property and derive at least 95 percent of their gross income (and 75 percent of net income) from real property.[23] A key characteristic of REITs is that they are publicly traded and required to pass on at least 90 percent of taxable income to their investors as dividends. TIMOs emerged in the 1960s and 1970s and timber REITs gained popularity following changes in the 1997 Real Estate Investment Trust Simplification Act.[24] By 2010, nearly all of the largest (and many of the smaller) forest corporations had restructured into either TIMOs or REITs.

The implications of this change are still being understood, but industry observers and academics have pointed to several TIMO/REIT transition outcomes that could change the ways that corporate forestlands are managed. There is some evidence that TIMOs and REITs tend to harvest more intensively than did their vertically integrated forebears.[25] In the South, TIMOs and REITs have converted hardwood and mixed hardwood-pine stands into pure stands of fast-growing pine.[26] At the same time, TIMOs and REITs were found to be more likely to reforest than family forest owners.[27] Some researchers have noted that TIMOs and REITs have substantially reduced their research and development expenditures—including both private in-house research and development as well as support for university-led research cooperatives. Although some individual firms continue to support these activities, the overall level of investment has declined in comparison to earlier periods.[28]

As part of being "blown up" and "reconfigured," forestland assets are increasingly put toward uses other than the production of trees for harvest, with residential real estate sales and resort development serving as important sources of income generation.[29] In some cases, proposals for large-scale development have triggered elected officials and local communities to respond. For example, the REIT Plum Creek Timber announced plans in 2002 to sell 20,000 acres of former industrial timberland in the Swan Valley of western Montana as recreational and residential real estate. A community-based organization (see Chapter 10) partnered with The Nature Conservancy, the Trust for Public Land, the Confederated Salish and Kootenai Tribes, and other entities to seek options for preventing runaway development. The group's early successes focused on conserving lands within the Swan Valley through a mix of land purchases, conservation easements, and other mechanisms. These efforts ultimately led to a much larger deal in which over 300,000 acres of Plum Creek timberlands across western Montana were purchased in a complex set of transactions that entailed use of a new forestry conservation bond championed by Montana Senator Max Baucus.[30] Chapter 10 provides additional details on the role of community-led forest conservation efforts in areas of corporate forestland divestment.

Additionally, many financialized firms have adopted a strategy of combining revenue streams from multiple sources on those lands that remain dedicated to growing and harvesting trees. Alternative revenue streams include hunting or other recreational leases, payments for ecosystem services (PES) such as carbon sequestration or water provisioning, wetland mitigation credits, and the sale of development rights via conservation easements (discussed in Chapter 8).[31] Lands identified for continued wood fiber production are often enrolled in third-party certification schemes as well. The sale of conservation easements, recreation leases, or ecosystem services provides short-term cash flow for investors and complements revenue streams from timber sales. While recreation leases and conservation easements are well-established tools, PES programs are still very much in a growth and experimentation phase as of the date of this writing (Box 9.1).

### BOX 9.1 PAYMENT FOR ECOSYSTEM SERVICES PROGRAMS

The concept of "ecosystem services" refers to any tangible or intangible benefit provided by natural processes. Ecosystem services are often conceptualized as falling into four categories: provisioning (for example, providing water or wood products), regulating (for example, climate regulation or erosion control), cultural (such as spiritual or recreational benefits), and supporting (for example, protecting the biodiversity upon which other ecosystem services depend).[32] In many cases, these are benefits that have long been taken for granted within prevailing social and economic systems; proponents of payment for ecosystem services (PES) mechanisms argue that putting a price on these benefits will help conserve them through the creation of economic incentives. PES schemes are distinct from other voluntary, incentive-based conservation programs primarily in the emphasis on what is called "conditionality": the payment is conditional upon verifiable actions by the seller that, at least theoretically, produce or protect specified ecosystem services.[33] Addressing the need for conditionality requires that a PES program make its payments only upon the enactment of a specified input (such as a change in land use activity) or outcome (such as production of a valued ecosystem service). It also requires initial reconnaissance for inventory followed by ongoing monitoring and verification; due to economies of scale, corporate forest owners may be especially well-poised to participate in PES marketplaces.

Beyond conditionality, PES schemes also must confront questions of additionality, leakage, and permanence. Additionality refers to the idea that any benefits derived from implementation of a PES scheme are above and beyond what would have occurred absent the scheme. This means that establishing credible counterfactuals (in essence, determining what would have happened if the PES scheme were not in place) is a key part of the process. Leakage, which refers to positive changes in one place being offset by negative changes somewhere else, is perhaps the thorniest issue of all for PES schemes to address. To provide a perhaps overly simplistic example, if one forest owner delays harvesting in order to sequester more carbon in standing trees, the existing demand for timber may simply drive another owner to harvest their own stand earlier. Although the first owner made a change and received a payment, the overall carbon benefit in this hypothetical case is nonexistent. Permanence refers to the durability of a positive change: a wood product that will decompose in five years provides less of a carbon sequestration benefit in comparison to a product that will last for 50 years.

To date, PES programs have largely focused on attempting to structure payment programs that drive the production of valued services through practices like afforestation, avoided deforestation, and forest management improvement. Confronting the problems of additionality, leakage, and permanence in this process has been neither simple nor uncontested. Questions have been raised by academics and others regarding the extent to which carbon credits in major offset programs are properly calculated.[34] One major timberland owner reported that at least some of the early forest carbon offsetting programs in which they participated provided little to no additionality because they were applied to forests that would not have been managed any differently in the absence of the program.[35]

The demand drivers for ecosystem service protection on forestlands come from a mix of sources, with corporate sustainability pledges and state and regional greenhouse gas regulation schemes (see Chapter 6) serving as some of the most important sources. Carbon markets are sufficiently developed that in late 2021 a joint venture backed by $500 million in investment capital was established with a focus on purchasing and managing forests primarily for carbon sequestration. The Adirondack region of New York was one of the first landscapes targeted for acquisition, with around 50,000 acres purchased in early 2022. The plan for these forestlands, as well as for future acquisitions, is to harvest only 10 to 20 percent of the available timber volume while generating salable carbon credits from the difference between these activities and "business as usual." The goal of the firm is to eventually acquire more than one million acres of forestland to manage primarily for carbon benefits.

## New Technologies, Products, and Markets

From the woods to the mills, new technologies have driven transformations in the forest industry, changing demands for human labor and opening new forest products markets. For example, in-woods harvesting, delimbing, and cutting to specified log lengths once required a crew of several loggers with chainsaws; now these same tasks can be carried out in seconds by a single individual operating a very expensive piece of machinery. Mills are increasingly automated as well, using diagnostic technology to optimize the value of products sawn from each log. Changes such as these mean that desired job skills and demand for labor in forestry-oriented communities are also in transition (see Chapter 10).

New technologies have also created opportunities for innovations in wood products and corresponding changes in forest management practices. Two especially important markets are those in bioenergy and structural building materials. Forest-based bioenergy includes relatively simple technologies such as woody biomass and wood pellets as well as more advanced applications including ethanol and jet fuel produced from woody material. Woody biomass generally consists of chipped or shredded forest residues (such as limbs and smaller trees) that are burned to generate heat, electricity, or both in combination. Pellets can be made of similar materials, although they can also be manufactured from larger trees. Chipped wood is pressed into relatively uniform shapes and sizes to make pellets, which burn more evenly than cordwood and are used to generate heat via pellet stoves scaled for residential, municipal, or industrial uses. Like biomass, pellets can also be used to generate electricity; co-generation facilities are commonly used to make maximum use of the heat generated in the process of combusting wood fuel for electricity. Renewable energy policies passed at the state level have generated a modest demand for wood-based bioenergy, but international policies are the more potent driver of US pellet production.[36] The US exported 7.5 million metric tons of pellets in 2021, with the United Kingdom, European Union countries, Japan, and Korea as major destinations. Pellet mills are heavily concentrated in the southern US from east Texas to Virginia and collectively have the capacity to produce over 10 million tons of pellets annually.

The growing market for sustainable transportation fuels has also sparked research on and investment in technologies that convert wood to products like ethanol and jet fuel. Federal renewable energy policies have attempted to both support and mandate the production of cellulosic ethanol since 2007, but actual production has lagged far behind expectations. As of 2019, there was only a single operational cellulosic ethanol plant in the country.[37] To date, the primary demand driver for sustainable aviation fuels has come from voluntary commitments on the part of individual airlines and airline consortia rather than from federal or state mandates. Comprehensive state climate change policies in some places provide another motivation to develop liquid fuel alternatives, including from wood (see Chapter 6).

Forest-based bioenergy presents the promise of a win-win scenario, given that it generates energy from a plant-based source while also providing market support for harvesting and removing the kind of low-value material (smaller diameter trees) that are abundant in landscapes that have experienced decades of fire exclusion. For this reason, federal and state policies proliferated after 2000 to support biomass harvesting and processing and to spur innovation in biomass technology (see Chapter 5). However, both biomass- and pellet-based bioenergy have come under criticism from conservation advocates and climate change activists. One concern is that market demand for low-value trees and residues may lead to harvest practices that degrade high conservation value forests or that leave little residual woody material on site to replenish soil fertility.[38] United Kingdom and European Union governments have grappled with these issues as part of their attempts to assure the benefits of their own energy policies, and several countries have adopted sustainability standards for bioenergy—including for the wood pellets they import from the US. Forest certification is the primary pathway to demonstrate compliance with many of these standards.[39]

A second key public policy question has to do with the carbon emissions associated with wood-based bioenergy. As the production of energy from woody biomass and wood pellets began to ramp up, dueling scientific analyses emerged that seemed to alternately endorse and reject wood-based bioenergy as a low-carbon feedstock. The primary argument in favor of classifying wood pellets and biomass as carbon-neutral energy sources is that the carbon dioxide released during combustion of these materials will be reabsorbed by future tree growth as harvested forests are replanted.[40] There are complementary arguments that the increased economic demand associated with this use of wood products will encourage additional tree planting and thereby maintain forest cover and, further, that the active management prompted by bioenergy markets will reduce the risk of wildfires and insect outbreaks.[41] Those arguing against the wider adoption of wood bioenergy typically point to the time lag between the carbon emissions associated with biomass burning and the eventual reuptake of carbon in new forests—a time lag that may be on the order of decades.[42] They argue that confronting the climate crisis requires maintaining carbon stocks in the tissue of living trees and developing low-carbon energy alternatives. Determining whether forest biomass-based energy truly comes with climate benefits appears to depend heavily upon the "perspectives, baselines, feedstocks, and forest management practices" that are included in the carbon analysis.[43] Beginning in 2018, Environmental Protection Agency appropriations bills have included language requiring the agency to develop regulations that "reflect the carbon-neutrality of forest bioenergy and recognize biomass as a renewable energy source, provided the use of forest biomass for energy production does not cause conversion of forests to non-forest use." The EPA has not finalized regulations implementing this guidance as of the date of this writing.

A somewhat less contentious set of policy issues surrounds the question of engineered wood products for residential and commercial construction, commonly called "mass timber" construction. Unlike conventional wood construction, mass timber takes advantage of novel products like cross-laminated timber, laminated veneer lumber, mass plywood panels, and other building materials.

What makes these products unique is that they are engineered from smaller pieces of wood joined in ways that give them the structural properties of larger solid materials. According to mass timber proponents, buildings made with these materials are both more sustainable and more fire-resistant than steel or concrete buildings. The largest mass-timber structure in the US is under construction in Milwaukee, Wisconsin, at the time of this writing and will measure 284 feet in height when complete. Some of the major barriers to their wider adoption are state and municipal building codes, which often assume that wood is only appropriate for relatively small structures. The 2021 iteration of the International Building Code included three types of construction utilizing mass timber components, and the Code has already been adopted by numerous states and municipalities.

## Conclusion

Corporate forestry has been part of the US forest landscape for well over a century. Since the first vertically integrated firms were assembled to harvest the Upper Midwest pineries, corporate forestry has undergone numerous organizational and technological changes and adapted to a shifting political and economic operating environment. The policy and governance influences on this class of forest owners and producers are multiple and complex, encompassing federal tax policies, corporate accounting practices, non-governmental certification schemes, voluntary ecosystem service markets, and varied policies related to issues as diverse as energy generation and building codes. Corporate forestlands are more "liquid" than they have been at any time since the emergence of the first vertically integrated firms: they change ownership more frequently and are put to a wider array of uses than was typical for most of the twentieth century. Improved planting stock and intensive fertilization and thinning practices continue to increase yields from the most productive terrain, even as alternative income streams related to carbon sequestration, habitat mitigation, and real estate development are pursued in places where they provide the greatest economic returns. The constant turnover in corporate forestland ownership has created new opportunities for some communities to take ownership of important nearby forests, a topic covered in greater depth in Chapter 10.

## Questions for Discussion

1. Why did corporate forestry face a "crisis of legitimacy" in the late twentieth century, and how was this crisis ultimately resolved?
2. What are the drivers of the transition from vertically integrated to financialized corporate organization in the timber industry? What are the implications of this transition for forests and people?
3. Why does there continue to be debate and controversy regarding the development of wood-based bioenergy as an alternative to fossil fuels? How might this debate ultimately be resolved?

4. When you purchase paper or wood products, do you seek out products that have been certified? Are you willing to pay more for products that have been certified? Why or why not?
5. Are payments for ecosystem services the right way to ensure that forestlands continue to provide the kinds of benefits that are usually taken for granted? What more needs to be done to address issues of additionality, leakage, and permanence?

## Notes

1 Highberger, 2001.
2 Fox et al., 2007.
3 Hillard, 2020.
4 Zhang, 2021, p. 15.
5 Fernholz et al., 2007; Zhang, 2021; note that the US Forest Service's definition of "corporate forests" includes a substantial amount of small corporate organizations with limited landholdings; the estimate of total forestland owned by TIMOs, REITs, and large corporate owners based on 2017 data is 65 million acres (Sass et al., 2021).
6 J. Bernstein, 2018.
7 Krill, 2001, p. 16.
8 Bartley, 2007; S. Bernstein, 2002.
9 For more on the background and history of forest certification, see Bartley, 2007; Cashore et al., 2004; Cashore & Auld, 2012; Gulbrandsen, 2010.
10 Krill, 2001; Sasser et al., 2006.
11 Moog et al., 2015.
12 https://ecoactions.homedepot.com/wp-content/uploads/Wood-Purchasing-Policy-May-2022.pdf.
13 Judge-Lord et al., 2020.
14 https://www.justice.gov/opa/pr/lumber-liquidators-inc-pleads-guilty-environmental-crimes-and-agrees-pay-more-13-million.
15 https://subscriber.politicopro.com/article/eenews/2022/05/20/home-depot-fails-to-derail-shareholder-push-on-rainforests-00033836.
16 The exception to this general trend is Alaska's Tongass National Forest, which continued a timber program focused on harvesting older stands and was at times exempted from the Roadless Rule through actions by the George W. Bush and Donald Trump administrations.
17 Zhang, 2021.
18 Epstein, 2005; Gunnoe, 2016; Gunnoe & Gellert, 2011; Meyfroidt, 2018.
19 Baral & Mei, 2022; Clutter et al., 2005.
20 Hillard, 2020, p. 171.
21 Zhang, 2021, pp. 43–44.
22 Kelly, 2010; Zhang, 2021.
23 Zhang, 2021, p. 141.
24 Baral & Mei, 2022.
25 Belair & Ducey, 2018; Jin & Sader, 2006; Zhao et al., 2020.
26 Zhang et al., 2012.
27 Sun et al., 2015.
28 Clutter et al., 2005, pp. 10–11; Zhang, 2021, pp. 135–136.
29 Bliss et al., 2010.
30 Parker, 2014.
31 Sass et al., 2021.
32 Millennium Ecosystem Assessment, 2003.

33 Wunder, 2015.
34 Badgley et al., 2022; Marino et al., 2019.
35 https://www.bloomberg.com/news/articles/2022-03-17/timber-ceo-wants-to-reform-flawed-carbon-offset-market.
36 Masum et al., 2019.
37 Masum et al., 2019.
38 Berger et al., 2013.
39 Barnett, 2016; Kittler et al., 2020.
40 Dale et al., 2017.
41 Cowie et al., 2021; Dale et al., 2017.
42 Booth, 2018; Norton et al., 2019; Searchinger et al., 2009.
43 Dwivedi et al., 2019.

## References


Badgley, G., Freeman, J., Hamman, J. J., Haya, B., Trugman, A. T., Anderegg, W. R. L., & Cullenward, D. (2022). Systematic over-crediting in California's forest carbon offsets program. *Global Change Biology*, *28*(4), 1433–1445.

Baral, S., & Mei, B. (2022). Development and performance of timber REITs in the United States: A review and some prospects. *Canadian Journal of Forest Research*, *52*(1), 1–10.

Barnett, B. (2016). An evaluation of the UK's use of SFM standards to procure solid woody biomass for electricity generation using sustainable bioenergy criteria. *Biofuels*, *7*(1), 1–11.

Bartley, T. (2007). Institutional emergence in an era of globalization: The rise of transnational private regulation of labor and environmental conditions. *American Journal of Sociology*, *113*(2), 297–351.

Belair, E. P., & Ducey, M. J. (2018). Patterns in forest harvesting in New England and New York: Using FIA data to evaluate silvicultural outcomes. *Journal of Forestry*, *116*(3), 273–282.

Berger, A. L., Palik, B., D'Amato, A. W., Fraver, S., Bradford, J. B., Nislow, K., King, D., & Brooks, R. T. (2013). Ecological impacts of energy-wood harvests: Lessons from whole-tree harvesting and natural disturbance. *Journal of Forestry*, *111*(2), 139–153.

Bernstein, J. (2018). Beyond "owls versus jobs": A twenty-year retrospective of the Headwaters Forest controversy. *Humboldt Journal of Social Relations*, *40*, 54–67.

Bernstein, S. (2002). Liberal environmentalism and global environmental governance. *Global Environmental Politics*, *2*(3), 1–16.

Bliss, J. C., Kelly, E. C., Abrams, J., Bailey, C., & Dyer, J. (2010). Disintegration of the US industrial forest estate: Dynamics, trajectories, and questions. *Small-Scale Forestry*, *9*(1), 53–66.

Booth, M. S. (2018). Not carbon neutral: Assessing the net emissions impact of residues burned for bioenergy. *Environmental Research Letters*, *13*(3), 035001.

Cashore, B., & Auld, G. (2012). Forestry review. In Steering Committee of the State-of-Knowledge Assessment of Standards and Certification (Ed.), *The roles and limitations of certification* (pp. A88–A124). RESOLVE, Inc.

Cashore, B., Auld, G., & Newsom, D. (2004). *Governing through markets: Forest certification and the emergence of non-state authority*. Yale University Press.

Clutter, M., Mendell, B., Newman, D., Wear, D., & Greis, J. (2005). *Strategic factors driving timberland ownership changes in the US South*. USDA Forest Service Southern Research Station.

Cowie, A. L., Berndes, G., Bentsen, N. S., Brandão, M., Cherubini, F., Egnell, G., George, B., Gustavsson, L., Hanewinkel, M., Harris, Z. M., Johnsson, F., Junginger, M., Kline, K. L., Koponen, K., Koppejan, J., Kraxner, F., Lamers, P., Majer, S., Marland, E., … Ximenes, F. A. (2021). Applying a science-based systems perspective to dispel misconceptions about climate effects of forest bioenergy. *GCB Bioenergy*, *13*(8), 1210–1231.

Dale, V. H., Kline, K. L., Parish, E. S., Cowie, A. L., Emory, R., Malmsheimer, R. W., Slade, R., Smith Jr, C. T., Wigley, T. B., Bentsen, N. S., Berndes, G., Bernier, P., Brandão, M., Chum, H. L., Diaz-Chavez, R., Egnell, G., Gustavsson, L., Schweinle, J., Stupak, I., … Wellisch, M. (2017). Status and prospects for renewable energy using wood pellets from the southeastern United States. *GCB Bioenergy*, 9(8), 1296–1305.

Dwivedi, P., Khanna, M., & Fuller, M. (2019). Is wood pellet-based electricity less carbon-intensive than coal-based electricity? It depends on perspectives, baselines, feedstocks, and forest management practices. *Environmental Research Letters*, *14*(2), 024006.

Epstein, G. A. (2005). Introduction: Financialization and the world economy. In G. A. Epstein (Ed.), *Financialization and the world economy* (pp. 3–16). Edward Elgar.

Fernholz, K., Bowyer, J., & Howe, J. (2007). *TIMOs and REITs: What, why, and how they might impact sustainable forestry*. Dovetail Partners.

Fox, T. R., Jokela, E. J., & Allen, H. L. (2007). The development of pine plantation silviculture in the southern United States. *Journal of Forestry*, *105*(7), 337–347.

Gulbrandsen, L. H. (2010). *Transnational environmental governance: The emergence and effects of the certification of forests and fisheries*. Edward Elgar.

Gunnoe, A. (2016). The financialization of the US forest products industry: Socio-economic relations, shareholder value, and the restructuring of an industry. *Social Forces*, *94*(3), 1075–1101.

Gunnoe, A., & Gellert, P. K. (2011). Financialization, shareholder value, and the transformation of timberland ownership in the US. *Critical Sociology*, *37*(3), 265–284.

Highberger, M. (2001). *The town that was Maxville*. Bear Creek Press.

Hillard, M. G. (2020). *Shredding paper: The rise and fall of Maine's mighty paper industry*. Cornell University Press.

Jin, S., & Sader, S. A. (2006). Effects of forest ownership and change on forest harvest rates, types and trends in northern Maine. *Forest Ecology and Management*, *228*(1–3), 177–186.

Judge-Lord, D., McDermott, C. L., & Cashore, B. (2020). Do private regulations ratchet up? How to distinguish types of regulatory stringency and patterns of change. *Organization & Environment*, *33*(1), 96–125.

Kelly, E. C. (2010). *Forest industry restructuring and emerging forest tenures in Deschutes and Klamath Counties, Oregon* [Doctoral dissertation]. Oregon State University.

Kittler, B., Stupak, I., & Smith, C. T. (2020). Assessing the wood sourcing practices of the US industrial wood pellet industry supplying European energy demand. *Energy, Sustainability and Society*, *10*(1), 23.

Krill, J. (2001). Felling the lumbering giants. *Multinational Monitor*, *22*(1/2), 16.

Marino, B. D. V., Mincheva, M., & Doucett, A. (2019). California Air Resources Board forest carbon protocol invalidates offsets. *PeerJ*, *7*, e7606.

Masum, M. F. H., Sahoo, K., & Dwivedi, P. (2019). Ascertaining the trajectory of wood-based bioenergy development in the United States based on current economic, social, and environmental constructs. *Annual Review of Resource Economics*, *11*, 169–193.

Meyfroidt, P. (2018). Financialization and the forestry sector. In C. Farcy, E. Rojas-Briales, & I. Martinez de Arano (Eds.), *Forestry in the midst of global changes* (pp. 307–316). Chapman and Hall/CRC.

Millennium Ecosystem Assessment (2003). *Ecosystems and human well-being: A framework for assessment*. Island Press.

Moog, S., Spicer, A., & Böhm, S. (2015). The politics of multi-stakeholder initiatives: The crisis of the Forest Stewardship Council. *Journal of Business Ethics*, *128*(3), 469–493.

Norton, M., Baldi, A., Buda, V., Carli, B., Cudlin, P., Jones, M. B., Korhola, A., Michalski, R., Novo, F., Oszlányi, J., Santos, F. D., Schink, B., Shepherd, J., Vet, L., Walloe, L., & Wijkman, A. (2019). Serious mismatches continue between science and policy in forest bioenergy. *GCB Bioenergy*, *11*(11), 1256–1263.

Parker, M. (2014). Swan story. In S. Charnley, T. E. Sheridan, & G. P. Nabhan (Eds.), *Stitching the West back together: Conservation of working landscapes* (pp. 123–135). University of Chicago Press.

Sass, E. M., Markowski-Lindsay, M., Butler, B. J., Caputo, J., Hartsell, A., Huff, E., & Robillard, A. (2021). Dynamics of large corporate forestland ownerships in the United States. *Journal of Forestry*, *119*(4), 363–375.

Sasser, E. N., Prakash, A., Cashore, B., & Auld, G. (2006). Direct targeting as an NGO political strategy: Examining private authority regimes in the forestry sector. *Business and Politics*, 8(3), 1–32.

Searchinger, T. D., Hamburg, S. P., Melillo, J., Chameides, W., Havlik, P., Kammen, D. M., Likens, G. E., Lubowski, R. N., Obersteiner, M., Oppenheimer, M., Philip Robertson, G., Schlesinger, W. H., & David Tilman, G. (2009). Fixing a critical climate accounting error. *Science*, *326*(5952), 527–528.

Sun, X., Zhang, D., & Butler, B. J. (2015). Timberland ownerships and reforestation in the Southern United States. *Forest Science*, *61*(2), 336–343.

Wunder, S. (2015). Revisiting the concept of payments for environmental services. *Ecological Economics*, *117*, 234–243.

Zhang, D. (2021). *From backwoods to boardrooms: The rise of institutional investment in timberland*. Oregon State University Press.

Zhang, D., Butler, B. J., & Nagubadi, R. V. (2012). Institutional timberland ownership in the US South: Magnitude, location, dynamics, and management. *Journal of Forestry*, *110*(7), 355–361.

Zhao, J., Daigneault, A., & Weiskittel, A. (2020). Forest landowner harvest decisions in a new era of conservation stewardship and changing markets in Maine, USA. *Forest Policy and Economics*, *118*, 102251.

# 10 Communities and Forests

In the mid-1990s, Wallowa County leaders were struggling to manage the multiple crises sparked by Endangered Species Act listings and the subsequent shutdown of timber harvests on the Wallowa-Whitman National Forest: shuttered sawmills, the loss of hundreds of jobs within a remote county of only 7,000 residents, and deep community social divisions. County Commissioner Ben Boswell convened a series of monthly meetings in the back room of the Cloud 9 Bakery in the county seat of Enterprise to chart a path forward among the varied interests and organizations involved in natural resource management in the county. He invited the director of Portland-based Sustainable Northwest to help facilitate these conversations. Wallowa County leaders already had some experience bringing diverse interests together to attempt to solve complex problems: county commissioners, the Nez Perce Tribe, and state and federal agency staff had previously collaborated on a plan to proactively restore salmon populations, and the Natural Resources Advisory Committee was formed in 1994 to bring diverse perspectives together to seek collaboratively designed solutions to difficult natural resource management issues. The result of the Cloud 9 Bakery meetings was the founding of Wallowa Resources, a community-based organization with a mission to help the county's families and businesses retain access to the benefits from natural resources even as the community navigated changes in forest and environmental policy that were out of their control.

Wallowa Resources moved quickly to set up an administrative apparatus and began to work on diverse economic development projects: taking on contracts for noxious weed control and subcontracting to local businesses; investing in local sawmill infrastructure; helping contractors and mills transition to utilizing smaller-diameter trees; and convening a collaborative restoration assessment in the Upper Joseph Creek watershed (see Chapter 5). Because of the capacity built in part through Wallowa Resources, the Wallowa-Whitman National Forest was chosen as one of the pilot sites for the then-novel stewardship contracting authorities. Over time, Wallowa Resources expanded to take on other tasks such as environmental education and wood-based biomass energy development.[1] The organization came to be recognized nationwide as a model for community-based natural resource management despite operating in a challenging social, political, and economic environment.

DOI: 10.4324/9781003043669-10

In 2011, Wallowa Resources co-founded the Wallowa Lake Moraines Partnership together with the Wallowa Land Trust, Wallowa County, and the Oregon Parks and Recreation Department. In 2020 the Partnership successfully established the East Moraine Community Forest on a parcel of land near Wallowa Lake that was slated for residential development (Figure 10.1). Funding to purchase the community forest came from a mix of federal Forest Legacy Program dollars (see Chapter 8), state park funds, private donations, and a sizeable contribution from the Nez Perce Tribe. Throughout these varied initiatives, Wallowa Resources maintained its focus on generating benefits for the local community while encouraging stewardship of the county's natural resources. Even as some other remote rural places responded to recent federal forest policy changes by reorienting their economies around tourism and retirement—or by slipping into depopulation and economic decline—the people of Wallowa County made it clear that they wanted to retain their cultural ties to a "working landscape."

This chapter focuses on communities—people, organizations, and institutions in specific places—that have developed unique relationships with surrounding forests. These relationships include economic aspects (forests as a source of work in sectors like wood products or recreation), direct livelihood aspects (forests as

*Figure 10.1* Wallowa County's East Moraine Community Forest was established on former industrial lands that were slated to be subdivided and converted to homesites. The forest will be managed for community benefits with input from the representatives of key local and regional governmental, tribal, and non-governmental organizations. Photograph by the author.

a source of water, firewood, and edible plants, animals, and fungi), and cultural aspects (forests as components of personal and community identity and as sources of emotional and spiritual renewal). We examine some of the varied ways that communities participate in forest governance, including through both formal and informal means.

## From Logging Camps to Forest Communities

In 1918, a US Forest Service investigator by the name of Samuel Trask Dana reported on his nationwide review of current and former logging towns. Titled "Forestry and Community Development," Dana's report was unequivocal about the social impacts of an industry that liquidated standing forests without consideration for the future: not only did typical logging practices devastate the land, but they also left shattered communities in their wake. Dana warned that "deserted villages are signposts that too often mark the trail of lumbering operations."[2] In contrast to the conditions found in settled agricultural communities, centers of timber production typically experienced economic and population booms followed by busts a few years later as local timber supplies were exhausted and logging operations moved on to the next untapped forest. Beyond serving as an exposé of prevailing timbering practices, Dana's report also made a case for the importance of the National Forest System and its management by the US Forest Service. Dana argued that, unlike private owners of the time who were trapped in cycles of speculation, debt, and timber liquidation, the federal government could afford to institute more rational and socially beneficial forestry practices—including on the eastern lands recently purchased under the Weeks Act (Chapter 2).

The importance to forestry of human communities was not lost on Gifford Pinchot; echoing the guidance of his mentor Sir Dietrich Brandis, Pinchot observed that "in the long run forestry cannot succeed unless the people who live in and near the forest are for it and not against it."[3] National forest policies included specific consideration for the needs of local communities beginning in 1906, when a new law directed ten percent of all timber receipts from forest reserves to be distributed to the county in which harvests occurred; the payment was increased to 25% two years later. This mechanism generated funding for counties hosting national forests and helped to foster local support for the federal bureaucracy and its mission.

Another example of community-focused national forest policy was the 1944 Sustained Yield Forest Management Act, which allowed for the creation of both "cooperative" and "federal" long-term sustained yield units on which federal timber would be offered noncompetitively to individual mills. Under the cooperative model (of which only one was ever established, on Washington's Olympic Peninsula), the participating timber company would agree to coordinate its own harvest levels with federal harvests within the unit, thereby providing local economic stability. While attractive in theory, this idea never spread beyond a few initial units.[4] One of these, the Vallecitos Federal Sustained Yield Unit on the Carson National Forest in northern New Mexico, was established in 1948 under the premise that a single timber operator with obligations to hire local workers would generate both environmental and local economic benefits. However, the unit was consistently plagued by unmet promises of local benefits, resulting in ongoing labor conflicts and even an act of arson at the Vallecitos sawmill.[5]

The US Forest Service was keenly aware that many communities' economic fates were tied to federal timber harvests; for much of the twentieth century, the agency made an explicit commitment to support "community stability"—generally measured in terms of a sustained output of timber from national forests.[6] Studies going back to the 1940s, however, cautioned that well-being in forest-dependent communities meant more than a steady supply of logs,[7] and this fact became increasingly apparent as the decades advanced. The production of national forest timber is only one variable in a complex scenario that includes private timber supplies, national and international timber markets, the presence of and competition between mills, patterns of worker productivity and technological advancements, the social and economic dynamics of communities (not all of which are tied to timber harvest levels), and societal demands for recreation and preservation on federal lands.[8]

A consistent supply of logs from national forests also did little to address longstanding tensions between capital and labor in the woods and in timber towns. Northwest timber workers had been organizing and striking in protest of dangerous working conditions and unhealthy living conditions as early as the first decades of the twentieth century, and unions continued to play an important role in community life and in forest policy dynamics throughout the twentieth century. As historian Erik Loomis observes, some woodworker unions embraced expansive visions of well-being that included concerns beyond a steady supply of timber. The International Woodworkers of America, for example, allied with environmental interests in support of park and wilderness protection proposals and called for federal regulation of private forestry practices to achieve community stability. The IWW came to support these positions out of a concern with the recreational, health, and spiritual needs of timber workers, who performed (and continue to perform) some of the country's most dangerous jobs.[9]

Woodworker union membership declined throughout the 1980s and 1990s as a result of changes in technologies and markets that led to lowered demand for domestic labor and due to the relocation of timber processing from the Northwest to the South.[10] Nationwide attention to forest community well-being experienced a resurgence, however, in the wake of the timber wars of the 1990s. Of particular concern were those places where loggers, haulers, millworkers, and the communities in which they lived and worked suffered job losses as the result of declining federal harvests, agency budget cuts, and other factors.[11] Federal support programs were established with the goal of retraining displaced forest industry workers to carry out work in forest restoration, road decommissioning, wildlife surveying, and other elements of a nascent "restoration economy." Although the federal programs themselves were insufficient to compensate for the jobs and wages lost, a restoration economy did emerge in some communities; in these areas it can represent a substantial element of economic activity.[12] However, federal contracting protocols often allow large regional businesses to capture disproportionate amounts of available restoration work, thereby depriving rural communities of the opportunity to benefit from new programs and practices in their own backyards.[13]

With some localized exceptions,[14] 1990s-era federal forest policy changes had less severe impacts on forest-dependent communities east of the Rocky Mountains, but technological changes, industry financialization, and changing markets nevertheless left their mark on these timber towns as well. In forest communities nationwide, the rapid transformations in land ownership associated with financialization—in addition to the threat of parcelization and conversion—alarmed residents and catalyzed community-led efforts to retain forests in local ownership (Chapter 9). At the same time, forest communities nationwide were also undergoing important changes in their populations and economies, and these changes carried implications for forest governance.[15] Although forest-based communities had long been interested in more than just a stable timber supply, by the end of the twentieth century they were more economically and demographically diverse than ever, and this diversity was reflected in their values and orientations toward local forests.

## The Rise of Community-Based Forestry in the US

The timber wars of the 1980s and 1990s exacted a heavy economic and social toll on rural forest communities, particularly those in which peoples' livelihoods and identities were closely tied to federal lands. Despite the seemingly insurmountable divides separating environmentalists, timber industry workers, federal land managers, local government leaders, and other stakeholders, peoples' common ties to their communities and surrounding forests provided at least a glimmer of hope for reconciliation.[16] In the wake of the divisive policy battles of the 1990s, some local leaders attempted to bring various interests together to explore whether there was any room for agreement on a common vision for nearby federal forestlands. Rather than fight battles in the courtrooms, in the woods, and through the bureaucratic processes of the federal land management agencies, early experiments in community-based public lands collaboration brought traditional adversaries face to face in an attempt to identify alternatives to the bureaucratic gridlock and negative socioeconomic outcomes that had become the status quo. Conversations among long-time adversaries were difficult, and progress often required the presence of trained mediators. Some powerful players accustomed to advancing their interests through political connections (including some national-level environmental and forest products representatives) refused to participate and occasionally tried to scuttle emerging efforts in collaboration and alternative dispute resolution. Many communities struggled to reconcile competing pressures for environmental protection and continued economic utilization of forests. Yet, perhaps miraculously, some managed to forge tentative agreements and even build a fragile trust among traditional rivals.

What eventually emerged was a new model of forest governance that came to be known as the US community-based forestry movement (distinct in important ways from community forestry as practiced in other countries).[17] The movement took hold in a variety of settings nationwide but was especially pronounced in the western US, where public land policy battles had left many rural communities facing linked social, economic, and ecological crises.[18] The community forestry movement was one element of a larger nationwide trend toward collaborative environmental management that went beyond forests to include watersheds, wildlife, and other concerns. The community-based forestry movement in the US, still very much active today, is represented by a broad diversity of communities, organizations, and institutions engaged in addressing a variety of social, ecological, and economic issues. In spite of this diversity, we can characterize the movement by examining the following key elements: 1) collaborative deliberation processes; 2) community-based organizations; 3) complementary formal policies; and 4) larger-scale support networks.

### *Collaborative Deliberation Processes*

In contrast to prior forest policy changes driven largely by new legislation (such as the Wilderness Act), litigation (such as the 1980s- and 1990s-era northern spotted owl lawsuits), or a combination of the two (such as the *Monongahela* lawsuit that catalyzed passage of the National Forest Management Act), the community-based forestry movement developed by way of an emergent, bottom-up process led primarily by people with no official authority over federal land management. For example, one of the earliest community-based efforts arose in the midst of the timber wars in the rural community of Quincy in California's Sierra Nevada range. The Quincy Library Group formed after years of legal and social conflicts over timber harvesting had left the community divided and its residents feeling disempowered. A small group of timber workers, environmentalists, county commissioners, timber industry leaders, and national forest managers took the radical step of sitting down at the table together in an attempt to forge agreement on a vision for managing the national forests in their backyard. Although the discussions were often strained, the group eventually agreed on a plan that would protect the California spotted owl and other species at risk, preserve roadless areas, restore older forest conditions, and reduce forest fuels while still providing jobs and income for local people. This breakthrough suggested that alternatives to litigation, division, and anger were possible, and the Quincy Library Group came to serve as a model of what an alternative, community-based approach to forest governance might look like.[19]

As national-level political and legal battles over forestry continued to rage in many public lands-dominated landscapes, other communities experimented with collaborative deliberation processes in hopes of arriving at something other than zero-sum solutions to forest conflicts. Additional early examples include the Applegate Partnership on the Oregon-California border, the Grand Canyon Forest Partnership in northern Arizona, the Northeast Washington Forestry Coalition, the Flathead Forestry Project in northwestern Montana, the Catron County Citizens Group in New Mexico, the Ponderosa Pine Forest Partnership in southwestern Colorado, and the Natural Resources Working Group in Arizona's White Mountains (Box 10.1). Many of these entities are still active as of the date of this writing, and dozens of other initiatives have been founded since the early 1990s. Importantly, there is no singular template for how these efforts are organized and run: each collaborative group sets its own rules regarding membership, organizational structure, decision-making process, and topical and geographic scope. Many groups attempt to make decisions on the basis of consensus; others employ simple majority or near-consensus voting processes. Some focus exclusively on federal forests and others consider a wider range of ownerships and ecosystems. What these varied efforts have in common is a commitment to a broadly inclusive process that seeks "win-win" solutions to steward the environment and support rural economies and to do so by empowering local community members as leaders.

Collaborative forums often begin by building relationships among traditional adversaries and attempting to identify a "zone of agreement"—the overlap, however small, of interests, values, and acceptable management approaches among participants.[20] Through continued learning and relationship-building over time, this zone of agreement can expand to encompass a broader range of activities and approaches. Throughout much of the country, the restoration of fire-adapted forests serves as a natural starting point for identifying a zone of agreement given shared concerns regarding wildfire hazard and the recognition that a century or more of fire exclusion has altered forest conditions and potential wildfire behavior (see Chapter 5).[21] Some of these treatments are guided by site-specific reconstructions of historical tree distributions;[22] others are guided by a focus on processes (such as fire) rather than structure or are built upon conservation of the largest contemporary trees in order to rebuild the kind of older forest structure that has largely been lost due to past harvesting activity.[23] Restoration-focused projects hold at least the promise of meeting diverse interests simultaneously: the reduction of fuels that can contribute to uncharacteristically intense wildfires; the creation of ecologically resilient forest conditions; the improvement of habitat for species that depend upon older or more open forests; and the creation of economic activity from the trees that are removed as a part of thinning operations. As participants in collaborative processes experiment with restoration treatments, learn from experience, and build trust, they are often able to take on more ambitious projects over time (Box 10.1).

## BOX 10.1 AN EXAMPLE OF COMMUNITY-BASED FORESTRY FROM ARIZONA

The Natural Resources Working Group formed in the mid-1990s in the White Mountains of east-central Arizona in an attempt to manage ongoing social conflict regarding forest management on and near the

Apache-Sitgreaves National Forests. The region found itself in the throes of legal and political battles over timber harvests on federal forestlands after a federal court ruling in 1995 put a stop to harvesting in order to plan for protection of the Mexican spotted owl, cutting off an important source of wood for area mills. Key members of the National Resources Working Group included county commissioners, wood products representatives, environmental organizations, wildlife advocates, homeowners, fire managers, and national forest staff. The group's early deliberations focused on building agreement to treat the Blue Ridge demonstration area, a 7,000-acre ponderosa pine forest in the wildland-urban interface. The pilot project included implementation of a variety of different approaches to thinning, providing opportunities for members of the collaborative to learn from experience. By the time the 467,000-acre Rodeo-Chediski Fire burned across eastern Arizona in 2002, the Natural Resources Working Group had already built sufficient trust and goodwill among its members to quickly broker a larger agreement. The result was the White Mountain Stewardship Project, a ten-year, 150,000-acre plan to reduce forest densities and restore more fire-resilient conditions in the wildland-urban interface across the Apache-Sitgreaves National Forest. The project was the largest and longest-term stewardship contract to date and included dedicated funding for a multiparty monitoring team to promote collaborative learning. A network of new and existing local processing firms helped to take the small-diameter material that came out of the project and turn it into wood pellets, dimensional lumber, biomass for electricity generation, and other products. The project was supported by numerous intermediary organizations, including those focused on supporting the local wood products industry, and received financial support through federal programs to stimulate innovation in woody biomass utilization. Although the project ultimately treated less than half of the acreage that was originally anticipated, it nevertheless helped to advance the state of practice on forest restoration and small-diameter timber utilization; project treatments were credited with helping to change fire behavior and protect homes and other structures when the Wallow Fire burned through the region in 2011.[24]

### *Community-Based Organizations*

Even as early collaborative efforts were showing promise in resolving seemingly intractable conflicts over forest management, rural communities found themselves facing myriad issues related to declining employment and capacity. As local tax rolls declined, county and local governments lost staffing and budgets. With mills shutting down and Forest Service ranger stations being closed or downsized in the wake of harvest reductions, many communities struggled to pivot toward new economic activities after decades of dependence on a few primary industries. The emerging opportunities generally focused on forest restoration, fuel reduction, and other non-traditional activities in the woods, requiring new skills, new equipment, and new contracting mechanisms that were unfamiliar to both Forest Service staff and to rural businesses. Community-based organizations (CBOs) were founded in many places as "bridging organizations" capable of filling in for missing governmental, civil society, and private sector capacity and helping to secure economic benefits for their respective communities.[25] Although CBOs often lead or otherwise participate in collaborative deliberation processes, their role goes beyond facilitating collaboration to implementing practical activities that improve local social, economic, and ecological conditions. They are typically registered nonprofit organizations, making them eligible to do such things as receive and administer public and philanthropic grants, take out loans, serve as contractors, and enter into agreements with other governmental and non-governmental organizations. By fulfilling the practical side of community-based forestry, CBOs help make innovative ideas a reality and allow their respective communities to access the benefits of forest stewardship.

Early examples of CBOs include Wallowa Resources, profiled at the beginning of this chapter, as well as the Watershed Research and Training Center in Hayfork, California; Framing Our Community in Elk City, Idaho; the Siuslaw Institute in Deadwood, Oregon; the Lake County Resources Initiative in Lakeview, Oregon; the Lomakatsi Restoration Project in Ashland, Oregon; the Swan Ecosystem Center in Seeley, Montana; and many others. Examples of typical CBO activities include investing in small-diameter wood processing facilities, providing workforce training, engaging in economic development planning, implementing monitoring activities, applying for and managing grants, and providing educational programming for youth and adults. In some cases, CBOs have gone as far as to participate in interdisciplinary teams for national forest planning (a role traditionally limited to Forest Service staff), administer restoration contracts on federal lands, create wholly owned for-profit subsidiaries, and contract out their own fire and forestry crews to accomplish work in the woods.[26]

CBOs are often more flexible than the governmental and private sector entities with which they partner, allowing them to creatively combine capacities, resources, and policy tools to achieve practical outcomes. For example, several CBOs have emerged as leaders of cross-boundary prescribed fire implementation by coordinating across agencies, landowners, and government entities.[27] Mt. Adams Resource Stewards combined state and federal funding to support the use of its own prescribed burn crews to reintroduce fire across national forests, national wildlife refuges, and community-owned forestlands in south-central Washington. The Forest Stewards Guild began leading burns on private and state land in New Mexico and was eventually able to assemble a multi-partner prescribed burning team with support from the Rio Grande Water Fund (see Chapter 4). Despite these important and wide-ranging roles, most CBOs are relatively small organizations. Research conducted in 2016 showed that western CBOs averaged three full-time employees and an approximately $500,000 annual budget; median values were one full-time employee and a $280,000 annual budget.[28]

### *Formal Policies*

Federal lawmakers slowly provided new policy tools to support community-based forestry as the model came to demonstrate a pathway forward through political and social gridlock. Stewardship end result contracting authorities were first piloted in the 1990s and then expanded in the early 2000s (Chapter 5). Stewardship contracting allows federal forest managers to bundle timber sale and restoration projects under a single contract and to consider the "best value" to the government—which can include considerations of local community benefits—rather than solely considering cost when awarding contracts. These administrative tools were intended to break down economic and bureaucratic barriers to accomplishing the kind of integrated restoration projects that were regularly being promoted by collaborative organizations, and many of the initial pilots of stewardship contracting authorities took place in forests where collaborative capacity was already established. The Secure Rural Schools and Community Self-Determination Act of 2000 allowed counties to receive their national forest and Bureau of Land Management "county payments" based upon historical—rather than current—timber sales. This temporary change, which has been extended several times since 2000, was designed to soften the economic blow to counties traditionally reliant upon federal timber receipts to fund public services. A policy specific to the state of New Mexico, authorized as part of the Secure Rural Schools legislation, provides annual funding for collaboratively developed forest restoration projects on public lands—including federal, state, tribal, or Hispano Land Grant forests. To be eligible for funding, the projects must be collaboratively designed and monitored. A similar model was adopted in 2009 for a nationwide policy of competitive grants on national forestlands: the Collaborative Forest Landscape Restoration Program funds collaboratively designed projects of at least 50,000 acres in size;[29] this program is described in more detail in Chapter 5.

Collaborative processes have been recognized in various federal policies, but the emergent and grassroots nature of community-based collaboration makes it an uneasy fit with federal mandates.[30] A collaborative group chartered by a federal entity could easily run afoul of the Federal Advisory Committee Act, which prohibits federal agencies from establishing advisory committees except under specified procedures. Federal policy tends to define collaboration broadly and to leave wide discretion in terms of how and where it is applied. For example, the 2003 Healthy Forests Restoration Act requires a collaborative process for the creation of Community Wildfire Protection Plans (see Chapter 5) and requires collaboration in forest restoration project planning more broadly. The law does not specifically define collaboration, however. Legislative language simply refers to it as a form of public participation involving "State and local governments and Indian tribes, and participation of interested persons." Other policies are more prescriptive; for example, the Secure Rural Schools Act calls for the creation of Resource Advisory Committees to recommend the allocation of funding under Title II of the Act. These committees are required to be composed of a mix of interests that include forest industry, organized labor, environmental interests, recreationists, and state, county, and tribal governments (Box 10.2).

### *Support Networks*

Forest collaborative groups and CBOs share in common a grounding in their respective communities of place. However, they are also forced to contend with policies and institutions at higher scales: federal policies matter in all forested landscapes, and state policies are important in places with high proportions of nonfederal land (see Chapter 6). During the early stages of the community-based forestry movement, many organizations and initiatives had to construct new procedures, institutions, and innovations from scratch without the benefit of established guiding models. Collectively, these factors drove the creation of larger-scale networks designed to help advocate for policy change and to facilitate mutual learning among collaborative and community organizations. For example, a Communities Committee was chartered at the Seventh American Forest Congress in 1996; it was led by the founder and then-director of the Hayfork, California-based Watershed Research and Training Center. The Communities Committee serves as a clearinghouse of information on the community forestry movement as well as a forum to advocate for policy changes and new programs at state and federal levels to support the movement.[31] A separate policy network, the Rural Voices for Conservation Coalition, was established as a forum for developing policy proposals that could be supported by diverse rural community, conservation, and natural resource interests and delivering them to agency decision-makers and elected officials in Washington, DC.[32] Several of the most important forest policy innovations since the 1990s—including many of the formal policies listed above—were innovated, tested, and advocated by the Coalition or its members.

Community forestry participants also worked to develop mutual support and learning networks such as the Collaborative Learning Circle and Lead Partnership Group, both focused on northern California and southern Oregon, and the National Network of Forest Practitioners. These and related initiatives found financial support through the Ford Foundation's Community-Based Forestry Program as well as organizational support from regional and national organizations such as Sustainable Northwest, American Forests, The Forest Trust (since renamed Forest Stewards Guild), The Nature Conservancy, National Forest Foundation, the Pinchot Institute for Conservation, and others. The movement also benefited from research support by social scientists at various colleges, universities, and US Forest Service research stations. Collectively, these networks of CBOs, community leaders, funders, regional and national NGOs, universities, and other entities helped to cultivate the community-based forestry movement, provide it with visibility and legitimacy, critically assess its accomplishments and shortcomings, and realize its capacity to transform the policy and governance landscape for western forests.

### *The Legacy of the Community-Based Forestry Movement*

An important commonality among community-based forestry efforts in the US is that they tend to lack formal authority to decide how the public forests in their respective backyards will be managed.[33] Despite lacking formal decision-making power, collaborative governance entities and CBOs often develop another kind of influence: "practical authority," in the words of political scientists Rebecca Abers and Margaret Keck.[34] Their power comes from the ability to forge agreements seen as legitimate by a wide range of interests and to demonstrate results on the ground. Given the combination of decreasing US Forest Service budgets and increasing opportunities for stakeholders to slow or halt proposed management projects, achieving any sort of forest management may be considered an accomplishment in some places; this was especially true during the most contentious period of the 1990s and early 2000s. At the same time, it can be challenging for informal and voluntary community-based forestry efforts to overcome the many legal requirements for federal forest management. With very few exceptions, the federal policies and programs supporting community-based forestry have not altered the scope or requirements of influential laws such as NEPA, the National Forest Management Act, or the Endangered Species Act. In some cases, this has led to unmet expectations and burnout among those who have devoted time and energy to advancing community-based forestry efforts.[35]

Despite these limitations, the concept of collaboration as a complement to (rather than a substitution for) conventional bureaucratic planning processes is now firmly established in national forest policy and practice. An estimated 97% of national forests had experimented with at least some kind of community-based collaboration by the late 1990s,[36] and collaboration as a practice is recognized in various formal forest policies as described above. Public land managers are sometimes themselves participants in these groups, and managers often look to collaboratives for guidance on how to proceed when faced with potentially controversial situations.[37] Although federal forest managers are still required to follow standard procedures dictated by laws like NEPA and the National Forest Management Act in any case, the inclusion of collaborative input as part of the decision-making process is often believed to result in fewer legal or administrative challenges and greater community support.[38] Still, not all forest-based communities have well-functioning collaborative groups, and some federal forest managers may choose not to engage or not to follow the lead of these organizations even where they do exist.[39]

Collaborative efforts also continue to be opposed by some forest stakeholder groups. There is often a particular concern among a segment of the environmental community who believes that local compromises could undermine expectations to strictly implement federal environmental protection laws or federal court rulings.[40] Numerous collaborative agreements have been derailed by legal or administrative challenges from environmental or other organized interests.[41] Concerns regarding the institutionalization of collaboration have also been raised by advocates for forest workers, non-timber forest product harvesters (see Chapter 11), and other forest stakeholders who may not have the availability or resources to participate in collaborative processes, who may not feel comfortable doing so due to language or cultural divides, or who may be excluded from participation by membership rules.[42] Indeed, the structure of these collaborative processes carries important implications for equity and representation. Although several community-based forestry efforts explicitly included social equity concerns from the beginning, others have tended to focus on ecological and economic concerns without specifically addressing questions of equity.[43] Important questions remain about who is included in definitions of "community" and which voices are able to participate in collaborative forums (Box 10.2).

**BOX 10.2 WHAT IS COLLABORATION? WHAT IS COMMUNITY?**

Community-based forestry efforts include a strong emphasis on the practice of collaboration—but meanings of the concepts of "collaboration" and "community" are far from obvious.

*Collaboration* is often conceptualized using the definition of Barbara Gray, who calls it "a process through which parties who see different aspects of a problem can constructively explore their differences and search for solutions that go beyond their own limited vision of what is possible."[44] Richard Margerum defines it as "an approach to solving complex environmental problems in which a diverse group of autonomous stakeholders deliberates to build consensus and develop networks for translating consensus into results,"[45] and Philip Brick and Sarah Van de Wetering describe collaborative conservation as emphasizing

> the importance of local participation, sustainable natural and human communities, inclusion of disempowered voices, and voluntary consent and compliance rather than enforcement by legal and regulatory coercion...to find working solutions to intractable problems that will surely languish unresolved for decades in the existing policy system.[46]

"Collaboration" is defined within National Forest Management Act regulations finalized in 2012 as "a structured manner in which a collection of people with diverse interests share knowledge, ideas, and resources while working together in an inclusive and cooperative manner toward a common purpose." According to collaboration scholars Judith Innes and David Booher, a genuine collaborative process includes a diversity of interdependent interests operating under a process that provides for authentic dialogue and aims toward establishing consensus. They contend that the "collaborative rationality" resulting from such processes is capable of building enduring relationships, promoting collective learning processes, building shared understandings, and moving past zero-sum solutions to achieve outcomes that meet multiple interests simultaneously. However, they concur with many other governance researchers that collaboration is not appropriate in all situations and should not be expected to solve every kind of natural resource conflict.[47]

*Community* refers to a collection of people that interact and that share something in common; the combination of these two features is believed to create at least the basic enabling conditions for collective action. In the environmental governance context, the term is most often used in reference to communities of place: the "something in common" for these communities of place is a physical location (including the built environment), local relationships, institutions, and elements of the landscape such as forests. Other varieties of the term include communities of interest and communities of identity. Although the word carries connotations of harmony and constructive interaction, communities are also typically characterized by conflict, inequality, and social divisions.[48] Community-based forestry derives its legitimacy in part from its claim to represent the interests of people who live closest to the forest; the way in which "community" is defined therefore carries implications for the empowerment or disempowerment of various people and groups through their ability to participate in community-based governance processes. Several questions complicate any simple conceptualization of community. For example, should community-based collaborative groups include representatives that do not recognize collaboration as a legitimate governance process or that do not recognize current forest managers as having legitimate rights to make decisions? How should non-local interests be included, if at all? Should regional, state-level, or national organizations and agencies be represented in local deliberations? How should dispossessed Indigenous nations with historical ties to the land be represented, and how does participation in collaboration affect tribes' government-to-government relationship with federal entities? Does the community include representatives of national or multinational corporations that own land in the area? Should recently arrived or part-time residents be considered part of the community? How well are community interests represented by volunteer or professional collaborative group members, given that they tend to be self-appointed or selected by higher-level officials rather than selected through a democratic process? How should the interests of "unseen" forest users (Chapter 11) be represented, especially when language or cultural barriers are present?[49]

## Community-Owned Forests

The community-based forestry movement in the US West has focused much of its energy on changing the formal and informal governance of federally owned forests. A separate nationwide movement focuses on securing community benefits through taking formal ownership of forests. The concept of community-owned forests has a long legacy in the United States; New England villages retained ownership of nearby forests important for fuelwood and timber as early as the seventeenth century, and states in the region established mechanisms to convert abandoned and tax-delinquent lands to town ownership in the early twentieth century.[50] The city of Seattle, Washington, began its purchases of land in the Cedar River watershed in 1899 and continues to own and manage the forested drainage for municipal water supplies. Forests on Native American reservations in many parts of the country have long been managed in ways that emphasize benefits to the tribal community as a whole (Chapter 7). Forest scholar Jill Belsky distinguishes between three different categories of community-owned forests: Indigenous community forests, such as those owned by tribes and by southwestern Hispano Land Grant communities; town or municipal county forests, owned by local governments; and community-based conservation organization community forests, owned by CBOs, land trusts, and other non-profit or limited liability entities.[51] What defines a community-owned forest, beyond formal property title, is that it is managed specifically to generate both material and nonmaterial benefits for a defined community. Members of that community typically have access to various means of participating in decision-making for the forest—and these differ from their options to participate in setting management direction for private or public forestlands.

Community-owned forests are more than just scaled-down versions of national forests, state parks, or other public lands. What makes them unique is their orientation toward a defined community of place: representatives of the community directly participate in the governance of the forest and the community as a whole can be expected to benefit from its management.[52] In this sense, community-owned forests represent one variation on the commons tenure whose utility was disparaged by Garrett Hardin and later redeemed by Elinor Ostrom. In the case of Indigenous forests (such as tribal or Hispano Land Grant forests), community membership is formally defined and non-members typically possess few to no access rights to community-owned forests. Town or municipal community-owned forests may also be oriented toward a formally defined community, in this case determined by residence. Many town forests in New England are governed through decisions made at town meetings, reflecting a tradition of "direct democracy" that runs deep in that region of the country. Not all town or municipal forests are necessarily open for public access, as they may be managed for watershed protection or other specified uses for the benefit of town residents. Community forests owned by community-based conservation organizations may be governed under less formal definitions of community, but local interests are still paramount in their management, and community members often participate in decision-making.[53] For example, the draft management plan for the East Moraine Community Forest in Wallowa County includes open space protection, public access, and contributions to the local economy in its vision statement; the proposed management committee for the forest includes representatives from the Nez Perce Tribe, Wallowa County, the Wallowa Land Trust, Wallowa Resources, and four state-level entities (Oregon Department of Forestry, Oregon Department of Fish and Wildlife, Oregon State University Extension, and Oregon Parks and Recreation Department).

Advancing community benefits and empowerment through ownership of forests is one response to recent changes in the corporate forest landscape. The property ownership turnover and the search for alternative income streams that are characteristic of the TIMO/REIT transition (Chapter 9) can present opportunities for tribes, municipalities, and other communities to acquire title to important properties. Taking advantage of these opportunities requires more than access to financing; it also requires the leadership of community advocates with the vision and creativity to imagine ways to reconcile the social, ecological, and economic dimensions of community-owned forestlands. For example, Chapter 7 describes the establishment of a community-owned forest by the Yurok Tribe in California that included multiple partner organizations and financing mechanisms along with the sale of carbon credits under state climate change legislation. As the Yurok case illustrates, acquisition of a community-owned forest can be a complex undertaking involving governmental and non-governmental organizations at multiple scales and the creative combination of various sources of funding.

The U.S. Forest Service supports community-owned forest acquisition through its Community Forest Program, managed by State and Private Forestry. Land trusts (introduced in Chapter 8 for their role in establishing conservation easements) can be centrally important players in acquisition because of their knowledge of property law and of various financing mechanisms for land conservation. Some larger national organizations such as The Nature Conservancy, The Conservation Fund, and Trust for Public Land have provided critical support to community-owned forest acquisition by temporarily purchasing forestlands that are at risk of being subdivided or sold to other owners and holding these lands until they can be sold to other conservation owners (such as municipalities and CBOs).[54] For example, The Conservation Fund purchased a 532-acre parcel of forestland in Hoke County, North Carolina in 2006 and held it until the county was able to raise enough funding to secure title and establish the first community-owned forest in the South. Private companies can also facilitate community ownership through arranging financing and brokering deals between forest sellers and buyers.

## Urban Forests

The majority of the US population lives in urban and suburban settings, and it is in these settings that most of us come into regular contact with forests. The urban forest is defined as "all trees within urban areas,"[55] including trees in residential yards, city parks, vacant lots, parking lots, schoolyards, cemeteries, and along streets, sidewalks, and other elements of urban infrastructure. US Forest Service researchers have estimated that the country's urban areas contain 5.5 billion trees and that the urban forest covers 141 million acres in total—nearly equivalent to the acreage of the entire National Forest System. Even as the area of US urban forest is expanding due to continued residential development of forested areas, the urban forest is also experiencing an overall loss of proportional tree cover due to trees dying or being removed and replaced with non-forest cover such as grass or impervious surfaces.[56]

Urban forests provide both economic and non-economic benefits to residents. Examples of economic benefits include moderating temperatures (resulting in lower heating and cooling bills) and contributing to property values.[57] Non-economic benefits include forests' ability to ameliorate air pollution, reduce noise, and attract birdlife in addition to their numerous documented positive effects on physical and mental health.[58] There is also strong evidence that urban tree cover is associated with lower crime rates.[59] These benefits are not evenly distributed, however. Urban forest cover tends to be greater in more affluent neighborhoods, as does access to greenspaces such as city parks.[60] A legacy of residential segregation based on race and ethnicity is reflected in patterns of tree cover and proximity to greenspaces.[61] This has created unequal access to the urban forest as well as unequal exposure to harms such as the urban heat island effect, leading to calls for greater attention to "tree equity."[62] As surprising as it may seem, trees are inseparable from patterns of urban segregation: their presence, diversity, and condition both reflect and reproduce inequality.

These uneven patterns result from a complex legacy of both historical and contemporary practices. Some urban tree-planting programs, for example, have focused on homeowners (rather than renters) as the recipients of street trees, further exacerbating the discrepancy in tree cover between more and less affluent neighborhoods.[63] Wealthier property owners may be targeted by city tree-planting programs and they may also be more likely to both participate in these programs and to maintain existing trees, while renters are often passed over for tree planting and also less able to provide care for trees.[64] Landlords of rental properties sometimes choose not to invest in tree planting and maintenance, particularly for properties in lower-income neighborhoods.[65] At the same time, neighborhoods with greater tree cover may also be more prone to gentrification, further exacerbating class and racial/ethnic disparities.[66]

The urban forest presents one of the most complex tenure scenarios in the country due to its ownership domination by millions of individual residents and due to the constellation of laws, ordinances, NGOs, community organizations, municipal agencies, private and public funding programs, and myriad other organizations and institutions that influence urban forest governance (Figure 10.2). Urban tree ordinances, common nationwide, regulate the removal and maintenance of trees and may require planting or other compensation when trees are removed as part of development activities. The regulation of tree-disturbing activities is founded upon a recognition of the many public benefits provided by urban forests. Landowners and developers may need permits to remove trees depending on the ordinances in place and the conditions of the site under consideration. For example, some municipalities and states only regulate the removal of trees above a certain height or diameter, some only regulate certain species, and some only regulate urban forest alteration when the number of trees,

*Figure 10.2* The urban forest presents some of the country's most complex tenure scenarios, exemplified by this white oak tree in Athens, Georgia, that owns itself. The text reads, "For and in consideration of the great love I bear this tree and the great desire I have for its protection for all time, I convey entire possession of itself and all land within eight feet of the tree on all sides – William H. Jackson." Photograph by author.

percent of canopy, or area disturbed exceeds a threshold value. Maryland's Forest Conservation Act, for example, requires an inventory of existing trees and a plan for their protection before certain development activities of 40,000 square feet or more are conducted. Many localities have adopted afforestation programs designed to meet canopy cover goals through a combination of planting on public property and encouraging residents and landowners to plant on private properties. The US Forest Service supports these and other urban forestry programs through the Urban and Community Forestry Program, active in all states and territories.

Social norms can also influence the planting and maintenance of urban trees in individual neighborhoods.[67] Whether on private or public property, urban trees and patches of urban forest may be widely perceived to be community assets, and there are sometimes strong social expectations for residents, landowners, and municipalities to ensure their continued survival. On the other hand, public pressure can also be brought to bear to remove trees or stands associated with nuisances—for example, those that threaten neighboring structures or that are believed to attract behaviors considered undesirable, such as dumping, drug use, or encampments for unhoused people. Social norms also extend to expectations for interactions between urban forest professionals and city residents. A study of a tree-planting program in Detroit, Michigan, for example, found that many of those who rejected offers to receive street trees did so because they wanted more information, more maintenance support, or more direct engagement from the city's urban forestry organization and not because they were opposed to hosting street trees. In some cases, past negative experiences with municipal service providers led residents to be more skeptical of tree-planting projects.[68]

Questions of risk loom large in both public and private decision-making for the urban forest given the potential damage to homes, people, and other values in the case of tree failure.[69] For example, a Florida law passed in 2019 preempts certain municipal tree protection and replanting ordinances; the purpose of the law is to facilitate the removal of trees that could pose risks in the case of a hurricane or tropical storm.[70] Liability laws vary from state to state and may influence the level of risk tolerance in both public and private decisions to prune or remove urban trees. The urban forest can also present fire risks to nearby residents; this dilemma grows along with the expanding wildland-urban interface. Events such as the 1991 Oakland Hills Fire (Oakland and Berkeley, California), the 2016 Chimney Tops Fire (Gatlinburg, Tennessee), and the 2017 Tubbs Fire (Santa Rosa, California) are just a few examples of wildfires that caused the loss of both human life and property in urban and suburban environments. Yet homeowners typically appreciate the scenery, privacy, and other benefits provided by surrounding trees and may be reluctant to remove them even when faced with the threat of wildfire.[71] In other cases, residents may not have the financial means to implement appropriate fire hazard reduction measures (see Chapter 5).

The growing presence and influence of homeowners' associations (HOAs) add another layer of complexity to urban forest governance. The requirement to comply with HOA compacts, covenants, and restrictions is written into property titles in HOA neighborhoods—meaning that rules established by HOA boards effectively carry the weight of law. These rules emphasize the protection of property values, including by promoting the maintenance of residential trees. In fire-prone landscapes, HOA rules requiring residents to retain trees or other vegetation within their yard—and rules making it difficult to remove vegetation—can have the unintended consequence of elevating residential fire risk.[72] At the same time, HOAs can be highly effective at hazard reduction by mandating changes that reduce the vulnerability of homes to fires and other natural events.

Urban environments also contain relatively large proportions of land that blends public and private tenure. For example, the linear strips of land between sidewalks and streets, sometimes called "parking strips," "tree lawns," or "hellstrips," are typically included in rights-of-way controlled by municipalities—yet adjacent residents or property owners are responsible for their upkeep. These areas are often targeted for tree planting under city afforestation programs. In selecting where and what species to plant, urban foresters might consider how the tree could affect aboveground or underground infrastructure (sidewalks, power lines, sewage lines, etc.), how well the species thrives in an urban environment, what kind of maintenance it will require, how likely the tree is to fall or break apart under high winds or ice, and how long the tree is likely to live. There is also a need to consider who will provide long-term care and maintenance. Adjacent renters or property owners may feel that the parking strip in front of their residence belongs to them even if the formal title is vested in the city, and their actions can determine the long-term success or failure of planted trees. As noted above, seemingly neutral decisions about where to plant and on whom to rely for continued maintenance may—wittingly or unwittingly—reproduce patterns of inequality by fostering changes in the urban canopy.

Conflict can occur even when the forest is under the relatively straightforward tenure category of municipal or county park. To some extent, conflict is inevitable given the large number and diversity of people making use of a limited supply of public parkland. It can be exacerbated, however, when municipal managers propose substantial changes to the visual quality of people's favored places. For example, the early years of the Chicago Wilderness project were characterized by substantial conflict as some area residents felt that changes to their local environment were being made without sufficient justification. The plan to conserve a network of green spaces in the greater Chicago area stretching from Michigan to Indiana included an important element of ecological restoration. This meant conducting extensive tree removal, invasive species removal, application of prescribed fire, and other management activities designed to restore forest conditions to the mix of savannahs and open stands characteristic of the region prior to Euro-American settlement. Although Chicago-area residents shared a love for urban greenspaces, the plan generated sharp disagreements, with some people asking how much those parks needed to be changed from the conditions to which they were accustomed—and who should decide.[73] In California, proposals to replace non-native (and highly flammable) eucalyptus trees in urban areas have also been met with resistance on the part of residents who feel a personal attachment to the exotic species. Similar conflicts have arisen regarding invasive species like Australian pine in Florida and mangroves in Hawaii.

## Conclusion

People tend to form strong connections to the places in which they live—both to the people that make up their communities and to the physical and ecological landscapes in which they are situated. People's attachments to their communities of place and to the forests in which they "work, live, play, and pray" are complex, multidimensional, and not easily reduced to simple measures. Along with shifts in the policy and governance context for federal and corporate forestlands in recent decades have come a suite of new initiatives that aim to ground decision-making and even ownership in communities of place. The terrain of community-oriented forest governance includes a tapestry of actors, organizations, and institutions: collaborative decision-making forums, community-based organizations, regional and national support networks, land trusts, local governments, urban and community foresters, individual residents and landowners, funding opportunities, formal policies, and informal practices. The emergence of new community-oriented forest governance models raises important questions about how decisions are made, who has the power to participate in decision-making, and how concepts like "community" are defined. A renewed attention to the urban forest raises a unique set of questions regarding the distribution of forest benefits and risks in varied urban settings.

## Questions for Discussion

1. How have definitions of forest-based communities changed since the early twentieth century? What accounts for these changes?
2. What factors allowed historic forest policy adversaries to productively collaborate on forest management proposals? Why might federal forest managers decide to adopt those proposals (or not)?
3. What explains the rise of community-based organizations throughout much of the western US, and how do these organizations differ from collaborative decision-making groups?
4. What are the benefits of community ownership of forests? Why might different kinds of communities (such as tribal nations, Hispano land grant communities, rural municipalities, and urban governments) choose to govern and manage these forests in different ways?
5. What makes urban forest governance so complex? What is the relationship between urban tree cover and inequality in urban areas?

## Notes

1 Christoffersen, 2005.
2 Dana, 1918, p. 4.
3 Pinchot, 1947, pp. 17–18.
4 Hirt, 1994.
5 Correia, 2007; Wilmsen, 2001. Despite this history, there have been attempts to reimagine the possibilities of the Vallecitos Sustained Yield Unit since 2005 through collaborative efforts involving the Forest Service and land grant communities bordering the unit (Krasilovsky & Valdez, 2007).
6 Schallau & Alston, 1987.
7 Kaufman & Kaufman, 1946.
8 Schallau, 1989; Schallau & Alston, 1987.
9 Loomis, 2016.
10 Loomis, 2016.
11 Kusel, 2001.
12 Baker & Quinn-Davidson, 2011; Nielsen-Pincus & Moseley, 2013.
13 Moseley & Reyes, 2008; Moseley & Shankle, 2001; Moseley & Toth, 2004.
14 Che, 2003.
15 Fortmann & Kusel, 1990.
16 Baker & Kusel, 2003; Cheng, 2006.
17 Baker & Kusel, 2003; Davis et al., 2020; Glasmeier & Farrigan, 2005.
18 Danks & Everett, 2022.

19 Marston, 2001.
20 McDermott, 2009.
21 Charnley et al., 2015.
22 Covington & Moore, 1994.
23 Allen et al., 2002; Falk, 2006.
24 Abrams, 2011; Abrams & Burns, 2007; Mottek Lucas et al., 2017; Waltz et al., 2014.
25 Abrams, Davis, et al., 2015; Danks, 2008.
26 Abrams, Davis, et al., 2015; Abrams et al., 2017; Danks & Everett, 2022.
27 Rural Voices for Conservation Coalition, 2022.
28 Abrams et al., 2016.
29 Butler & Schultz, 2019.
30 Cheng, 2006.
31 Abrams, Davis, et al., 2015.
32 Enzer & Goebel, 2014.
33 An exception is the Quincy Library Group, whose recommendations were formally recognized in federal legislation.
34 Abers & Keck, 2013; Abrams et al., 2017.
35 Moote, 2008.
36 Selin et al., 1997.
37 Abrams, 2011.
38 Davis et al., 2017.
39 Maier & Abrams, 2018; Moote, 2008.
40 Coggins, 2001; Hibbard & Madsen, 2003.
41 Moseley & KenCairn, 2001.
42 Cestero & Belsky, 2003; Cheng & Mattor, 2010; Davis et al., 2020; Foster, 2002.
43 McDermott, 2009.
44 Gray, 1989, p. 5.
45 Margerum, 2011, p. 6.
46 Snow, 2001, p. 2.
47 Innes & Booher, 2010.
48 Agrawal & Gibson, 1999.
49 Foster, 2002.
50 McCullough, 1995.
51 Belsky, 2008.
52 Lyman et al., 2013.
53 McGinley et al., 2022.
54 Parker, 2014.
55 Nowak & Greenfield, 2018b, p. 164.
56 Nowak & Greenfield, 2018a.
57 Donovan & Butry, 2009.
58 Jennings & Gaither, 2015; Nowak & Dwyer, 2007.
59 Donovan & Prestemon, 2012; Kondo et al., 2017; Troy et al., 2012.
60 Watkins & Gerrish, 2018.
61 Hoffman et al., 2020; Nardone et al., 2021.
62 Hoffman et al., 2020; Jesdale et al., 2013; Leets et al., 2022; Mitchell & Chakraborty, 2015.
63 Heynen et al., 2006.
64 Donovan & Mills, 2014; Flocks et al., 2011.
65 Heynen et al., 2006.
66 Wolch et al., 2014.
67 Grove et al., 2014.
68 Carmichael & McDonough, 2018, 2019.
69 Judice et al., 2021.
70 Hilbert et al., 2019.

71 Brenkert–Smith et al., 2006; Hjerpe et al., 2016.
72 Abrams, Knapp, et al., 2015; Paveglio et al., 2016.
73 Almassi, 2021; Helford, 2000.

## References


Abers, R. N., & Keck, M. E. (2013). *Practical authority: Agency and institutional change in Brazilian water politics*. Oxford University Press.

Abrams, J. (2011). The policy context of the White Mountain stewardship contract. In D. Egan, E. Hjerpe, & J. Abrams (Eds.), *Human dimensions of ecological restoration: Integrating science, nature, and culture* (pp. 163–176). Island Press.

Abrams, J., & Burns, S. (2007). *Case study of a community stewardship success: The White Mountain Stewardship Contract*. Ecological Restoration Institute, Northern Arizona University.

Abrams, J., Davis, E. J., & Moseley, C. (2015). Community-based organizations and institutional work in the remote rural West. *Review of Policy Research*, *32*(6), 675–698.

Abrams, J., Davis, E. J., Moseley, C., Ellison, A., & Nowell, B. (2016). *Community-based organizations in the US West: Status, structure, and activities* (Ecosystem Workforce Program Working Paper #68).

Abrams, J., Davis, E. J., Moseley, C., & Nowell, B. (2017). Building practical authority for community forestry in and through networks: The role of community-based organisations in the US West. *Environmental Policy and Governance*, *27*(4), 285–297.

Abrams, J., Knapp, M., Paveglio, T. B., Ellison, A., Moseley, C., Nielsen-Pincus, M., & Carroll, M. S. (2015). Re-envisioning community-wildfire relations in the US West as adaptive governance. *Ecology and Society*, *20*(3), 34.

Agrawal, A., & Gibson, C. C. (1999). Enchantment and disenchantment: The role of community in natural resource conservation. *World Development*, *27*(4), 629–649.

Allen, C. D., Savage, M., Falk, D. A., Suckling, K. F., Swetnam, T. W., Schulke, T., Stacey, P. B., Morgan, P., Hoffman, M., & Klingel, J. T. (2002). Ecological restoration of southwestern ponderosa pine ecosystems: A broad perspective. *Ecological Applications*, *12*(5), 1418–1433.

Almassi, B. (2021). Value disputes in urban ecological restoration: Lessons from the Chicago Wilderness. *Studies in History and Philosophy of Science Part A*, *87*, 93–100.

Baker, J. M., & Quinn-Davidson, L. N. (2011). Jobs and community in Humboldt County, California. In D. Egan, E. E. Hjerpe, & J. Abrams (Eds.), *Human dimensions of ecological restoration: Integrating science, nature, and culture* (pp. 221–237). Island Press.

Baker, M., & Kusel, J. (2003). *Community forestry in the United States: Learning from the past, crafting the future*. Island Press.

Belsky, J. M. (2008). Creating community forests. In E. M. Donoghue & V. Sturtevant (Eds.), *Forest community connections: Implications for research, management, and governance* (pp. 219–242). Resources for the Future.

Brenkert–Smith, H., Champ, P. A., & Flores, N. (2006). Insights into wildfire mitigation decisions among wildland–urban interface residents. *Society & Natural Resources*, *19*(8), 759–768.

Butler, W. H., & Schultz, C. A. (2019). *A new era for collaborative forest management: Policy and practice insights from the Collaborative Forest Landscape Restoration Program*. Routledge.

Carmichael, C. E., & McDonough, M. H. (2018). The trouble with trees? Social and political dynamics of street tree-planting efforts in Detroit, Michigan, USA. *Urban Forestry & Urban Greening*, *31*, 221–229.

Carmichael, C. E., & McDonough, M. H. (2019). Community stories: Explaining resistance to street tree-planting programs in Detroit, Michigan, USA. *Society & Natural Resources*, *32*(5), 588–605.

Cestero, B., & Belsky, J. M. (2003). Collaboration for community and forest well-being in the upper Swan Valley, Montana. In J. Kusel & E. Adler (Eds.), *Forest communities, community forests* (pp. 149–170). Rowman and LIttlefield.

Charnley, S., Poe, M. R., Ager, A. A., Spies, T. A., Platt, E. K., & Olsen, K. A. (2015). A burning problem: Social dynamics of disaster risk reduction through wildfire mitigation. *Human Organization*, *74*(4), 329–340.

Che, D. (2003). The new economy and the forest: Rural development in the post-industrial spaces of the rural Alleghenies. *Social Science Quarterly*, *84*(4), 963–978.

Cheng, A. S. (2006). Build it and they will come? Mandating collaboration in public lands planning and management. *Natural Resources Journal*, *46*(4), 841–858.

Cheng, A. S., & Mattor, K. M. (2010). Place-based planning as a platform for social learning: Insights from a national forest landscape assessment process in western Colorado. *Society & Natural Resources*, *23*(5), 385–400.

Christoffersen, N. D. (2005). Wallowa Resources: Gaining access and adding value to natural resources on public lands. In B. Child & M. W. Lyman (Eds.), *Natural resources as community assets: Lessons from two continents* (pp. 147–180). The Sand County Foundation and The Aspen Institute.

Coggins, G. C. (2001). Of Californicators, quislings, and crazies: Some perils of devolved collaboration. In P. Brick, D. Snow, & S. Van de Wetering (Eds.), *Across the Great Divide: Explorations in collaborative conservation and the American West* (pp. 163–171). Island Press.

Correia, D. (2007). The sustained yield forest management act and the roots of environmental conflict in Northern New Mexico. *Geoforum*, *38*(5), 1040–1051.

Covington, W. W., & Moore, M. M. (1994). Postsettlement changes in natural fire regimes and forest structure: Ecological restoration of old-growth ponderosa pine forests. *Journal of Sustainable Forestry*, *2*(1–2), 153–181.

Dana, S. T. (1918). *Forestry and community development* (Bulletin 638). US Department of Agriculture.

Danks, C. (2008). Institutional arrangements in community-based forestry. In E. M. Donoghue & V. Sturtevant (Eds.), *Forest community connections: Implications for research, management, and governance* (pp. 185–204). Resources for the Future.

Danks, C., & Everett, Y. (2022). Community-based forestry in the western United States: Reimagining the role of communities in federal forest management. In J. Bulkan, J. Palmer, A. M. Larson, & M. Hobley (Eds.), *Routledge handbook of community forestry* (pp. 168–181). Routledge.

Davis, E. J., Hajjar, R., Charnley, S., Moseley, C., Wendel, K., & Jacobson, M. (2020). Community-based forestry on federal lands in the western United States: A synthesis and call for renewed research. *Forest Policy and Economics*, *111*, 102042.

Davis, E. J., White, E. M., Cerveny, L. K., Seesholtz, D., Nuss, M. L., & Ulrich, D. R. (2017). Comparison of USDA Forest Service and stakeholder motivations and experiences in collaborative federal forest governance in the western United States. *Environmental Management*, *60*(5), 908–921.

Donovan, G. H., & Butry, D. T. (2009). The value of shade: Estimating the effect of urban trees on summertime electricity use. *Energy and Buildings*, *41*(6), 662–668.

Donovan, G. H., & Mills, J. (2014). Environmental justice and factors that influence participation in tree planting programs in Portland, Oregon, US. *Arboriculture & Urban Forestry*, *40*(2), 70–77.

Donovan, G. H., & Prestemon, J. P. (2012). The effect of trees on crime in Portland, Oregon. *Environment and Behavior*, *44*(1), 3–30.

Enzer, M. J., & Goebel, M. (2014). Place-based conservation finds its voice: A case study of the Rural Voices for Conservation Coalition. In S. Charnley, T. E. Sheridan, & G. P. Nabhan (Eds.), *Stitching the West back together: Conservation of working landscapes* (pp. 101–117). University of Chicago Press.

Falk, D. A. (2006). Process-centred restoration in a fire-adapted ponderosa pine forest. *Journal for Nature Conservation*, *14*(3–4), 140–151.

Flocks, J., Escobedo, F., Wade, J., Varela, S., & Wald, C. (2011). Environmental justice implications of urban tree cover in Miami-Dade County, Florida. *Environmental Justice*, *4*(2), 125–134.

Fortmann, L., & Kusel, J. (1990). New voices, old beliefs: Forest environmentalism among new and long-standing rural residents. *Rural Sociology*, *55*(2), 214–232.

Foster, S. (2002). Environmental justice in an era of devolved collaboration. In K. M. Mutz, G. C. Bryner, & D. S. Kenney (Eds.), *Justice and natural resources: Concepts, strategies, and applications* (pp. 139–160). Island Press.

Glasmeier, A. K., & Farrigan, T. (2005). Understanding community forestry: A qualitative meta-study of the concept, the process, and its potential for poverty alleviation in the United States case. *Geographical Journal*, *171*(1), 56–69.

Gray, B. (1989). *Collaborating: Finding common ground for multiparty problems*. Jossey-Bass.

Grove, J. M., Locke, D. H., & O'Neil-Dunne, J. P. M. (2014). An ecology of prestige in New York City: Examining the relationships among population density, socio-economic status, group identity, and residential canopy cover. *Environmental Management*, *54*(3), 402–419.

Helford, R. M. (2000). Constructing nature as constructing science: Expertise, activist science, and public conflict in the Chicago wilderness. In P. H. Gobster & R. B. Hull (Eds.), *Restoring nature: Perspectives from the social sciences and humanities* (pp. 119–142). Island Press.

Heynen, N., Perkins, H. A., & Roy, P. (2006). The political ecology of uneven urban green space: The impact of political economy on race and ethnicity in producing environmental inequality in Milwaukee. *Urban Affairs Review*, *42*(1), 3–25.

Hibbard, M., & Madsen, J. (2003). Environmental resistance to place-based collaboration in the US West. *Society and Natural Resources*, *16*(8), 703–718.

Hilbert, D. R., Koeser, A. K., Roman, L. A., Hamilton, K., Landry, S. M., Hauer, R. J., Campanella, H., McLean, D., Andreu, M., & Perez, H. (2019). Development practices and ordinances predict inter-city variation in Florida urban tree canopy coverage. *Landscape and Urban Planning*, *190*, 103603.

Hirt, P. W. (1994). *A conspiracy of optimism: Management of the national forests since World War Two*. University of Nebraska Press.

Hjerpe, E., Kim, Y.-S., & Dunn, L. (2016). Forest density preferences of homebuyers in the wildland-urban interface. *Forest Policy and Economics*, *70*, 56–66.

Hoffman, J. S., Shandas, V., & Pendleton, N. (2020). The effects of historical housing policies on resident exposure to intra-urban heat: A study of 108 US urban areas. *Climate*, 8(1), 12.

Innes, J. E., & Booher, D. E. (2010). *Planning with complexity: An introduction to collaborative rationality for public policy*. Routledge.

Jennings, V., & Gaither, C. J. (2015). Approaching environmental health disparities and green spaces: An ecosystem services perspective. *International Journal of Environmental Research and Public Health, 12*(2), 1952–1968.

Jesdale, B. M., Morello-Frosch, R., & Cushing, L. (2013). The racial/ethnic distribution of heat risk–related land cover in relation to residential segregation. *Environmental Health Perspectives, 121*(7), 811–817.

Judice, A., Gordon, J., Abrams, J., & Irwin, K. (2021). Community perceptions of tree risk and management. *Land, 10*(10), 1096.

Kaufman, H. F., & Kaufman, L. C. (1946). *Toward the stabilization and enrichment of a forest community*. University of Montana.

Kondo, M. C., Han, S., Donovan, G. H., & MacDonald, J. M. (2017). The association between urban trees and crime: Evidence from the spread of the emerald ash borer in Cincinnati. *Landscape and Urban Planning, 157*, 193–199.

Krasilovsky, E., & Valdez, A. (2007). *Community owned forests: An example from the Southwest*. Forest Guild.

Kusel, J. (2001). Assessing well-being in forest dependent communities. *Journal of Sustainable Forestry, 13*(1–2), 359–384.

Leets, L., Sprenger, A., Hartman, R. O., Howard Jackson, J., Britt, M., Gulley, A., Simon Thomas, J., & Wijesinghe, S. (2022). Promoting tree equity in Washington, D.C. *Trees, Forests and People, 7*, 100209.

Loomis, E. (2016). *Empire of timber: Labor unions and the Pacific Northwest forests*. Cambridge University Press.

Lyman, M. W., Danks, C., & McDonough, M. (2013). New England's community forests: Comparing a regional model to ICCAs. *Conservation and Society, 11*(1), 46–59.

Maier, C., & Abrams, J. B. (2018). Navigating social forestry: A street-level perspective on national forest management in the US Pacific Northwest. *Land Use Policy, 70*, 432–441.

Margerum, R. D. (2011). *Beyond consensus: Improving collaborative planning and management*. MIT Press.

Marston, E. (2001). The Quincy Library Group: A divisive attempt at peace. In P. Brick, D. Snow, & S. Van de Wetering (Eds.), *Across the great divide: Explorations in collaborative conservation and the American West* (pp. 79–90). Island Press.

McCullough, R. (1995). *The landscape of community: A history of communal forests in New England*. University Press of New England.

McDermott, M. H. (2009). Equity first or later? How US community-based forestry distributes benefits. *International Forestry Review, 11*(2), 207–220.

McGinley, K., Charnley, S., Cubbage, F. W., Hajjar, R., Frey, G. E., Schelhas, J., Hovis, M., & Kornhauser, K. (2022). Community forest ownership, rights, and governance regimes in the United States. In J. Bulkan, J. Palmer, A. M. Larson, & M. Hobley (Eds.), *Routledge handbook of community forestry* (pp. 146–167). Routledge.

Mitchell, B. C., & Chakraborty, J. (2015). Landscapes of thermal inequity: Disproportionate exposure to urban heat in the three largest US cities. *Environmental Research Letters, 10*(11), 115005.

Moote, M. A. (2008). Collaborative forest management. In E. M. Donoghue & V. Sturtevant (Eds.), *Forest community connections: Implications for research, management, and governance* (pp. 243–260). Resources for the Future.

Moseley, C., & KenCairn, B. (2001). Problem solving or social change? The Applegate and Grand Canyon Forest Partnerships. In R. K. Vance, C. B. Edminster, W. W. Covington, & J. A. Blake (Eds.), *Ponderosa pine ecosystems restoration and conservation: Steps toward stewardship* (pp. 121–129). USDA Forest Service Rocky Mountain Research Station.

Moseley, C., & Reyes, Y. E. (2008). Forest restoration and forest communities: Have local communities benefited from Forest Service contracting of ecosystem management? *Environmental Management*, *42*(2), 327–343.

Moseley, C., & Shankle, S. (2001). Who gets the work? National forest contracting in the Pacific Northwest. *Journal of Forestry*, 99(9), 32–37.

Moseley, C., & Toth, N. A. (2004). Fire hazard reduction and economic opportunity: How are the benefits of the National Fire Plan distributed? *Society and Natural Resources*, *17*(8), 701–716.

Mottek Lucas, A., Kim, Y.-S., Greco, B., Becker, D. R., Hjerpe, E. E., & Abrams, J. (2017). Social and economic contributions of the White Mountain Stewardship Project: Final 10-year assessment: Lessons learned and implications for future forest management initiatives. *Journal of Forestry*, *115*(6), 548–558.

Nardone, A., Rudolph, K. E., Morello, -Frosch Rachel, & Casey, J. A. (2021). Redlines and greenspace: The relationship between historical redlining and 2010 greenspace across the United States. *Environmental Health Perspectives*, *129*(1), 017006.

Nielsen-Pincus, M., & Moseley, C. (2013). The economic and employment impacts of forest and watershed restoration. *Restoration Ecology*, *21*(2), 207–214.

Nowak, D. J., & Dwyer, J. F. (2007). Understanding the benefits and costs of urban forest ecosystems. In J. E. Kuser (Ed.), *Urban and community forestry in the Northeast* (pp. 25–46). Springer Netherlands.

Nowak, D. J., & Greenfield, E. J. (2018a). Declining urban and community tree cover in the United States. *Urban Forestry & Urban Greening*, *32*, 32–55.

Nowak, D. J., & Greenfield, E. J. (2018b). US urban forest statistics, values, and projections. *Journal of Forestry*, *116*(2), 164–177.

Parker, M. (2014). Swan story. In S. Charnley, T. E. Sheridan, & G. P. Nabhan (Eds.), *Stitching the West back together: Conservation of working landscapes* (pp. 123–135). University of Chicago Press.

Paveglio, T. B., Abrams, J., & Ellison, A. (2016). Developing fire adapted communities: The importance of interactions among elements of local context. *Society & Natural Resources*, *29*(10), 1246–1261.

Pinchot, G. (1947). *Breaking new ground*. Harcourt, Brace.

Rural Voices for Conservation Coalition. (2022). *Pathways to prescribed fire*. Rural Voices for Conservation Coalition.

Schallau, C. H. (1989). Sustained yield versus community stability: An unfortunate wedding? *Journal of Forestry*, *87*(9), 16–23.

Schallau, C. H., & Alston, R. M. (1987). Commitment to community stability: A policy or shibboleth? *Environmental Law*, *17*(3), 429–481.

Selin, S. W., Schuett, M. A., & Carr, D. S. (1997). Has collaborative planning taken root in the national forests? *Journal of Forestry*, 95(5), 25–28.

Snow, D. (2001). Coming home: An introduction to collaborative conservation. In P. Brick, D. Snow, & S. Van de Wetering (Eds.), *Across the great divide: Explorations in collaborative conservation and the American West* (pp. 1–11). Island Press.

Troy, A., Morgan Grove, J., & O'Neil-Dunne, J. (2012). The relationship between tree canopy and crime rates across an urban–rural gradient in the greater Baltimore region. *Landscape and Urban Planning*, *106*(3), 262–270.

Waltz, A. E. M., Stoddard, M. T., Kalies, E. L., Springer, J. D., Huffman, D. W., & Meador, A. S. (2014). Effectiveness of fuel reduction treatments: Assessing metrics of forest resiliency and wildfire severity after the Wallow Fire, AZ. *Forest Ecology and Management, 334*, 43–52.

Watkins, S. L., & Gerrish, E. (2018). The relationship between urban forests and race: A meta-analysis. *Journal of Environmental Management, 209*, 152–168.

Wilmsen, C. (2001). Sustained yield recast: The politics of sustainability in Vallecitos, New Mexico. *Society & Natural Resources, 14*(3), 193–207.

Wolch, J. R., Byrne, J., & Newell, J. P. (2014). Urban green space, public health, and environmental justice: The challenge of making cities 'just green enough.' *Landscape and Urban Planning, 125*, 234–244.

# 11 Unseen Forest Users and Uses

## Introduction

Responding to a national directive to reduce the resource impacts from unmanaged recreational vehicle use on public lands, Wallowa-Whitman National Forest administrators embarked on a travel management planning process in 2007. Forest Service staff from the supervisor's office and district offices utilized NEPA procedures regularly employed by the agency and held a series of community meetings over the next year to gather public input. They received thousands of written comments, email messages, and letters as well as signed petitions and some form letters, representing both those opposed to road closures and those who felt that reducing open road density was the only way to conserve wildlife and water quality. Hundreds of additional letters and comments poured in following the draft NEPA decision in 2009. When the final travel management plan proposing to close roughly half of the forest's 9,000 miles of roads was released in 2012, the backlash from Wallowa County residents was swift. A key concern voiced by local people was access to firewood. A survey administered by county commissioners found that 79% of residents in this cold, high-elevation terrain heated their homes with firewood and 73% cut some or all of their own wood. Residents also questioned whether they would be able to access berries, mushrooms, and game meat if their road access were reduced. Following a series of emotionally charged public meetings and threatened appeals by county governments, the Wallowa-Whitman National Forest withdrew the plan and reassigned the forest supervisor.

The previous year brought evidence of a very different kind of under-the-radar use of the Wallowa-Whitman National Forest when a group of bear hunters stumbled upon a clandestine cannabis grow site that extended for over a mile along a remote ravine.[1] A coalition of local, state, and national law enforcement agents subsequently raided the site, arresting six individuals with purported links to Mexican drug cartels. Materials found at the site included semi-automatic weapons and handguns, fertilizers, pesticides, camps stocked with weeks' worth of supplies, miles of plastic irrigation tubing, and an estimated 91,000 cannabis plants. It was the largest known illegal grow operation in the state's history.

Although firewood cutting, berry picking, and clandestine cannabis cultivation may appear to be wholly unrelated activities, what they have in common

DOI: 10.4324/9781003043669-11

is that they often occur "unseen"—that is, they take place on the peripheries of official governmental or market authorization and documentation. To say that a forest use or user is "unseen" is not to imply total invisibility; rather, it is to say that it represents an activity or a population that is difficult to study and that is often ignored or overlooked when we talk about forest policy and governance. As Sally Fairfax and colleagues explain, unseen forest users often

> do not seek agency recognition of their presence—in fact they are generally clandestine—but they do claim the land they access or upon which they operate, often in ways that effectively exclude other users, occasionally with violence.[2]

Informal and unsanctioned uses of forests have a long and storied history in the US. Chapter 1, for example, discusses how numerous African Americans made use of remote, inhospitable forests to establish "maroon" communities in the antebellum South as refuges from legal slavery. Contemporary unseen forest uses are not necessarily restricted to the most remote forest regions; in some cases, they take place on the fringes of urban areas and even within urban forests themselves. This chapter explores a variety of unseen uses and users and examines the policy and governance processes that influence them.

## Non-Timber Forest Products

The category of non-timber forest products (NTFPs) encompasses a wide range of goods produced by forests beyond timber, pulpwood, and woody biomass. The diversity of NTFPs harvested in the US is extensive; interviews with NTFP harvesters near eastern Kentucky's Daniel Boone National Forest revealed 105 distinct species harvested for uses such as food, medicine, crafts, dyes, and perfumes.[3] A study on Michigan's Upper Peninsula identified 140 products derived from over 100 NTFP species harvested by local Native American and Euro-American populations.[4] For some forest users, NTFP harvesting is a largely recreational pursuit. For others, it represents one of many activities that contribute to their livelihoods. In some cases, it is a primary or sole source of income or sustenance. Although it is difficult to collect reliable economic data on an activity characterized by its informality, conservative estimates of the economic activity associated with NTFP harvesting in the US are typically on the order of \$200-\$300 million annually, with wholesale market values approaching \$1 billion.[5]

Although NTFP harvesting is an activity that is growing in popularity, it often remains unseen due to a high level of participation from marginalized social groups: Native Americans, recent immigrants (some undocumented), and low-income populations of any race, ethnicity, or immigration status. Harvesters often avoid encounters with government forest managers and title forest owners; any conflict that arises between harvesters is often settled informally. Conflict can occur not only due to competition for finite resources but also because property boundaries often go unrecognized and unacknowledged in NTFP harvesting.[6] This can result, for example, in non-Native harvesters encroaching on territories to which enrolled tribal members have exclusive legal access or in the perceived violation of customary gathering places by new harvesters.[7]

Under some circumstances, NTFP harvesting can represent an example of what Garrett Hardin erroneously called the "tragedy of the commons" (but should have called the "tragedy of open access"): with weak enforcement of access limitations and tenure rights, some NTFPs are subject to overharvesting, leading to declining availability over time.[8] Other NTFP species appear to be relatively resilient in the face of harvest pressures. The most vulnerable species are those that grow slowly and that are harvested by taking whole plants or roots; less vulnerable are fast-growing, quickly reproducing species or those (such as mushrooms or berries) for which harvesting leaves a viable organism intact.[9] The patterns and governance processes for NTFP harvesting vary across jurisdictions and differ according to the kind of product being harvested.

### *Edible Fungi, Plants, and Plant Parts*

Mushrooms—the fruiting bodies of fungal mycelial networks that grow unseen beneath the soil or within decaying organic matter—are highly sought-after NTFPs nationwide (Figure 11.1). Those who scour the woods in search of chanterelles, morels, boletes, matsutakes, truffles, and other mushroom varieties are also typically unseen. Mushroom hunting in the US is a popular recreational pursuit, with many users only seeking to harvest enough for personal consumption (and perhaps gift to friends and family members). However, the wild mushrooms that appear on the shelves of domestic grocery stores and are shipped as far afield as Europe and Asia come from the effort of independent harvesters working in a cash-based system of daily forays into the woods and nightly sales to buyers. Many of these harvesters operate in an annual circuit, moving seasonally across multiple states to take advantage of prime harvesting opportunities in different environments.[10] Year-to-year differences in mushroom abundance can be substantial; for example, some morel species flourish in recently burned forests, meaning that wildfires one year will often attract mushroom hunters the next. Patterns of precipitation, drought, frost, and forest disturbances can mean the difference between a bumper crop and a bust in any given year.

Commercial-scale wild mushroom harvesting in many forests includes heavy representation by Indigenous, Latino, and Southeast Asian populations, although pickers of all ethnic and racial backgrounds participate to some extent. Anthropologist Anna Lowenhaupt Tsing described the rich cultural mixing zone of matsutake mushroom camps in central Oregon in her detailed ethnography:

> I seemed to have stepped into rural Southeast Asia. Mien wearing sarongs boiled water in kerosene cans over stone tripods and hung strips of game and fish over the stove to dry. Hmong all the way from North Carolina brought home-canned bamboo shoots for sale. Lao noodle tents sold not only *pho*

*Figure 11.1* A commercial chanterelle mushroom harvester sorting their harvest in the field in western Oregon. Photograph by the author.

> but also the most authentic *laap* I had eaten in the United States...Native Americans sometimes came to sell their mushrooms. There were also both whites and Latinos, although most avoided the official camp, staying in the woods alone or in small groups.[11]

Many of the southeast Asian matsutake hunters Tsing profiled were refugees from armed conflicts or the children of refugees, and many of the White harvesters were themselves veterans of these same conflicts.

Matsutake mushrooms can bring high prices to those who know where to find them: for a brief moment in 1993, pickers were offered $300 to $600 per pound; since then, prices have typically remained between five and twenty dollars per pound. Even at these prices, skilled matsutake hunters can earn thousands of dollars over several weeks of effort. Nearly all the matsutake harvested in Oregon and elsewhere in the Northwest is exported to Japan, where a sharp decline in the native matsutake has led to demand for related species from elsewhere around the world.

The seasonal mushroom camps in places like central Oregon have often been described as having a "Wild West" feel given the large amounts of cash changing hands and the fact that many pickers carry firearms (as much for signaling as for defense). The US Forest Service has developed systems for managing the resource impacts of large numbers of campers, mushroom harvesters, buyers, and associated individuals by designating places on national forests where mushroom buying and commercial harvester camping are allowed to take place. These decisions are not always welcomed or respected by the pickers and buyers themselves, however.[12] National forest managers also routinely utilize permit systems for commercial harvesters, but these typically do little to track the number of mushrooms removed or the consequent resource impacts.[13] It is widely acknowledged that many commercial-scale mushroom hunters operate without permits and, often, move freely between private and public lands in search of nature's bounty. It is important to point out, however, that mushroom harvesting in violation of formal legal restrictions is not an activity confined to commercial pickers, as described below.

In addition to mushrooms, many other wild-growing edible species are collected in forests across the country. Fiddlehead ferns, ramps, nettles, walnuts, hazelnuts, acorns, pawpaws, blueberries, huckleberries, and piñon nuts are just a few of the most common edible NTFPs collected by harvesters. Recreational harvest of berries for personal use is common in some parts of the country, for example, in New England, the Northern Rocky Mountains, and throughout much of Alaska. However, commercial harvesting also takes place, including for species like wild blueberries and huckleberries. In the Southwest, piñon nut harvesting represents a culturally and economically important practice across large areas dominated by piñon-juniper woodlands. The best stands can produce as much as 400 pounds of seeds per acre, and the retail value of piñon nuts can exceed $20 per pound—although, as in the mushroom economy, relatively little of that value may be captured by the harvesters themselves. Harvests have been dropping in recent decades as droughts and high temperatures driven by climate change have taken their toll on piñon pine populations and seed production across the Southwest.[14]

Research shows that unseen forest uses are abundant even in urban and suburban areas, with urban foraging of edible, decorative, and medicinal NTFPs an increasingly popular activity.[15] As with recreational NTFP harvesters in more remote areas, urban foragers are often motivated by interests in healthy lifestyles, engaging with the natural world, and continuing cultural traditions. For example, many urban and suburban residents of Italian, eastern European, and Russian heritage engage in mushroom harvesting and various Asian traditions emphasize nut and wild greens harvesting.[16] At the same time, some urban foragers may rely on harvests to help supplement their diets or to make ends meet during times of financial hardship.[17] Laws on NTFP harvesting in municipal and state parks commonly prohibit the collection of any plant or fungi material, with prohibitions sometimes extending even to abundant and invasive species such as dandelions.[18] However, this does little to stop some urban NTFP harvesters.[19] Mycologist David Arora, for example, described the commitment of some recreational mushroom enthusiasts in the San Francisco Bay Area to continue hunting for bolete (porcini) mushrooms in local parks despite a formal harvest prohibition: "Many mushroom hunters…were willing to risk being apprehended and fined in order to continue doing what they love doing."[20]

In contrast, some municipalities have established parks for the specific purpose of providing foraging opportunities. The country's first public "food forest" was established in Asheville, North Carolina, in 1997, and several other cities have since followed suit. Edible species in these parks tend to be dominated by apples, pears, and other imported cultivars but can also include native species such as walnuts, pecans, and hazelnuts. The city of Atlanta established the seven-acre Urban Food Forest at Browns Mill in 2020 in a predominantly African American neighborhood.[21] In addition to community garden plots, the property contains black walnut, pecan, and mulberry trees along with pawpaws and other native edible species. This food forest is significant not only because it is found within a neighborhood classified as a "food desert," but also because food production on residential lots was illegal within the city of Atlanta as recently as 2014.

### Floral Greens and Decorative and Weaving Materials

Dozens of wild plant, moss, and lichen species are used in floral arrangements, wreaths, and other decorative uses. Some of the most important species include beargrass, salal, Oregon grape, Douglas-fir, noble fir, and western red cedar from Pacific Northwest forests; galax, grapevine, and dutchman's pipe from southern Appalachia; and mountain laurel and winterberry holly from the Northeast. Spanish moss from places like Florida, Georgia, and Mississippi is widely used as a natural packing material, and decorative mosses and lichens from the Northwest and Appalachia make up a multimillion-dollar export industry.[22] Willow, ash, and oak species, along with palmetto, sweetgrass, beargrass, pine straw, and other plants and plant products, are sought after as ingredients for basketmaking. As with edible NTFPs, decorative and weaving NTFPs are gathered both for personal or family use as well as for formal or informal market exchange, with some floral and decorative materials being shipped to Europe and Asia.

Mirroring patterns in the mushroom trade, floral green and weaving material harvesters are a culturally and ethnically diverse group. In varied settings across the US, harvesters include Black urban and rural residents, Native Americans, recent immigrants (particularly from Latin America and Southeast Asia), and White rural populations, among other groups. One contrast with mushroom and edible plant harvesting is the prevalence of contract labor arrangements. In the Pacific Northwest, for example, floral green buyers will often enter into formal land leases and provide transportation for harvesters, who are then obligated to sell only to that buyer.[23] As is true for many other NTFPs, the management of floral greens and weaving materials by harvesters often entails the application of local knowledge.[24] This includes knowledge of where, when, and how to locate target species as well as how to harvest to maximize quality and minimize resource damage. Research has shown that many harvesters implement voluntary ethical practices designed to support the long-term sustainability of their resources. Techniques commonly practiced by harvesters include rotating gathering places, selective gathering, minimizing the impacts of harvest, intentionally propagating new generations, and protecting the knowledge of key site locations.[25]

### *Medicinal Species*

Numerous forest species are used for medicinal purposes and are harvested both commercially and non-commercially. For example, echinacea harvesting accelerated in the 1990s following the popularization of the plant as a possible immune system booster. Other important medicinal plants include black cohosh, slippery elm, goldenseal, saw palmetto berry, devil's club, and even common "weeds" like dandelion and red clover. For Indigenous and traditional populations, NTFP species may hold combined medicinal, nutritional, and spiritual values; the practice of harvesting, sharing, and making use of these species can represent an important element in the reproduction and transmission of culture.[26] For example, a team of Native and non-Native scholars explains,

> The "Strawberry Ceremony" or "berry fast" is still used for [Wabanaki] female adolescents' initiation into womanhood. The traditional medicinal uses of berries and certain portions of the plant strengthen women's reproductive organs and act as the primary medicine for endocrine system balancing, reproductive toning, birthing preparation, post-partum bleeding and toning as well as a preventive for post-partum infections and uses during menopause.[27]

No medicinal NTFP is more valuable or sought-after than American ginseng, a plant renowned for its medicinal qualities that grows in shady forests throughout the Appalachian and Ozark ranges and as far north as Minnesota. In some economically depressed rural communities, wild ginseng harvest has provided a lifeline for people with few alternative options. Ginseng harvest regulations vary by state and landowner, but typically include season restrictions, permit requirements for ginseng harvest on national forestlands, and written permission for harvesting on private lands. Some states require a license in order to buy and sell ginseng, and most states require plants to be at least five years old before they can be harvested (age can be determined by the number of leaves or the number of stem scars). However, as with many other NTFPs, the harvest of wild ginseng typically takes place deep in the forest away from oversight by public or private forest managers, and overharvesting has caused populations to dwindle. Some state and national forests have responded to this decline by temporarily or permanently prohibiting ginseng harvest. Because of its precarious status, American ginseng is included in Appendix II of the Convention on International Trade in Endangered Species of Wild Fauna and Flora (CITES), an international treaty ratified by the US. In practice, this means that the exportation of ginseng is regulated by the US Fish and Wildlife Service; the agency evaluates state and tribal ginseng conservation programs to determine whether they ensure the species' perpetuation. Nineteen states and the Menominee Tribe regulate the harvest of wild ginseng under rules approved by the US Fish and Wildlife Service.

### *Formal and Informal Governance of NTFPs*

Formal policies governing NTFP access vary greatly by jurisdiction and ownership category. For example, the state of Alaska has some of the country's strongest legal provisions protecting the rights of subsistence harvest; personal collection of a wide variety of plant, fungi, and fish and wildlife species is expressly allowed across Alaska state lands. The Alaska National Interest Lands Conservation Act (ANILCA) provides for subsistence activities on federal lands for both Native and non-Native rural residents of the state.[28] Native populations in Hawaii and US territories in the South Pacific have legally recognized rights to continue traditional gathering practices, including within national parks. NTFP harvesting policies vary on national park units across the rest of the country; a review in 2018 found that 46 parks allowed at least some level of foraging, whereas all foraging was prohibited on 13 other units.[29] Members of some tribes maintain specific treaty rights to gather important plant species on federal lands, including in national parks, but tribes have historically been restricted from hunting, gathering, or practicing resource management on traditional lands that became national parks. In 2016, the National Park Service finalized a rule that allows federally recognized tribes to participate in "traditional gathering" of vascular plants for "traditional purposes" within national parks, following a permit and NEPA analysis. The first use of this authority was by the Eastern Band of Cherokee Indians to harvest sochan and ramps within Great Smoky Mountains National Park.[30]

The rising value of some NTFPs has sparked political action by private landowners to crack down on trespass and unpermitted harvest. For example, the fallen needles of longleaf and slash pines are valuable enough that pine straw theft has become a major concern; some forest owners have gone as far as burying GPS trackers in bales of pine straw to aid in tracking and detaining thieves. In 1997, a coalition of North Carolina forest owners successfully lobbied the state to reclassify pine straw theft as a felony. As of 2018, Florida requires a state permit to harvest saw palmetto berries; Georgia followed with a similar law in 2020. As a means of controlling trespass and theft, Maine requires shippers moving more than 25 pounds of wild blueberries to operate under state permits. However, Maine continues to protect the longstanding tradition of allowing personal (noncommercial) harvesting of NTFPs on private lands.[31]

Forest management decisions can have impacts on the abundance and accessibility of NTFPs. Wildfire and prescribed fire, for example, can promote the growth of huckleberries and morel mushrooms. In some cases, tribal managers have been able to use cultural fire on tribal or public lands to stimulate specific kinds of ecological responses, such as the growth of tender new shoots and branches needed for basketmaking or to encourage the production of quality acorns. Depending on the details of stand conditions and harvest practices, timber harvest operations can increase the abundance of some NTFPs, diminish the abundance of others, and potentially facilitate or impede physical access to gathering sites.[32] For example, a series of proposed silvicultural operations in the lodgepole pine-dominated epicenter of matsutake gathering in central Oregon were ultimately canceled due to concerns about the effect of soil compaction and canopy changes on mushroom growth—though this is a rare example of a timber operation being modified due to possible effects on NTFPs.

Research has shown a relatively low overall level of engagement of NTFP harvesters in formal forest planning and decision-making processes. This results from a combination of factors including a lack of NTFP interest group organization and poor communication between land managers and NTFP harvesters.[33] The marginal position of many commercial NTFP harvesters may make them both less visible to formal decision-makers and less willing to assert their interests in public forums. Additionally, the rich and detailed traditional and local knowledges of many harvesters often go unrecognized or undervalued within the scientific tradition that guides professional forest management.[34] As a result, decision-making for national, state, county, and urban parks, as well as for national forests and other public lands, tends to exclude NTFP concerns and knowledges in favor of the interests of more powerful and well-organized stakeholder groups such as environmental preservationists and commercial timber harvesters.[35]

Despite the application of traditional and local knowledge held by many harvesters, the abundance and accessibility of some NTFPs have declined in many places, sometimes due to forces outside of harvesters' control and sometimes due to weak or nonexistent governance rules as numerous individuals compete to gather NTFPs under open-access conditions. The growing popularity of NTFP harvesting as both a recreational pursuit and as a livelihood practice has put pressure on some resources such as ginseng and huckleberries. Novice harvesters may lack the knowledge needed to harvest sustainably and may also be unaware of tenure patterns such as treaty language specifying exclusive NTFP harvest rights within an area to members of a particular tribe. Another phenomenon is the formalization or "hardening" of land tenure over time as existing or new landowners restrict traditional uses that had been informally permitted in the past. A key example of this phenomenon comes from the South Carolina Lowcountry, where exurban development has threatened access to plant species utilized by members of the Gullah-Geechee culture for basketmaking (Box 11.1).

### BOX 11.1 SWEETGRASS BASKETMAKING IN THE SOUTH CAROLINA LOWCOUNTRY

The Gullah-Geechee of the North Carolina, South Carolina, Georgia, and Florida "Lowcountry" have long relied on plant NTFPs as ingredients in an artisan basketmaking tradition with roots in West African culture. Sales of handmade baskets crafted with local plant species serve as an important source of income for Gullah-Geechee families, particularly in the greater Charleston, South Carolina, area. Although many Gullah-Geechee families possess private lands that have been in their families since shortly after the Civil War, the sweetgrass, longleaf pine needles, black rush, and sabal palmetto fronds essential to basketmaking are found across diverse properties that were traditionally "treated as de facto resource commons…embedded within a property regime characterized by privately owned land in a regional economy tied to agricultural and timber production."[36] As a growing amenity-seeking population and associated development have altered the physical and demographic character of the Lowcountry, access to these basketmaking materials has been put at risk. This is due to both physical changes to the landscape (for example, habitat alteration conducted as part of residential development) and changes to traditional patterns of access to NTFP gathering sites. Many sites that were once accessible for NTFP collection as part of commons-type land tenure arrangements have effectively been closed to entry as new landowners with more formal and exclusionary understandings of property have moved in. Researchers Patrick Hurley and Angela Halfacre observe that area developers, conservation advocates, and elected officials often celebrate the rich cultural heritage of the Lowcountry, yet relatively few have worked to ensure that cultural practitioners like Gullah-Geeche basketmakers maintain access to the raw ingredients of their cultural traditions.[37] Local changes have led many NTFP collectors to alter their strategies: some now travel to more distant collection sites, some have attempted to cultivate their own plant populations, some have negotiated new formal arrangements with landowners and neighborhood associations, and some have even started collecting in unexpected places like urban roadsides and parking lots where sweetgrass has been planted for ornamental landscaping purposes.[38]

## Precarious and Undocumented Laborers

In November 2005, *Sacramento Bee* journalist Tom Knudson and photographer Hector Amezcua published an exposé based on nine months of investigative work. Titled "Pineros: Men of the Pines," the award-winning series documented the dangerous and sometimes abusive working conditions experienced by the Latino forest workers charged with conducting fuel reduction, brush disposal, and other management activities on US national forestlands. The laborers profiled in the series included both undocumented workers and those legally in the US under the H2B guest worker program. Knudson and Amezcua described a system in which many workers were subjected to slavery-like conditions and paid pittances (some even going into debt) for arduous work in dangerous environments with little to no training, protective equipment, or access to basic resources like potable water. They reported on many workers who were left dead or permanently injured. In many cases, the contracting businesses that employed these workers had long histories of labor violations but continued to be offered contracts to manage national forestlands.

Forest workers may seem an odd fit for the category of unseen forest users given the deep cultural legacy of loggers as symbols of masculinity, epitomized in popular icons like Paul Bunyan and in the stars of reality television shows like *Ax Men*, *American Loggers*, and *Swamp Loggers*. Yet beyond the highly visible images of loggers and their increasingly advanced technology is what has been called the "Latinization" of the forestry workforce. Even as equipment-intensive work in the US forestry industry—such as that featured on reality television shows—is mostly conducted by White operators, manual labor-intensive work is now largely provided by migrant workers from Mexico and Central America—both documented (that is, working legally under guest worker or other visa guidelines) and undocumented (working without legal authorization).[39] These workers include both individuals for whom Spanish is a first language and those from Indigenous communities that may speak Mixtec, Nahuatl, Mam, or other Indigenous languages as a first language and Spanish as a second language (if at all).

The varieties of wage labor that are typically performed by migrant forest workers from Mexico and Central America include tree planting and herbicide application on private forestlands in the South plus stand thinning, seed collection, trail maintenance, brush piling, nursery work, and firefighting on public forestlands in the West.[40] Laborers often work for companies that bid on competitive contracts to perform specific tasks on private or public lands; some of these businesses are themselves owned and operated by Latinos. The work performed tends to be seasonal, poorly paid, and rarely offers health insurance or other benefits.[41] Although forest work has always been a dangerous and physically demanding field, the Latinization of the forest workforce magnifies risks due to the vulnerable position of the workers themselves. As Brinda Sarathy observes, "The undocumented status of many pineros renders them highly vulnerable to threats of deportation and job loss, and most avoid contacting federal officials to report workplace injustice."[42] These same factors also inhibit their active participation in formal governance processes (such as commenting on proposed forest policy or management plans) as well as in informal governance processes like place-based collaboration.

Policy and governance drivers of this pattern of forest worker vulnerability include intersections between immigration laws (including guest worker programs), private and public forestry contracting procedures, and the oversight—or lack of oversight—of labor conditions in the woods. For many years, the US Forest Service claimed that it was not responsible for investigating labor conditions among workers hired by contractors.[43] Within two weeks of the *Sacramento Bee*'s publication of "Pineros," however, the then-Chief of the US Forest Service announced changes to the agency's contracting policies; Congressional hearings followed soon after. Among other instituted changes was the passage of federal legislation allowing guest forest workers to be represented by the Legal Service Corporation. Yet these changes did little to improve the overall precarious position of immigrant woods workers.

## Clandestine Cultivation

The remoteness and deep visual cover of many forested landscapes in the US provide opportunities for clandestine cultivation: the planting, care, and harvesting of lucrative illegal crops. Of these, cannabis has become the most common since the 1980s, and concerns abound regarding its consequences for forest ecosystems. Southern Appalachia was a focus for federal eradication of illegal cannabis grow operations in the 1980s and early 1990s; since that time, the center of activity shifted to California and Oregon, where grow sites had become widespread across private, tribal, national forest, and even national park properties by the end of the century.[44] The thousands of illegal "trespass" grows on public lands bring with them worrying environmental impacts: abundant use of both legal and illegal pesticides (including some linked to the mortality of rare and sensitive species), unlicensed withdrawals of water from streams, wildlife poaching, increased wildfire starts, and the degradation of native tree and plant cover.[45]

Public land grow operations represent both sites of environmental degradation and potential violence given the practices in place and the steps taken to defend lucrative but illegal activities in the woods. Residents of areas where cannabis cultivation occurs often report threats and intimidation from grow site operators, and law enforcement agencies have struggled to respond to the scale of these operations. Additionally, the costs of cleaning and rehabilitating illegal grow sites can run from tens of thousands to over a hundred thousand dollars per site, diverting funds from limited agency budgets.[46] The federal government—in partnership with states, counties, and tribes—has prosecuted an aggressive campaign to eradicate illegal grow sites on public land. Some of the major federal enforcement operations focused on California's public lands included sites in Sequoia and Kings Canyon National Parks and the Sierra and Mendocino National Forests.[47]

The scenario for clandestine cultivation is complicated by the legal status of cannabis, which varies widely from state to state while remaining illegal at the federal level as of the date of this writing. This creates a legal grey zone or "quasi-legality" where cultivation in a particular location may be legal at one level of government and illegal at another.[48] A prime example comes from Mendocino County in California, which in 2008 passed ordinance 9.31 allowing for cooperatives to cultivate up to 99 plants for medicinal use under requirements for security measures, environmental compliance, and inspections. The fact that Mendocino County was one of the first jurisdictions to attempt to regulate the legal cultivation of cannabis was no coincidence. Rural northern California had attracted a counterculture population since the 1960s that included a substantial number of recreational cannabis enthusiasts. The region slowly became a center of commercial (not just personal) cannabis cultivation as a result of numerous factors that included the prosecution of the US-led drug war in places like Mexico, rising street prices for cannabis, and the abundance of remote forestland suitable for clandestine grow sites.[49]

California voters passed a ballot measure legalizing medicinal cannabis in 1996, and the law was modified through legislative approval of the Medical Marijuana

Program Act in 2003.[50] Under the latter policy, counties were given authority to regulate the cultivation and licensing of medical cannabis. Statewide legalization for recreational purposes came via a 2016 ballot measure. This framework of quasi-legality for cannabis cultivation and use (in which cannabis was legal at county and state levels but illegal at the federal level) helped to feed a boom in production, particularly in rural northern California counties like Mendocino, where some of the cultivation was taking place in small clearings scattered throughout public and private forestlands. Although Mendocino County attempted to regulate cultivation through its 9.31 program, the federal government ultimately refused to recognize any cannabis cultivation as legal, and the county was forced to revise its approach.[51] The federal crackdown on operators attempting to produce openly and in compliance with state and county law may also have further encouraged the spread of illegal cultivation on tribal and public lands.[52]

Under the quasi-legality regime, cannabis cultivation has created a substantial level of economic activity in places that experienced some of the highest unemployment and poverty rates in the region following the end of the timber wars. One estimate calculated the cannabis industry to represent over a quarter of Humboldt County, California's overall economic activity in 2010.[53] Public lands are often specifically targeted as cultivation sites under the quasi-legality regime because culpability cannot be easily traced to a particular individual. Mirroring patterns in some forestry and NTFP sectors, many undocumented laborers have been hired to perform the manual tasks associated with grow site clearing, cultivation, and harvesting—thereby taking on the immediate legal risk in the case of an enforcement operation (as well as the health and safety risks of working with dangerous chemicals in remote places). For example, Fredy Figueroa-Montes, one of the individuals arrested in the raid on the Wallowa-Whitman National Forest grow site described at the beginning of this chapter, claimed in court that he was recruited with promises of a job conducting forest thinning in Washington and only learned of his real duties once he was transported to the grow site camp deep within Wallowa County. In his testimony, he said that he felt he had no option but to participate because the organizers of the grow site were from a powerful cartel family from his home state of Jalisco, Mexico, and that his own family would be in danger if he fled.[54]

An important question is how the growing trend of state-level cannabis legalization will affect outcomes on private, public, tribal, and community forestlands. Research on the topic to date has found that state-level legalization can reduce clandestine cultivation on public lands within the state, but other factors such as the law enforcement approach and the level of taxation on legal cannabis also have an effect.[55] A study in the Pacific Northwest, for example, found that state legalization was associated with a reduction in the discovery of illegal national forest grow sites in Oregon but not in Washington. The study authors speculate that the relatively higher tax burden on legal cannabis in Washington may continue to encourage illegal cultivation.[56] The quasi-legality of cannabis cultivation raises other complications as well, especially regarding the administration of conservation programs that entail state-federal cooperation. For example, will landowners continue to receive federal tax benefits for conservation easements if they engage in cannabis cultivation in states where production is legal? Will the land trusts overseeing those easements risk legal consequences? How does the quasi-legality of cannabis affect cooperative federalism when a potential land use is legal at the state level but illegal at the federal level?[57]

## Other Unseen Uses

In 2019, a federal judge in Arizona ordered all national forests in New Mexico and one in Arizona to suspend timber management activities until they demonstrated compliance with their obligations to protect the Mexican spotted owl, a species listed as threatened under the Endangered Species Act. The ruling came in response to a lawsuit filed by the environmental organization WildEarth Guardians. In addition to suspending all commercial timber sales, the national forests also stopped issuing firewood cutting permits and notified existing permit holders not to cut any more trees. Some New Mexicans who depend on firewood for heating their homes may have felt a sense of *déjà vu*, as the same restrictions on national forest firewood cutting were imposed as the result of a Mexican spotted owl lawsuit filed by the same organization (under a slightly different name) in 1995. Although the 2019 firewood cutting ban was lifted just a few days later, the injunction shone a light on the importance of firewood in many southwestern households—especially but not exclusively on Native American reservations and in rural areas populated by multigenerational Hispano populations. As with piñon nuts, the harvest and sale of firewood represent a crucial element in the forest and domestic economy of the Southwest, even if much of that activity takes place outside of formal channels.

Another potentially unseen use of forests is unsanctioned or unmanaged recreation, ranging from hiking or mountain biking on unofficial "social trails" to the use of motorized vehicles in places where they can cause damage to resources, property, and competing uses. Recreation (in its various forms) is one of the predominant uses of forests in the US, and while much of it occurs within the parameters set by forest owners, the growing popularity of motorized, electrified, and even airborne recreational technologies has created new opportunities for unsanctioned and unregulated uses (Box 11.2). This unique challenge has been recognized by forest managers and policymakers since at least the 1970s, but solutions remain elusive even as recreational technology continues to evolve.

A distinct trend is the long-term unsanctioned inhabitation of forests, often referred to as "non-recreational camping." Public lands such as national and state forests can be attractive places for unhoused or transient populations to live, at least seasonally. Preliminary research on this phenomenon suggests that non-recreational campers on national forests are a diverse population that includes retirees living in recreational vehicles, asocial "separatists," families or single parents with children, homeless adults, nomadic intentional communities, college students, recreation industry workers, forest workers, fugitives from the law, and other categories of forest residents.[58] In many cases, remaining unseen may be a primary goal for these populations.

**BOX 11.2 SEEN AND UNSEEN RECREATIONAL USES OF FORESTS**

In April 1970, Forest Supervisor Claude Ferguson documented the damage to the Hoosier National Forest caused by the Buffalo 100, an unauthorized off-road motorcycle race organized by an adjacent landowner. Impacts included

> the tearing up of hiking trails, the destruction of fragile forest land not meant for trails, the trampling of young trees, the ripping down of branches, the destruction of wildlife habitat for both endangered and nonendangered species, littering, and excessive noise levels.[59]

What followed was a seven-year ordeal in which Ferguson struggled against his agency supervisors to prevent further expansion of off-highway vehicle (OHV) use within the forest. Ferguson went as far as supporting a lawsuit against his own agency filed by Indiana conservation organizations alleging violations of Forest Service requirements for trail-building on steep slopes. Despite being reassigned—and later fired—in retaliation for his efforts, Ferguson's efforts eventually paid off with a prohibition of OHV use on the Hoosier National Forest. Indiana in the early 1970s was far from an outlier in having to confront these issues. A growing concern with OHV use on public lands led to executive orders in 1972 by President Nixon and 1977 by President Carter directing federal land management agencies to minimize both OHV impacts and conflicts with other uses. The persistence of the problem three decades after the Nixon and Carter executive orders prompted the Forest Service to issue a rule in 2005 requiring all forests to identify which roads and trails would be open for OHV use and to close all others by default. Implementation of the Travel Planning Rule sparked the community outcry on the Wallowa-Whitman National Forest described at the beginning of this chapter. These kinds of concerns are not limited to OHVs; the growing popularity of mountain bikes starting in the 1980s led to conflicts with hikers and equestrians and raised questions of resource impacts among many park and forest managers.[60] More recently, park and forest managers have struggled to manage a new generation of recreational technology that includes devices like electric bicycles (e-bikes) and drones. Meanwhile, geolocated pictures of scenic locations shared on social media have driven crowds to places lacking appropriate infrastructure; the litter, resource damage, and human safety issues that are not widely shared on social media are some of the problems associated with a sudden interest in being "seen" in certain photogenic settings.[61]

## Conclusion

Unseen forest uses and unseen forest users are, ironically, ubiquitous. Nearly all public forests (and many private forests) show evidence of unofficial and unsanctioned recreation, and other unsanctioned activities such as long-term nonrecreational camping and even illicit crop cultivation are widespread. One study estimated that over 17% of the population in four New England states had engaged in some kind of NTFP harvesting in the previous year, making the activity more widely practiced than more visible outdoor activities such as hunting, backpacking, and horseback riding.[62] A study near Atlanta's Urban Food Forest at Browns Mill found that 20% of the neighborhood's largely African-American residents had harvested some kind of edible plant or mushroom product within the past five years.[63] Despite often being imagined as places of pure nature, public and private forests nationwide are sites of human labor, entailing practices such as harvesting foods and medicines, collecting materials for crafts both lasting and ephemeral, and planting, harvesting, thinning, and maintaining forests. In many cases, this labor is conducted by populations that remain on the peripheries of society and outside the official gaze. Understanding the governance of these activities requires looking beyond the design and intent of formal laws to issues such as the unintended consequences of the "quasi-legality" of cannabis, the blind spots in public land contracting protocols, and the cultural conservation practices enacted by individuals even when no one else is looking.

## Questions for Discussion

1. If the uses and users described in this chapter are so ubiquitous, why do you think they remain "unseen" by most forest managers, forest users, and researchers?
2. What factors motivate the activities of urban recreational NTFP harvesters? Are these motivations different from those of rural NTFP harvesters, both recreational and commercial?
3. How are unseen users' informal claims to forest access established and defended? What role do formal rules (such as laws and regulations) play in resolving conflicts regarding access to forests for activities like NTFP gathering, firewood cutting, motorized recreation, and long-term non-recreational camping?
4. What aspects of forest work make it especially dangerous? How might labor and immigration laws exacerbate the danger for some workers?
5. What are some of the factors that led northern California to become the nucleus for "trespass" cannabis grow operations? How did state legalization of cannabis change these practices, if at all?
6. What could the leadership of the Wallowa-Whitman National Forest have done differently to reduce conflict and increase community consensus in their travel management planning?

## Notes

1 This event took place prior to the statewide legalization of recreational cannabis in Oregon.
2 Fairfax et al., 1999, p. 637.
3 Hembram & Hoover, 2008.
4 Emery, 2001.
5 Frey et al., 2019.
6 Hansis, 1998; Tsing, 2015.
7 Hansis et al., 2001.
8 Pilz et al., 2007.
9 Frey et al., 2021.
10 Pilz et al., 2007.
11 Tsing, 2015, pp. 57–58.
12 McLain, 2008; Tsing, 2015.
13 McLain, 2008.
14 Redmond et al., 2012.
15 Charnley, McLain, et al., 2018; Hurley et al., 2015; McLain et al., 2014.
16 Arora, 2008; Hurley et al., 2015.
17 Poe et al., 2013.
18 Hurley et al., 2015; Linnekin, 2018.
19 McLain et al., 2014.
20 Arora, 2008, p. 370.
21 Gaither et al., 2020.
22 Chamberlain et al., 2018.
23 McLain & Lynch, 2010; Watson et al., 2018.
24 Ballard & Huntsinger, 2006; Emery, 2001.
25 Ballard & Huntsinger, 2006; Emery, 2001; Emery & Ginger, 2014; Watson et al., 2018.
26 Lake et al., 2018; Lynn et al., 2013.
27 Lynn et al., 2013, p. 549.
28 Emery & Pierce, 2005.
29 Linnekin, 2018.
30 Schrack, 2018.
31 Linnekin, 2018.
32 Pilz & Molina, 2002.
33 McLain & Jones, 2001.
34 Ballard & Huntsinger, 2006; Love & Jones, 2008.
35 Arora, 2008; McLain & Jones, 2001.
36 Hurley et al., 2015, p. 203.
37 Hurley & Halfacre, 2011.
38 Grabbatin et al., 2011; Hurley et al., 2013.
39 McDaniel & Casanova, 2005; Moseley, 2006; Sarathy, 2012.
40 McDaniel & Casanova, 2005; Sarathy, 2012.
41 Charnley, Kline, et al., 2018, p. 824.
42 Sarathy, 2012, p. 68.
43 Sarathy, 2008.
44 Polson, 2018; Spencer & Miller, 2018.
45 Bauer et al., 2015; Klassen & Anthony, 2019; Prestemon et al., 2019; Wiens et al., 2019.
46 Klassen & Anthony, 2019.
47 Polson, 2018.
48 Short Gianotti et al., 2017.
49 Corva, 2014; Irwin, 2018; Silvaggio, 2018.
50 Silvaggio, 2018.

51 Polson, 2018.
52 Silvaggio, 2018.
53 Meisel, 2017.
54 Irwin, 2018.
55 Prestemon et al., 2019.
56 Klassen & Anthony, 2019.
57 Owley, 2018.
58 Baur & Cerveny, 2019; Cerveny & Baur, 2020.
59 O'Leary, 2009, p. 1068.
60 Chavez, 1996; Lucas, 2020; Rothman, 2001.
61 Presley, 2020.
62 Robbins et al., 2008.
63 Gaither et al., 2020.

## References

Arora, D. (2008). California porcini: Three new taxa, observations on their harvest, and the tragedy of no commons. *Economic Botany*, *62*(3), 356–375.

Ballard, H. L., & Huntsinger, L. (2006). Salal harvester local ecological knowledge, harvest practices and understory management on the Olympic Peninsula, Washington. *Human Ecology*, *34*(4), 529–547.

Bauer, S., Olson, J., Cockrill, A., Hattem, M. van, Miller, L., Tauzer, M., & Leppig, G. (2015). Impacts of surface water diversions for marijuana cultivation on aquatic habitat in four northwestern California watersheds. *PLOS ONE*, *10*(3), e0120016.

Baur, J. W. R., & Cerveny, L. (2019). Social impacts of homelessness and long-term occupancy on national forests and grasslands: A national study of US Forest Service law enforcement officers. *Landscape and Urban Planning*, *184*, 69–77.

Cerveny, L. K., & Baur, J. W. R. (2020). Homelessness and nonrecreational camping on national forests and grasslands in the United States: Law enforcement perspectives and regional trends. *Journal of Forestry*, *118*(2), 139–153.

Chamberlain, M. J., Emery, M. R., & Patel-Weynand, T. (2018). *Assessment of nontimber forest products in the United States under changing conditions* (SRS-GTR-232; p. SRS-GTR-232). US Department of Agriculture, Forest Service, Southern Research Station.

Charnley, S., Kline, J., White, E. M., Abrams, J., McLain, R. J., Moseley, C., & Huber-Stearns, H. (2018a). Community well-being and forest management in the Northwest Forest Plan area. In *Synthesis of science to inform land management within the Northwest Forest Plan area* (Vol. 3, pp. 625–713). US Department of Agriculture, Forest Service, Pacific Northwest Research Station.

Charnley, S., McLain, R. J., & Poe, M. R. (2018b). Natural resource access rights and wrongs: Nontimber forest products gathering in urban environments. *Society & Natural Resources*, *31*(6), 734–750.

Chavez, D. J. (1996). *Mountain biking: Issues and actions for USDA Forest Service managers* (Research Paper PSW-RP-226). US Department of Agriculture, Forest Service, Pacific Southwest Research Station.

Corva, D. (2014). Requiem for a CAMP: The life and death of a domestic US drug war institution. *International Journal of Drug Policy*, *25*(1), 71–80.

Emery, M. R. (2001). Non-timber forest products and livelihoods in Michigan's Upper Peninsula. In I. Davidson-Hunt, L. C. Duchesne, & J. C. Zasada (Eds.), *Forest communities in the third millennium: Linking research, business, and policy toward a sustainable non-timber forest product sector* (pp. 23–30). USDA Forest Service North Central Research Station.

Emery, M. R., & Ginger, C. (2014). *Special forest products on the Green Mountain and Finger Lakes National Forests: A research-based approach to management*. USDA Forest Service Northern Research Station.

Emery, M. R., & Pierce, A. R. (2005). Interrupting the telos: Locating subsistence in contemporary US forests. *Environment and Planning A*, *37*(6), 981–993.

Fairfax, S. K., Fortmann, L. P., Hawkins, A., & Huntsinger, L. (1999). The federal forests are not what they seem: Formal and informal claims to federal lands. *Ecology Law Quarterly*, *25*, 630–646.

Frey, G. E., Alexander, S. J., Chamberlain, J. L., Blatner, K. A., Coffin, A. W., & Barlow, R. J. (2019). Markets and market values of nontimber forest products in the United States: A review, synthesis, and identification of future research needs. *Journal of Forestry*, *117*(6), 613–631.

Frey, G. E., Chamberlain, J. L., & Jacobson, M. G. (2021). Producers, production, marketing, and sales of non-timber forest products in the United States: A review and synthesis. *Agroforestry Systems*. https://doi.org/10.1007/s10457-021-00637-3.

Gaither, C. J., Aragón, A., Madden, M., Alford, S., Wynn, A., & Emery, M. (2020). "Black folks do forage": Examining wild food gathering in Southeast Atlanta communities. *Urban Forestry & Urban Greening*, 56, 126860.

Grabbatin, B., Hurley, P. T., & Halfacre, A. (2011). "I still have the old tradition": The co-production of sweetgrass basketry and coastal development. *Geoforum*, *42*(6), 638–649.

Hansis, R. (1998). A political ecology of picking: Non-timber forest products in the Pacific Northwest. *Human Ecology*, *26*(1), 67–86.

Hansis, R., Jones, E. T., & McLain, R. J. (2001). Recent trends: Non-timber forest product pickers in the Pacific Northwest. *Journal of Sustainable Forestry*, *13*(3–4), 141–146.

Hembram, D., & Hoover, W. L. (2008). Nontimber forest products in Daniel Boone National Forest region–economic significance and potential for sustainability. In D. F. Jacobs & C. Michler (Eds.), Proceedings, 16th Central Hardwood Forest Conference: Vol. Gen. Tech. Rep. NRS-P-24. (pp. 148–156). US Department of Agriculture, Forest Service, Northern Research Station, West Lafayette, IN.

Hurley, P. T., Emery, M. R., McLain, R., Poe, M., Grabbatin, B., & Goetcheus, C. L. (2015). Whose urban forest? The political ecology of foraging urban nontimber forest products. In C. Isenhour, G. McDonogh, & M. Checker (Eds.), *Sustainability in the global city: Myth and practice* (pp. 187–212). Cambridge University Press.

Hurley, P. T., Grabbatin, B., Goetcheus, C., & Halfacre, A. (2013). Gathering, buying, and growing sweetgrass (Muhlenbergia sericea): Urbanization and social networking in the sweetgrass basket-making industry of lowcountry South Carolina. In R. Voeks & J. Rashford (Eds.), *African ethnobotany in the Americas* (pp. 153–173). Springer.

Hurley, P. T., & Halfacre, A. C. (2011). Dodging alligators, rattlesnakes, and backyard docks: A political ecology of sweetgrass basket-making and conservation in the South Carolina Lowcountry, USA. *GeoJournal*, *76*(4), 383–399.

Irwin, A. (2018). Double bind: The intractability of undocumented immigrant trespass marijuana grow operations in US national forests. In C. Miller (Ed.), *Where there's smoke: The environmental science, public policy, and politics of marijuana* (pp. 109–126). University Press of Kansas.

Klassen, M., & Anthony, B. P. (2019). The effects of recreational cannabis legalization on forest management and conservation efforts in US national forests in the Pacific Northwest. *Ecological Economics*, *162*, 39–48.

Lake, F. K., Emery, M. R., Baumflek, M. J., Friday, K. S., Kamelamela, K., Kruger, L., Grewe, N., Gilbert, J., & Reo, N. J. (2018). Cultural dimensions of nontimber products. In J. L. Chamberlain, M. R. Emery, & T. Patel-Weynand (Eds.), *Assessment of nontimber forest products in the United States under changing conditions* (Vol. 2018, pp. 84–99). USDA Forest Service Southern Research Station.

Linnekin, B. J. (2018). Food law gone wild: The law of foraging. *Fordham Urban Law Journal*, *45*(4), 995–1050.

Love, T., & Jones, E. T. (2008). Why Is non-timber forest product harvesting an "issue"? *Journal of Sustainable Forestry*, *13*(3–4), 105–121.

Lucas, E. (2020). A review of trail-related fragmentation, unauthorized trails, and other aspects of recreation ecology in protected areas. *California Fish and Wildlife Journal, Recreational Special Issue*, 95–125.

Lynn, K., Daigle, J., Hoffman, J., Lake, F., Michelle, N., Ranco, D., Viles, C., Voggesser, G., & Williams, P. (2013). The impacts of climate change on tribal traditional foods. *Climatic Change*, *120*(3), 545–556.

McDaniel, J., & Casanova, V. (2005). Forest management and the H2B guest worker program in the southeastern United States: An assessment of contractors and their crews. *Journal of Forestry*, *103*(3), 114–119.

McLain, R. J. (2008). Constructing a wild mushroom panopticon: The extension of nation-state control over the forest understory in Oregon, USA. *Economic Botany*, *62*(3), 343–355.

McLain, R. J., Hurley, P. T., Emery, M. R., & Poe, M. R. (2014). Gathering "wild" food in the city: Rethinking the role of foraging in urban ecosystem planning and management. *Local Environment*, *19*(2), 220–240.

McLain, R. J., & Jones, E. T. (2001). Expanding non-timber forest product harvester/buyer participation in Pacific Northwest forest policy. *Journal of Sustainable Forestry*, *13*(3–4), 147–161.

McLain, R. J., & Lynch, K. (2010). Managing floral greens in a globalized economy: Resource tenure, labour relations and immigration policy in the Pacific Northwest, USA. In S. A. Laird, R. J. McLain, & R. P. Wynberg (Eds.), *Wild product governance: Finding policies that work for non-timber forest products* (pp. 265–286). Earthscan.

Meisel, J. (2017). Hidden in plain sight: Cannabis cultivation in the Emerald Triangle. *California Geographer*, 56, 3–26.

Moseley, C. (2006). Ethnic differences in job quality among contract forest workers on six national forests. *Policy Sciences*, 39(2), 113–133.

O'Leary, R. (2009). When a career public servant sues the agency he loves: Claude Ferguson, the Forest Service, and off-road vehicles in the Hoosier National Forest. *Public Administration Review*, 69(6), 1068–1076.

Owley, J. (2018). Unforeseen land uses: The effect of marijuana legalization on land conservation programs. *U.C. Davis Law Review*, *51*, 1673–1716.

Pilz, D., McLain, R., Alexander, S., Villarreal-Ruiz, L., Berch, S., Wurtz, T. L., Parks, C. G., McFarlane, E., Baker, B., Molina, R., & Smith, J. E. (2007). *Ecology and management of morels harvested from the forests of western North America*. (PNW-GTR-710; p. PNW-GTR-710). US Department of Agriculture, Forest Service, Pacific Northwest Research Station.

Pilz, D., & Molina, R. (2002). Commercial harvests of edible mushrooms from the forests of the Pacific Northwest United States: Issues, management, and monitoring for sustainability. *Forest Ecology and Management*, *155*(1), 3–16.

Poe, M. R., McLain, R. J., Emery, M., & Hurley, P. T. (2013). Urban forest justice and the rights to wild foods, medicines, and materials in the city. *Human Ecology*, *41*(3), 409–422.

Polson, M. (2018). Making marijuana an environmental issue: Prohibition, pollution, and policy. *Environment and Planning E: Nature and Space*, *2*(2), 229–251.

Presley, M. (2020). *You belong among the wildflowers… unless you're an Instagrammer: The political economy of the Instagram Effect in the news* [PhD Thesis]. The Florida State University.

Prestemon, J. P., Koch, F. H., Donovan, G. H., & Lihou, M. T. (2019). Cannabis legalization by states reduces illegal growing on US national forests. *Ecological Economics*, *164*, 106366.

Redmond, M. D., Forcella, F., & Barger, N. N. (2012). Declines in pinyon pine cone production associated with regional warming. *Ecosphere*, *3*(12), 1–14.

Robbins, P., Emery, M., & Rice, J. L. (2008). Gathering in Thoreau's backyard: Nontimber forest product harvesting as practice. *Area*, *40*(2), 265–277.

Rothman, H. K. (2001). The war for the future: Mountain bikes and Golden Gate National Recreation Area. *The George Wright Forum*, *18*(1), 34–47.

Sarathy, B. (2008). The marginalization of Pineros in the Pacific Northwest. *Society & Natural Resources*, *21*(8), 671–686.

Sarathy, B. (2012). *Pineros: Latino labour and the changing face of forestry in the Pacific Northwest*. University of British Columbia Press.

Schrack, A. (2018). The shifting landscape of ancestral lands: Tribal gathering of traditional plants in national parks. *Arizona Journal of Environmental Law and Policy*, 9(1), 1–24.

Short Gianotti, A. G., Harrower, J., Baird, G., & Sepaniak, S. (2017). The quasi-legal challenge: Assessing and governing the environmental impacts of cannabis cultivation in the North Coastal Basin of California. *Land Use Policy*, *61*, 126–134.

Silvaggio, A. (2018). Cannabis agriculture in California: The environmental consequences of prohibition. In C. Miller (Ed.), *Where there's smoke: The environmental science, public policy, and politics of marijuana* (pp. 13–28). University Press of Kansas.

Spencer, H., & Miller, C. (2018). Southern exposure: Marijuana and labor in the Appalachians. In C. Miller (Ed.), *Where there's smoke: The environmental science, public policy, and politics of marijuana* (pp. 127–135). University Press of Kansas.

Tsing, A. L. (2015). *The mushroom at the end of the world: On the possibility of life in capitalist ruins*. Princeton University Press.

Watson, K., Christian, C. S., Emery, M. R., Hurley, P. T., McLain, R. J., & Wilmsen, C. (2018). Social dimensions of nontimber forest products. In J. L. Chamberlain, M. R. Emery, & T. Patel-Weynand (Eds.), *Assessment of nontimber forest products in the United States under changing conditions* (Vol. 2018, pp. 102–117). USDA Forest Service Southern Research Station.

Wiens, J. D., Dilione, K. E., Eagles-Smith, C. A., Herring, G., Lesmeister, D. B., Gabriel, M. W., Wengert, G. M., & Simon, D. C. (2019). Anticoagulant rodenticides in Strix owls indicate widespread exposure in west coast forests. *Biological Conservation*, *238*, 108238.

# 12 Policy Trends, Controversies, and Proposals for Change

The basic ownership patterns of Wallowa County have remained mostly unchanged since the late nineteenth and early twentieth centuries when the Wallowa Band of the Nez Perce was expelled by force; the Wallowa, Imnaha, Wenaha, and Chesnimnus Forest Reserves were carved out from the public domain; and the remainder of the landscape "disposed of" to timber companies and Euro-American settlers. There have been some important changes since that time, including the repatriation of lands under title to the Nez Perce Tribe, the designation of Hells Canyon National Recreation Area, land purchases by conservation NGOs such as The Nature Conservancy, the creation of state parks, and the establishment of the East Moraine Community Forest, but the basic patterns of land tenure have remained mostly steady. This is not to say that alternatives have not been raised, however. Believing that the US Forest Service was degrading the ecological quality of area landscapes, a regional conservation organization pushed (unsuccessfully) for many years to transfer much of the county's national forestland to the National Park Service for the creation of a Chief Joseph National Park and Preserve. Many local landowners and forest users, in contrast, have criticized the Forest Service for not managing aggressively enough, especially following the 1994 timber harvesting injunction. A familiar refrain is that the boundary between federal and private forests is easily identified because the federal side of the fence is overcrowded with dead and sickly trees whereas the adjacent private side of the fence contains healthy, vigorous stands. Correctly or incorrectly,[1] federal forestlands are often perceived by landowners as sources of fire, insect, and disease outbreaks that then spread onto adjacent nonfederal-lands. Many residents of Wallowa County also point to their shuttered sawmills as a sign that the federal agencies have abandoned their obligations to support local economic development. Some Wallowa County residents and leaders have called for the federal government to cede control over the national forests in the interest of local economic benefits. Some suggested the forests be privatized; others have advocated for the county government to take ownership.

Federal lands and federal policies have been at the center of many of Wallowa County's most divisive conflicts in recent decades, from timber management on national forests to sheep grazing within Hells Canyon to federal protections for gray wolves. However, other tenure categories have also drawn critical scrutiny.

DOI: 10.4324/9781003043669-12

The Nature Conservancy was criticized by some Wallowa County residents for its purchase of large expanses of rangeland and forestland. Although the conservation NGO continues to run cattle on the preserve—and to pay property taxes despite its tax-exempt status—some residents claimed that the purchase contributed to rising local land prices that put property out of reach for local families. Along similar lines, there has been concern regarding the purchase of productive land by private buyers from outside of Wallowa County who were attracted by the county's scenery, isolation, and low land prices. According to some multigenerational residents, these newer, "amenity"-oriented landowners drive up the price of land and fail to manage in ways considered acceptable to long-time locals.[2] However, erecting any kind of policy barriers to the free buying and selling of land would conflict sharply with the county's prevailing embrace of private property rights.[3]

At a national scale, various proposals have been put forward in recent decades to change, sometimes in radical ways, the governance of the country's forests. Many of these have come from organizations opposed to the management approaches adopted by federal multiple-use agencies. Some campaigns have called for more local control of federal forests, claiming that federal managers are out of touch with local needs and constrained to the point of paralysis by layers of laws and regulations. Others have called for an end to commercial activities on federal lands altogether. As described in Chapter 6, some state forests have likewise been caught between counterposing campaigns for more timber production and more nature conservation. Private forests have not entirely escaped these debates; even as some comprehensive conservation policies have been undermined in places like Oregon (see Chapter 6), there have been calls for greater regulatory scrutiny on forest practices in states where a hands-off approach has historically prevailed. Even as the most radical proposals have failed to gain traction, less obvious dynamics have led to subtle but important changes in the governance of forests across tenure categories. In this chapter, we take a look at these contemporary debates and controversies and examine what they might mean for forest policy and governance trends going forward.

## Federal Policies and Forestlands

The multiple-use federal agencies (the US Forest Service and Bureau of Land Management) have long been targeted for reform by activist campaigns representing all points along the political spectrum. This is not surprising, given the complex and diverse missions of these agencies and the relatively limited guidance for site-specific decision-making provided by laws like the Multiple Use-Sustained Yield Act, National Forest Management Act, and Federal Land Policy and Management Act. Most federal forest managers are trained as scientists, yet in their daily work they are tasked with making what are essentially political decisions when they prioritize one use (or set of uses) over another on a particular piece of land. The presence of conflict over federal land management itself is nothing new; as described in Chapter 2, a variety of western interests were opposed to the expansion of forest reserves under Theodore Roosevelt and ultimately forced the president to rescind his powers to declare new reserves. Even as the first forest reserves were established, conservation leaders like John Muir advocated for them to be preserved like national parks. The first campaigns attempting to transfer forest reserves and national forests to state governments began shortly after these federal lands were designated.

### *Regulatory Reform*

With the passage of environmental legislation in the 1970s came new environmental and procedural obligations for federal land management agencies along with an expansion of opportunities for non-agency publics to comment, file administrative appeals and objections, and bring lawsuits to force compliance. A major topic of policy debate since that time is whether these requirements go "too far": whether they provide too much latitude for non-agency publics to contest proposed activities, create unreasonable inefficiencies in planning and management, and complicate or prevent urgently needed treatments to address hazardous or unhealthy forest conditions. The debate, at its core, is about the tension between accountability and flexibility. A highly accountable agency—one that finds its decisions under constant scrutiny and challenge—is one that quickly loses the flexibility to make decisions in line with managers' judgments. On the other hand, an agency with maximal flexibility and minimal accountability can easily drift into making decisions that fail to reflect the interests and values of diverse publics. This was precisely the critique leveled against the US Forest Service during the post-World War II era when intensive timber production objectives dominated management of the national forests (see Chapter 3).

The balance between federal agency accountability and flexibility moved sharply in favor of accountability as a result of the judicial decisions and executive actions that ultimately resolved the timber wars of the 1990s. However, calls for more flexibility grew louder in the face of the growing wildfire dilemma. For example, the US Forest Service itself published an analysis titled "The Process Predicament" in 2002, identifying excessive analysis, ineffective public involvement, and management inefficiencies as contributors to an inability to conduct needed management of overstocked and fire-prone forests. Passage of the Healthy Forests Restoration Act in 2003 (see Chapter 5) was significant in part because of the policy narrative it endorsed: specifically, that the requirements of NEPA, the Endangered Species Act, and other environmental laws were interfering with actions needed to reduce wildfire and insect risks to people and forests.[4] Various legislative and regulatory changes since that time have likewise taken aim at the numerous checks and balances built into federal land governance by amending procedures related to NEPA and administrative appeals and objections. The debate continues, however, with timber industry and professional forestry advocates often calling for more flexibility while environmental preservation advocates call for more accountability.

Whether flexibility or accountability should prevail in the case of post-fire timber harvest, or "salvage" as it is known, is an especially controversial topic

(Box 12.1). Environmental advocates often argue that the use of heavy equipment in recently burned landscapes threatens to exacerbate erosion while removing ecologically valuable snags. At the same time, the fact that trees start to decay after they die means that the longer the time lapse between fire and harvest, the lower the quality (and economic value) of the remaining timber. This has led some to argue for expedited processes that allow timber harvest to move forward immediately after a fire; advocates of this approach argue that moving quickly to conduct salvage operations would recover the economic value in the remaining trees while removing hazardous dead fuel from the landscape.[5] The policy response to date has been to allow for flexibility on a small scale in the post-fire environment while retaining accountability across the larger landscape. The US Forest Service can use a categorical exclusion from full NEPA analysis to remove burned trees from along roadways that could present hazards or block roads. In 2020, the Bureau of Land Management expanded to 3,000 acres the size of salvage operations that it can conduct under a categorical exclusion. The use of a categorical exclusion requires a nominal public notice process and foregoes environmental analysis and opportunities for administrative appeals or objections, allowing management to move forward quickly. Bills have been introduced into Congress that would expedite or eliminate some environmental analysis and public involvement processes to allow salvage to occur more rapidly following a fire, but none have been passed into law as of the date of this writing.

As the salvage debate illustrates, questions of regulatory reform are inseparable from questions of whether and how federal forests should be managed—a debate going back at least as far as Gifford Pinchot's and John Muir's divergent views on the proper use of federal forests. What is new is the recognition of the complexities introduced by climate change and decades of fire exclusion. Many of the changes to federal forest policies since the 1990s have been framed as necessary to address fire and restoration concerns; place-based collaborative processes have often found common ground by focusing on these same issues.[6] Yet not everyone agrees that these actions are necessary or prudent. For example, in 2022 the singer Carole King testified before Congress on issues related to wildfires and forest management in the West. In her testimony, she argued that "Wildfires ARE getting worse, not because we have too many trees, but because of climate-driven weather events accelerated by removing trees under the guise of 'thinning' or 'fuel reduction.'"[7] She concurred with some in the environmental advocacy community who have called for forest treatments to focus on the immediate home environment rather than on fuels and stand density management across the wider landscape.

### BOX 12.1 THE BISCUIT FIRE AND SALVAGE HARVESTING

The 2002 Biscuit Fire burned nearly half a million acres of rugged forested terrain in southern Oregon and northern California, with a designated wilderness area and inventoried roadless areas constituting most of the burned area. A salvage harvesting plan was proposed by the Rogue River-Siskiyou National Forest, justified in part as a means to help reduce fuels that could contribute to future wildfires and in part as a revenue source to fund reforestation efforts. This plan was hotly contested as it involved logging in areas previously protected under the 2001 Roadless Area Conservation Rule and in areas designated as Late Successional Reserves under the 1994 Northwest Forest Plan. The roadless area element of the plan was particularly complex because a series of federal court decisions in 2002 and 2003 provided conflicting interpretations of the scope and validity of the Roadless Rule. To plan the salvage sale, forest staff conducted an environmental impact statement, the most detailed of the various analysis pathways under NEPA. Upon completion of the NEPA process, the Regional Forester used emergency authorities to allow harvesting to begin immediately rather than providing the standard 105-day appeal process in which outside publics can challenge decisions or elements of the analysis. No fewer than seventeen separate lawsuits were filed to contest various elements of the plan, and some of these resulted in temporary injunctions on harvest. As salvage activities proceeded, two Northwest congressmen introduced a bill designed to expedite post-fire salvage activities in response to what they saw as an unreasonable delay in the NEPA process associated with the Biscuit Fire. Around this same time, university researchers published findings that undermined some of the assumptions of both the Biscuit salvage plan and the proposed legislation. One study found that salvage logging within the Biscuit area negatively affected natural regeneration and left more burnable fuel on the ground compared to unlogged areas.[8] Federal funding was subsequently withdrawn from the lead author (a graduate student at the time), and his study was criticized by the co-sponsors of the bill to expedite salvage harvests. Another study showed that salvaged areas from a previous fire in southern Oregon burned hotter than neighboring non-salvaged areas when the Biscuit fire moved through.[9] These events demonstrate the complex intersections between science, policy, and politics in the case of post-fire salvage harvesting.

### *Divestment*

Critics of both procedural and substantive policies guiding federal land management have led a decades-long effort to transfer formal authority over the federal forest estate to state, county, or private decision-makers. The federal land divestment campaign has found support from a variety of communities and causes: rural residents that feel left behind as federal land management has shifted away from commodity production and toward ecosystem restoration; elected officials and candidates running on anti-federal government platforms; libertarian think tanks and funders. Within the last half century, movements to transfer federal multiple-use lands to state or local control have been most active during the administrations of conservation-minded Democrats, only to die down again under the next Republican presidency. The "Sagebrush Rebellion" that flared up under President Jimmy Carter was a reaction to the raft of new environmental laws of the 1970s, particularly those that appeared to portend stronger federal control over land and water in the West.[10] During this time, several states passed legislation calling for the transfer of federal land to the states. However, these bills had no authority to effect the change they sought, and the rebellion largely dissipated once Ronald Reagan took office. The Wise Use movement that arose during the presidency of Bill Clinton supported a similar agenda (see Chapter 3). This movement went dormant under the George W. Bush years, only to emerge again during the Barack Obama administration as a series of coordinated efforts to transfer federal lands to states and counties. The latter campaign saw some of its greatest political support in Idaho and Utah, two states with especially large proportions of public land. Despite spending substantial amounts of money on studies, lawyers, and lobbying organizations, these and other states arrived at the end of the Obama years without any change in the federal footprint within their boundaries.[11]

Proponents of federal land transfers to states, counties, and private owners point to the large percentages of western states that are effectively controlled by the federal government: 62% of Idaho and 63% of Utah, for example. In Nevada, the state with the highest proportion of federal ownership, the federal government owns fully 80% of land within state boundaries. Those advocating for federal forest divestment often echo the refrain that the federal government has locked up valuable resources and prevented their use for local economic benefit. As Chapter 6 describes, state and county forestlands typically produce more timber per acre and operate under simpler legal mandates than do federal lands; transfer proponents point to these patterns as evidence that nonfederal entities would be more efficient and effective managers of lands currently under federal control. Those who argue for maintaining federal land ownership, conversely, say that what appears to be "inefficiency" simply reflects the more complex mandate of federal multiple-use agencies and the additional obligations they face to examine the possible impacts of their actions and involve the public before moving forward with management. They argue that timber production is only one of several uses for these lands and, therefore, that commodity output metrics fail to capture the spectrum of values provided by federal management. They also point out that the escalating costs associated with wildfire prevention, suppression, and response threaten to outweigh any additional timber or other commodity revenues realized by states and counties under a transfer scenario.[12]

Some have suggested a stronger tribal role in federal land governance as an alternative to federal land divestment to state, county, or private ownership. Ojibwe author David Treuer argued in *The Atlantic* magazine, for example, that transferring national parks to Native tribal control would simultaneously meet restorative justice and land stewardship objectives while also opening possibilities for creative natural resource stewardship approaches grounded in Indigenous knowledge.[13] Chickasaw Nation enrolled member and Dean of the University of Iowa College of Law Kevin Washburn observed that many tribes have already built substantial capacity through the administration of federal programs as "self-governance tribes."[14] Greatly expanding co-management arrangements, he argues, would not necessarily require a new policy framework given the existence of laws like the Tribal Self-Governance Act, Tribal Forest Protection Act, and Good Neighbor Authority.

### *Collaborative and Networked Governance*

Although states, counties, and private interests have been unsuccessful in acquiring formal title to federal forestlands, their influence over federal lands has steadily expanded in other ways as governance approaches have evolved. The US Forest Service still maintains formal authority over decision-making for the lands it manages, but it increasingly depends upon other entities—especially collaborative and community-based organizations as well as NGOs, states, and local governments—to help forge consensus on programs of action and to help fund management activities.[15] This shift is a product of the various changes the agency has experienced during the past several decades: the decreasing deference to federal foresters as a result of controversies surrounding proper forest management; legal and policy changes that have created numerous options for non-agency publics to contest agency decisions; the diversification of social values both interior and exterior to the agency; and the growing financial burden of wildfire suppression and wildfire risk reduction.[16] In this context, community-based collaborative efforts have helped to build legitimacy for programs of work within an identified "zone of agreement," community-based organizations and larger NGOs have provided capacity for planning and implementation, tribes have expanded their role through the various mechanisms detailed in Chapter 7, and states and counties have contributed financial and non-financial resources.

This evolution in national forest governance opens opportunities for various outside entities to play a larger role in setting management direction on public lands, but state governments are especially well-poised to benefit. In part, this is because states tend to have financial resources that few other entities can muster, and in part because they have a unique status in some recent federal policy changes. For example, the George W. Bush administration allowed states to petition for changes to roadless area designations under the Roadless Area Conservation Rule announced in the waning days of the Clinton administration. The Good Neighbor Authority, described in Chapter 5, allows states, counties, and federally recognized tribes to conduct management on federal lands under agreement with the relevant federal agency (the US Forest Service or Bureau of Land Management). The 2014 Farm Bill created an option for governors to request a declaration of national forest areas experiencing insect or disease epidemics, and the agency is required to respond with a designation (Chapter 5).

Although this authority did not change any laws or provide any additional resources, it represents one example of a larger pattern: the slowly growing formal role of states in governing federal forests. States were identified as especially important partners in the 2018 Shared Stewardship initiative (see Chapter 5), and many states had already established programs of funding for federal forest-related work prior to the initiative's launch. In at least one state—Utah—the Shared Stewardship initiative was seen by key stakeholders as an opportunity to assert more state control over federal forestlands and possibly defuse some of the energy that had been funneled into federal land divestment efforts.[17]

At the date of this writing, the opportunities for states to directly influence federal land management are modest at best, but there has been a clear growing trend toward greater integration of state resources and authorities with federal forestlands. This trend raises questions regarding where the trajectory is headed. As states contribute more financial and human capacity resources to federal land management, will they take on a more active role in setting management direction as well? Will they also be asked to pick up greater proportions of federal firefighting budgets? How would a growing state role affect the ability of other entities to participate in federal land governance—especially those lacking the financial resources at states' disposal?

### *Resilience and Adaptation*

Even as federal land managers increasingly operate in partnership and collaboration with various non-agency governments, organizations, and publics, they are being tasked with addressing highly complex social and ecological issues on and near federal lands. This points toward another important long-term trend, one that arguably began with the first forest plans created after the passage of the National Forest Management Act. Since then, a model focused on producing outputs of measurable commodities (like timber, forage, and recreation user-days) has slowly been eclipsed by a model focused on reducing the risks of high-intensity wildfire and other disturbances under a changing climate and altered forest conditions—and doing so while recovering threatened and endangered species, protecting water quality, providing for outdoor recreation, and providing economic opportunities to local communities. While the basic framework for federal forest management has remained relatively unchanged since the 1970s, a series of new concepts have emerged since that time to provide guidance for prioritizing and integrating activities: "ecosystem management" under the Clinton administration in the 1990s, "healthy forests" under the George W. Bush administration in the 2000s, and "ecological integrity" and "resilience" starting with the Obama administration.

The concepts of ecological integrity and resilience were both included in the 2012 revision to the administrative regulations implementing the National Forest Management Act. This revision represented the first successful update of the Act's administrative rules since 1982; whereas earlier regulations framed forest plans largely around timber harvest planning, the updated rules emphasize restoration and sustainability in the face of climate change. Ecological integrity is defined in the 2012 regulations as "the quality or condition of an ecosystem when its dominant ecological characteristics occur within the natural range of variation and can withstand and recover from most perturbations imposed by natural environmental dynamics or human influence." The concept of resilience to climate change is included in the regulations and in various other forms of policy guidance for the US Forest Service, including the 2008 Strategic Framework for Responding to Climate Change, the National Roadmap for Responding to Climate Change, and Strategic Plan.[18]

As examples from throughout this text illustrate, climate change is already making itself felt in forests nationwide, in many cases interacting in complex ways with other environmental change drivers like introduced species and the forest landscape alterations stemming from past timber and fire management practices. Numerous scholars have endorsed the idea of reconfiguring forest governance toward a concern with resilience.[19] The concept of resilience implies the ability to constructively adapt to change and incorporate disturbances without losing valued system elements, such as human lives, rare species, or drinkable water. Rather than attempting to restore forests to match the conditions associated with a particular time period or to precisely recreate a particular disturbance regime, a resilience-oriented approach attempts to set up both human and nonhuman systems to be able to withstand future challenges, whatever those may be. With fires taking ever-greater tolls on ecosystems and human communities—and occupying ever-greater proportions of agency budgets—promoting the resilience of both forests and communities would appear to be a pathway out of the dilemma.

Yet this pivot raises its own questions as well. Various 1970s-era policies, such as NEPA and the National Forest Management Act, reflect a paradigm that sits uneasily with the ideals of resilience. The resilience perspective emphasizes the need for flexibility and continual adaptation within a complex, integrated socio-ecological context that can never be fully understood or predicted in advance. In contrast, laws like NEPA were designed under the assumption that specialists would be able to predict the outcomes of actions on specific resources or systems (for example, individual species, discrete resources like water, or processes like wildfire). NEPA expects planners to commit to a course of action in advance and not to deviate from the plan as a project proceeds. In short, the flexibility needed for adaptative governance conflicts with expectations of strict accountability to commitments made before the project begins.[20]

Recent attempts to make the NEPA process more flexible include an "outcome-based" approach in which forest managers adjust their management actions based upon changes in key indicators as the project proceeds. The challenge in implementing approaches like these often comes back to trust: forest managers often find it difficult to escape from decades-old planning and management models because a more flexible approach requires skeptical publics and partners to trust them.[21] Distrust regarding federal land management has proliferated since the late twentieth century. This distrust comes in several dimensions:

distrust of various publics toward each other, distrust of the intentions and motivations of federal forest managers and the agencies they represent, even distrust of basic facts regarding the drivers of forest and wildfire dynamics.

Trust is more than interpersonal; it is also institutional.[22] The incentives driving institutional behavior can influence interactions with non-agency partners and publics, often in ways that create conflict or misunderstanding or that erode trust. As management projects continue to grow more complex and integrated, federal forest managers are still guided by a budget process that treats resource areas as separate concerns.[23] They are also frequently under pressure to meet simplistic quantitative targets for these budget lines—targets such as "acres treated for hazardous fuels" and "board feet of timber sold"—to demonstrate that they are using taxpayer dollars wisely.[24] This national-scale bureaucratic accountability can, ironically, undermine the accountability to local partners and stakeholders that is needed to rebuild trust and provide latitude for more flexible, innovative management.

Federal forest agencies find themselves in an especially vexing position in the first decades of the twenty-first century, facing forest dilemmas that are more complex than anything Gifford Pinchot could have imagined. Federal forest management is no longer a question of producing the maximum sustained yield of specified outputs; rather, it is oriented toward integrated landscape-scale projects that attempt to undo the effects of a century of management guided by visions of control and simplicity in the woods. Yet management decisions are driven by a legacy of policies and performance metrics that often run counter to the capacities and approaches needed to tackle these dilemmas. How can federal forest managers remain accountable to multiple publics at multiple scales while also maintaining the flexibility needed to manage creatively and flexibly? Is creative and flexible management possible given the agency's prevailing planning, budgetary, and incentive structures? Do agencies need an update of the fifty-year-old federal laws guiding their management? What would the consequences be of overhauling those laws? What other changes might be needed to encourage more nimble and adaptive approaches to federal forest management without sacrificing "too much" accountability?

## Nonfederal Policies and Forestlands

The situation appears no less complex if we shift our focus away from federal policies and federal forests. State forestlands in places like Washington, Oregon, California, and Massachusetts have become embroiled in the same kinds of controversies that were common on federal lands three decades ago. Changes in markets and technologies mean that even where timber supplies are relatively secure, leaders of forest-dependent communities continue to search for alternative economic activities. Meanwhile, millions of acres of corporate and family-owned private forestlands are poised to be lost to development in the near future. With these dynamics as a backdrop, some broad trends are apparent in the nonfederal forest policy landscape. Among the most important is the growing dominance of voluntary and market-based solutions. One set of tools is represented by the various optional policy carrots meant to incentivize forest conservation, including policy mechanisms such as current use taxation, conservation easements, state and federal cost-share programs, support for wood-based bioenergy, and conservation land purchases from willing sellers. A second broad category includes the emerging emphasis on payments for ecosystem services (PES), tradeable mitigation and conservation credits, and other tools designed to address forest conservation problems through buying, selling, and trading. A separate but related set of tools includes the various NGOs and programs that are increasingly called upon to provide assurances of responsible forest management on private lands, especially on corporate lands where technological change continues to drive increases in productivity in the woods and innovation in the mills.

### *Voluntary Policy Carrots and Market-Based Solutions*

The suite of market-based policy tools are diverse, but they tend to have in common two characteristics: first, they are purely voluntary, meaning that landowners can choose whether or not to participate in them; second, they typically work by making a payment (either a direct payment or something more indirect such as a tax credit or privileged access to markets) to private actors in exchange for actions believed to produce benefits for the broader public. These benefits include things like restricting or deferring subdivision and development, maintaining forest cover, providing habitat for species in decline, and protecting water quality. The voluntary and incentive-based approach to forest regulation has only continued to grow in popularity in the decades since passage of some of the nation's bedrock environmental laws in the 1970s. Importantly, many of the key policies from the "environmental decade" at least initially began by employing a more punitive approach toward environmental quality assurance, in some cases leading to the unintended consequences detailed in Chapter 4.

Voluntary and incentive-based policy instruments are often lauded for meeting conservation objectives without provoking fear, resentment, or the creation of perverse incentives for private landowners. As this model of policy design continues to evolve, the lines between governmental and non-governmental programs are growing increasingly blurry: for example, conservation easements typically come with state and federal tax benefits but are often administered by non-governmental organizations such as land trusts (Chapter 8). Forest certification is a non-governmental model, but by certifying their forests, landowners may be able to show compliance with Clean Water Act requirements (Chapter 4) or maintain favorable tax treatment (Chapter 8). California's Global Warming Solutions Act and some regional greenhouse gas initiatives (Chapter 6) have helped to develop a nationwide marketplace for carbon-smart forestry programs, most of which are operated by private firms and NGOs. Indeed, many of the existing programs for ecosystem service payments and other "market-based" instruments are not based on true markets of private buyers and sellers but rather on payments by entities like governments, utilities (such as water providers), or NGOs.[25]

There is currently a great deal of enthusiasm for market-based instruments such as PES programs. In recent decades this model has been eagerly adopted and promoted by private, governmental, and non-governmental stakeholders who tout its superior efficiency and effectiveness as well as its attractiveness to private landowners. However, the model has its detractors as well. In the latter category are those arguing that PES program implementation does not always work as well in practice as it does in theory. In part, their argument is that PES schemes to date have not fully addressed issues of conditionality, additionality, leakage, and permanence (see Chapter 9). Some have argued that contemporary forest-based carbon markets are built upon a series of shared, but potentially problematic, assumptions among carbon buyers, sellers, and traders regarding issues like additionality.[26] Calculating the precise carbon benefits of specific actions—and the carbon consequences of their counterfactual alternatives—is a highly challenging exercise and one that, according to some critics, has failed to meet the exacting standards of prevailing carbon sequestration policies.

Another concern regarding PES programs and other market-based approaches is that they may shift the blurry line dividing the foundational responsibilities of landowners from actions that go above and beyond the basics and, therefore, merit compensation. The phenomenon of "crowding out" altruistic actions for conservation occurs when landowners or other ecosystem service providers "come to expect financial assistance to undertake activities they otherwise would have seen as their own responsibility."[27] This is not a new dilemma; Aldo Leopold wrote about something very similar when he considered the role of government conservation programs in promoting private land stewardship. After his own experiences administering subsidy and technical assistance programs for farm conservation, Leopold walked away dispirited with the results, leading him to conclude that "a system of conservation based solely on economic self-interest is hopelessly lopsided."[28] He came to this realization after seeing that many ecologically important elements of the biotic community lacked commercial value and were too numerous and integral to be managed by government programs. For Leopold, "an ethical obligation on the part of the private owner is the only visible remedy for these situations."[29]

Beyond questions of efficacy, there is a question of equity at play in the choice of policy instruments as well: choosing to pay a landowner to take a socially beneficial action, rather than to require it through regulation, means that money will be transferred to landowners from taxpayers, ratepayers, or other general publics in order to secure the public benefits of forests. Conversely, prescriptive regulation exacts costs on landowners while delivering benefits to the public. Whether conserving and providing ecosystem services on private land is more effectively, efficiently, and equitably achieved through making the landowner pay or making the public pay is an important and complex public policy debate, and one that extends well beyond questions of PES programs. For example, Chapter 9 describes a private forest conservation effort in Montana organized in response to the planned sale of former industrial timberlands for residential real estate development. While the community-led effort was ultimately successful, the final bill to achieve this outcome came to hundreds of millions of dollars in public, philanthropic, and NGO funds; some critics argued that similar conservation results could have been achieved through the proper use of zoning.[30]

On the other hand, advocates for market-based instruments can point to numerous examples of apparently successful market-based programs. Well-designed PES programs can provide new funding for conservation or restoration activities and protect valuable services like water quality protection and wildlife habitat production that might remain at risk in the absence of an economic incentive. PES advocates often argue that if we leave ecosystem services outside of the system of market exchange, they will be treated as if they are worth nothing; they say that only by putting a price tag on these services will they be seen as valuable, and treated as such, within capitalist society. Market-based instrument defenders also point out that many of these tools are relatively new; some mistakes and learning processes in the early years are inevitable and improvement over time can be expected. They also argue that market-based instruments may be imperfect, but so are all policy tools—and that tools favored by landowners are more likely to succeed than those loathed by landowners.

The PES model is primarily focused on attaching economic values to ecological processes and outputs that have traditionally been "outside" of market exchange. A separate but related policy approach is to further develop demand for goods already "inside" of markets that have associated conservation benefits. Conservation strategies for areas dominated by private forests consistently stress the importance of strong markets for forest products: the argument is that pressures to convert forests to other uses will go down as markets for forest products go up. The development of new wood-based technologies such as mass timber construction and advanced wood-based biofuels (see Chapter 9), along with the expansion of markets for older products like dimensional lumber and wood pellets, have been touted by some for creating demand drivers to prevent the conversion of forests to other uses.[31] The carbon benefits of forest-based products are increasingly emphasized in policies and programs encouraging their use and development: using wood-based energy in place of fossil fuels and wood-based building materials in place of concrete and steel is argued to result in net benefits for both the climate and for forest conservation. Chapter 9 details some of the dilemmas of carbon accounting when substituting wood-based bioenergy for fossil fuels. There is also a growing debate about the carbon tradeoffs between using forests to produce wood products (including those that replace more carbon-intensive materials) and allowing existing forests to continue growing and sequestering carbon.[32] Older trees in particular store substantial amounts of carbon and may be able to continue sequestering carbon well after reaching their economic maturity.[33] How, then, should policy be designed to conserve the carbon benefits of standing trees while also encouraging the use of renewable building and energy materials? What role should governments play in providing economic incentives to discourage conversion of forests to residential, agricultural, or other uses? More broadly, to what extent should our policy frameworks rely on economic drivers rather than (or in addition to) required standards of practice to achieve effective, efficient, and equitable outcomes on private lands? What role should governments play when those economic drivers are not strong enough to stem patterns of forest conversion and loss?

## Beyond the Public–Private Divide

The debates and controversies described above speak to the continuing need to critically examine policies and practices in the face of accelerating environmental changes, especially those driven by climate change. As paradigms for forest management have evolved from conservation of timber supplies to the consideration of broader social-ecological systems, concepts such as resilience and adaptation have become more prominent across ownership categories. Yet it is not entirely clear what it means for a forest—or for communities living in and near forests—to be resilient. It is also unclear what changes to policy and governance will be needed to achieve this goal. Prevailing policy direction tends to emphasize fuel reduction on federal lands, and debates over policy reforms needed to accelerate these treatments have been central to national forest policy reforms for more than two decades. However, addressing only the federal side of the wildland-urban interface will not achieve resilience under almost any definition. Very few jurisdictions have placed limits on the scale, speed, or pattern of wildland-urban interface expansion on private lands, meaning that the amount of built infrastructure and number of human lives in this zone will only continue to increase, as will the watershed-level impacts of forest conversion and development.

As public, private, tribal, and community forest owners confront the many challenges associated with the contemporary era, they will increasingly be forced to confront questions of how to plan for a future climate. Whereas earlier paradigms looked to concepts like "natural range of variability" or "historic range of variability" for guidance, climate change and other elements of a changing global environment complicate the value of previous forest patterns for informing future-oriented decision-making. Many scientists have warned that we will increasingly need to think in terms of "no-analog" ecosystems characterized by combinations of species, patterns, and processes for which there is no historical precedent.[34] To what extent will governance actors begin to plan for future climates when conducting forest management activities? What might the role be for things like offsite seedling stock or genetically modified trees that are believed to have better chances at survival under changing climatic conditions? How will the "assisted migration" of species play into our forest policy and governance deliberations? How will these kinds of issues be addressed in cost-share support initiatives, certification schemes, payment for ecosystem service programs, or collaborative decision-making?

Another common thread tying together public and private land policies is the growing importance of coordinated and networked approaches to forest governance. Forest governance is increasingly characterized by the participation of multiple governmental, non-governmental, and private actors working through both formal and informal means. Tribal, county, and state resource managers may help to plan and implement projects on federal lands, and governmental and non-governmental entities may hold stewardship agreements to provide long-term benefits to those lands. On nonfederal forestlands, decision-making might be driven not only by the landowner and applicable state or federal regulatory agencies but also by non-regulatory agencies that administer voluntary incentive programs, representatives of third-party certification programs, buyers and sellers of tangible or intangible ecosystem services, land trusts that hold distinct property rights, and cooperating NGOs that provide resources, advice, and capacity-building. Landowners within the range of certain important wildlife or tree species may find themselves invited to participate in landscape-scale conservation and recovery programs that involve coordination with neighboring landowners and land managers. In short, the forest governance landscape is increasingly one of diverse entities at multiple scales working in partnership toward common goals. This trend raises questions about which interests and organizations are included in collaborative and networked approaches and which interests and organizations are not. Will there continue to be sufficient funding for the NGOs that increasingly serve as the bridging organizations connecting these varied participants? Do our public agencies and policies need to be reformed so that they take on these roles instead of NGOs? Will the private, public, and NGO entities with the deepest pockets come to dominate the informal spaces of forest governance? What role will there be—and what role should there be—for less powerful interests and less well-funded organizations?

## Conclusion

Top-down edicts and programs controlled solely by government agencies have fallen out of favor since the 1990s; instead, prevailing governance approaches promote the cooperative inclusion of non-governmental partners and the use of creative incentives (including, but not limited to, market-based instruments) in place of inflexible regulations. Landscape-scale restoration approaches associated with issues like red-cockaded woodpecker recovery (Chapter 4) exemplify this trend, as do policies and programs associated with watershed conservation (Chapter 4), the reduction of uncharacteristic wildlife and insect risk across ownerships (Chapter 5), voluntary water quality standards (Chapter 6), tribal co-management (Chapter 7), voluntary incentive programs (Chapter 8), PES schemes (Chapter 9), third-party certification (Chapter 9), place-based collaborative processes (Chapter 10), and many other examples. These kinds of approaches have been widely celebrated by many scholars and practitioners who argue that they produce better results by drawing upon local knowledge and local motivations for stewardship; foster cooperation in place of conflict; support innovation instead of simply encouraging compliance; increase capacity by creatively aligning resources and opportunities; and allow for decisions that are less politically contentious and therefore more sustainable over the long term.[35]

Even as these varied approaches have become institutionalized nationwide, questions remain about the degree to which they are efficient, effective, and equitable as well as whether they are adaptable. The efficiency criterion has to do with the balance between costs and benefits, with each of these terms possibly being defined and measured in multiple ways. Effectiveness is a related concern that has to do with the degree to which policies and programs produce lasting, real-world outcomes; controversies regarding additionality, leakage, and permanence are ultimately about effectiveness. Equitable policies are those that are designed around justice and fairness—concepts that might encompass issues like historical dispossession, modern-day inequities in the distribution of the benefits of forests, opportunities to participate in governance processes, and the ability of varied actors to contribute resources. Adaptability refers to the ability of policies and governance approaches to take in new information and change in response, a quality that will only grow in importance as forests continue to experience the pressures of environmental change.

## Questions for Discussion

1. Why do some observers feel that the policy framework for federal land management agencies impedes efficient management?
2. Are changes to federal policies or performance metrics needed? What changes, if any, would you recommend? What would be the consequences of such changes for the balance between flexibility and accountability?
3. What does it mean for forests to be resilient in the face of climate change and other processes of environmental change? What does it mean for policies and programs to be adaptable? Do adaptable policies and programs lead to resilient forests?
4. What are the controversies regarding voluntary and market-based instruments? How do these approaches perform in terms of the criteria of efficiency, effectiveness, equity, and adaptability?
5. Given the rise of collaborative, partnership-based, and networked governance models, what is the unique role for governmental agencies and policies going forward? Can the complex problems of forest conservation under environmental change be solved without a capable and competent government presence?

## Notes

1 Downing et al., 2022.
2 Abrams & Bliss, 2013.
3 Abrams et al., 2013; Abrams & Bliss, 2013.
4 Vaughn & Cortner, 2005.
5 Lindenmayer et al., 2008.
6 Charnley et al., 2015; Schultz et al., 2021; Vaughn & Cortner, 2005.
7 Emphasis in original. https://oversight.house.gov/sites/democrats.oversight.house.gov/files/Carole%20King%20written%20testimony%20FINAL.pdf.
8 Donato et al., 2006.
9 Thompson et al., 2007.

10 Freemuth & Babcock, 2016.
11 Freemuth & Babcock, 2016.
12 Hjerpe, 2013.
13 Treuer, 2021.
14 Washburn, 2022.
15 Abrams, 2019; Maier & Abrams, 2018.
16 Abrams et al., 2020.
17 Kooistra et al., 2022.
18 Timberlake et al., 2017.
19 Hessburg et al., 2021; Stephens et al., 2016.
20 Benson & Garmestani, 2011a, 2011b.
21 Coleman et al., 2018; Stern & Coleman, 2015.
22 Stern & Coleman, 2015.
23 Schultz et al., 2018.
24 Schultz et al., 2016, 2019.
25 Wunder, 2015.
26 Gifford, 2020.
27 Lockie, 2013, p. 96.
28 Leopold, 1949, p. 214.
29 Leopold, 1949, p. 214.
30 Parker, 2014; note that Montana laws at that time would have made any actual use of zoning for these purposes nearly impossible.
31 Cabiyo et al., 2021; Dale et al., 2017; Nepal et al., 2016; Parish et al., 2018.
32 Hudiburg et al., 2009, 2011; Law et al., 2018.
33 Luyssaert et al., 2008.
34 Hobbs et al., 2009; Williams & Jackson, 2007.
35 Sheridan et al., 2014.

## References

Abrams, J. (2019). The emergence of network governance in US National Forest administration: Causal factors and propositions for future research. *Forest Policy and Economics*, *106*, 101977.

Abrams, J., & Bliss, J. C. (2013). Amenity landownership, land use change, and the re-creation of "working landscapes." *Society & Natural Resources*, *26*(7), 845–859.

Abrams, J., Bliss, J. C., & Gosnell, H. (2013). Reflexive gentrification of working lands in the American West: Contesting the 'middle landscape.' *Journal of Rural and Community Development*, 8(3), 144–158.

Abrams, J., Huber-Stearns, H., Gosnell, H., Santo, A., Duffey, S., & Moseley, C. (2020). Tracking a governance transition: Identifying and measuring indicators of social forestry on the Willamette National Forest. *Society and Natural Resources*, *33*(4), 504–523.

Benson, M. H., & Garmestani, A. S. (2011a). Can we manage for resilience? The integration of resilience thinking into natural resource management in the United States. *Environmental Management*, 48(3), 392–399.

Benson, M. H., & Garmestani, A. S. (2011b). Embracing panarchy, building resilience and integrating adaptive management through a rebirth of the National Environmental Policy Act. *Journal of Environmental Management*, 92(5), 1420–1427.

Cabiyo, B., Fried, J. S., Collins, B. M., Stewart, W., Wong, J., & Sanchez, D. L. (2021). Innovative wood use can enable carbon-beneficial forest management in California. *Proceedings of the National Academy of Sciences*, *118*(49), e2019073118.

Charnley, S., Poe, M. R., Ager, A. A., Spies, T. A., Platt, E. K., & Olsen, K. A. (2015). A burning problem: Social dynamics of disaster risk reduction through wildfire mitigation. *Human Organization*, *74*(4), 329–340.

Coleman, K., Stern, M. J., & Widmer, J. (2018). Facilitation, coordination, and trust in landscape-level forest restoration. *Journal of Forestry*, *116*(1), 41–46.

Dale, V. H., Kline, K. L., Parish, E. S., Cowie, A. L., Emory, R., Malmsheimer, R. W., Slade, R., Smith Jr, C. T., Wigley, T. B., Bentsen, N. S., Berndes, G., Bernier, P., Brandão, M., Chum, H. L., Diaz-Chavez, R., Egnell, G., Gustavsson, L., Schweinle, J., Stupak, I., … Wellisch, M. (2017). Status and prospects for renewable energy using wood pellets from the southeastern United States. *GCB Bioenergy*, *9*(8), 1296–1305.

Donato, D. C., Fontaine, J. B., Campbell, J. L., Robinson, W. D., Kauffman, J. B., & Law, B. E. (2006). Post-wildfire logging hinders regeneration and increases fire risk. *Science*, *311*(5759), 352–352.

Downing, W. M., Dunn, C. J., Thompson, M. P., Caggiano, M. D., & Short, K. C. (2022). Human ignitions on private lands drive USFS cross-boundary wildfire transmission and community impacts in the western US. *Scientific Reports*, *12*(1), 1–14.

Freemuth, J., & Babcock, E. (2016). Governance, science, and the federal lands. In J. Freemuth & Z. A. Smith (Eds.), *Environmental Politics and Policy in the West* (pp. 55–78). University Press of Colorado.

Gifford, L. (2020). "You can't value what you can't measure": A critical look at forest carbon accounting. *Climatic Change*, *161*(2), 291–306.

Hessburg, P. F., Prichard, S. J., Hagmann, R. K., Povak, N. A., & Lake, F. K. (2021). Wildfire and climate change adaptation of western North American forests: A case for intentional management. *Ecological Applications*, *31*(8), e02432.

Hjerpe, E. (2013). *Fiscal impacts to the State of Idaho from HR 22 implementation*. Idaho Conservation League.

Hobbs, R. J., Higgs, E., & Harris, J. A. (2009). Novel ecosystems: Implications for conservation and restoration. *Trends in Ecology & Evolution*, *24*(11), 599–605.

Hudiburg, T., Law, B., Turner, D. P., Campbell, J., Donato, D., & Duane, M. (2009). Carbon dynamics of Oregon and Northern California forests and potential land-based carbon storage. *Ecological Applications*, *19*(1), 163–180.

Hudiburg, T., Law, B., Wirth, C., & Luyssaert, S. (2011). Regional carbon dioxide implications of forest bioenergy production. *Nature Climate Change*, *1*(8), 419–423.

Kooistra, C., Schultz, C. A., Abrams, J., & Huber-Stearns, H. (2022). Institutionalizing the United States Forest Service's Shared Stewardship strategy in the western United States. *Journal of Forestry*. https://doi.org/10.1093/jofore/fvac010.

Law, B. E., Hudiburg, T. W., Berner, L. T., Kent, J. J., Buotte, P. C., & Harmon, M. E. (2018). Land use strategies to mitigate climate change in carbon dense temperate forests. *Proceedings of the National Academy of Sciences*, *115*(14), 3663–3668.

Leopold, A. (1949). *A Sand County almanac, and sketches here and there*. Oxford University Press.

Lindenmayer, D. B., Burton, P. J., & Franklin, J. F. (2008). *Salvage logging and its ecological consequences*. Island Press.

Lockie, S. (2013). Market instruments, ecosystem services, and property rights: Assumptions and conditions for sustained social and ecological benefits. *Land Use Policy*, *31*, 90–98.

Luyssaert, S., Schulze, E.-D., Börner, A., Knohl, A., Hessenmöller, D., Law, B. E., Ciais, P., & Grace, J. (2008). Old-growth forests as global carbon sinks. *Nature*, *455*(7210), 213–215.

Maier, C., & Abrams, J. B. (2018). Navigating social forestry: A street-level perspective on National Forest management in the US Pacific Northwest. *Land Use Policy*, *70*, 432–441.

Nepal, P., Skog, K. E., McKeever, D. B., Bergman, R. D., Abt, K. L., & Abt, R. C. (2016). Carbon mitigation impacts of increased softwood lumber and structural panel use for nonresidential construction in the United States. *Forest Products Journal*, *66*(1–2), 77–87.

Parish, E. S., Herzberger, A. J., Phifer, C. C., & Dale, V. H. (2018). Transatlantic wood pellet trade demonstrates telecoupled benefits. *Ecology and Society*, *23*(1), 28.

Parker, M. (2014). Swan story. In S. Charnley, T. E. Sheridan, & G. P. Nabhan (Eds.), *Stitching the West back together: Conservation of working landscapes* (pp. 123–135). University of Chicago Press.

Schultz, C. A., Abrams, J. B., Davis, E. J., Cheng, A. S., Huber-Stearns, H. R., & Moseley, C. (2021). Disturbance shapes the US forest governance frontier: A review and conceptual framework for understanding governance change. *Ambio*, *50*(12), 2168–2182.

Schultz, C. A., Mattor, K. M., & Moseley, C. (2016). Aligning policies to support forest restoration and promote organizational change. *Forest Policy and Economics*, *73*, 195–203.

Schultz, C. A., Moseley, C., Mattor, K. M., McIntyre, K., & Ellison, A. (2018). Key findings and recommendations based on the USDA Forest Service Integrated Resource Restoration Pilot third-party review. *Journal of Forestry*, *116*(1), 5–12.

Schultz, C. A., Thompson, M. P., & McCaffrey, S. M. (2019). Forest Service fire management and the elusiveness of change. *Fire Ecology*, *15*(1), 13.

Sheridan, T. E., Nabhan, G. P., & Charnley, S. (2014). Conclusions and policy implications. In S. Charnley, T. E. Sheridan, & G. P. Nabhan (Eds.), *Stitching the West back together: Conservation of working landscapes* (pp. 301–310). University of Chicago Press.

Stephens, S. L., Collins, B. M., Biber, E., & Fulé, P. Z. (2016). US federal fire and forest policy: Emphasizing resilience in dry forests. *Ecosphere*, *7*(11), e01584.

Stern, M. J., & Coleman, K. J. (2015). The multidimensionality of trust: Applications in collaborative natural resource management. *Society and Natural Resources*, *28*(2), 117–132.

Thompson, J. R., Spies, T. A., & Ganio, L. M. (2007). Reburn severity in managed and unmanaged vegetation in a large wildfire. *Proceedings of the National Academy of Sciences*, *104*(25), 10743–10748.

Timberlake, T., Schultz, C. A., & Abrams, J. (2017). *Resilience in land management planning: Policy mandates, approaches, and resources* (Working Paper 77). Ecosystem Workforce Program, Institute for a Sustainable Environment.

Treuer, D. (2021). Return the national parks to the tribes. *The Atlantic*, *327*, 30–45.

Vaughn, J., & Cortner, H. (2005). *George W. Bush's healthy forests: Reframing the environmental debate*. University Press of Colorado.

Washburn, K. K. (2022). Facilitating tribal co-management of federal public lands. *Wisconsin Law Review*, *2022*(2), 263–328.

Williams, J. W., & Jackson, S. T. (2007). Novel climates, no-analog communities, and ecological surprises. *Frontiers in Ecology and the Environment*, *5*(9), 475–482.

Wunder, S. (2015). Revisiting the concept of payments for environmental services. *Ecological Economics*, *117*, 234–243.

# Conclusion

Property lines dividing the distinct categories of forest ownership are difficult if not impossible to discern with the naked eye from atop the east moraine of Wallowa Lake. Wallowa County's forests appear to transition seamlessly from federal to state to community-owned to tribal to corporate to family ownership before disappearing into treeless farmland and rangeland. Yet this apparent seamlessness is something of an optical illusion; forests are managed in very different ways on varied ownerships, guided by different policies and programs and incorporating different processes for the inclusion of outside perspectives. Unseen beneath the verdant canopy is a vast diversity of histories, owners, uses, and interests, all of them influenced by and connected to varied formal policies, informal practices, modes of participation, and animating paradigms.

It could be easy for the uninitiated to imagine that the system of forest policy and governance in the US is itself seamless, that it exists as a rationally designed framework reflecting a unified vision of the proper tradeoffs between rights and responsibilities, freedoms and constraints. Like a structure designed and constructed by a visionary architect, this ideal structure of forest policies and institutions would create a cohesive whole that blends into and complements the surrounding landscape. The preceding twelve chapters suggest that no such cohesive system exists in the US. Instead of a structure designed and constructed out of a unitary vision, what we have is perhaps more like an old house that started out small and has been built up to something much larger over the years, with each expansion making use of the technologies, styles, and architectural standards of its time. This is not to say that US forest policy and governance is chaotic, simply that it contains legacies that reflect various historical periods and notions of good governance, all cobbled together into a still-evolving structure. Despite contrasts in style and substance, we can identify some consistent themes running throughout the legacy of forest policy and governance institutions by returning to the four categories of property introduced at the beginning of this text. We consider where each of them might be found in the US forest landscape and where they might play a role as we look forward. In doing so, we reflect on the role of formal policies, informal practices, participation, and paradigms in providing a kind of messy coherence to the US policy and governance system.

DOI: 10.4324/9781003043669-13

## Open Access

Open-access arrangements are characterized by a lack of rules regulating users and uses. Open-access systems have long been recognized as problematic because they tend to encourage overexploitation of existing resources and underinvestment in the future. There are, fortunately, relatively few real-world examples of open-access regimes on US forestlands. Some contemporary examples of NTFP harvesting show characteristics of an open-access arrangement, particularly where the resource in question is exhaustible and where oversight and enforcement are weak. Despite formal policies in place at state, tribal, and federal levels to protect wild ginseng, for example, populations of the prized medicinal species continue to decline in some places because of difficulties enforcing restrictions on access and harvesting. Whereas informal practices and relatively low harvest pressures appear to have helped control the exploitation of many other NTFPs, the combination of high demand for ginseng on the global market, the plant's slow growth rate, and its distribution across millions of acres of mostly remote forestland have led to unique dilemmas in its conservation. Other examples of weak access control leading to resource degradation come from uncontrolled recreation (especially motorized recreation) and the proliferation of illegal cannabis grow operations on public forestlands. Where open access-type arrangements do currently exist, they tend to be associated with a lack of effective enforcement rather than with a lack of rules themselves (the ginseng, recreation, and cannabis cases are fitting examples of this). In these instances, proper stewardship hinges on the capacity of private, tribal, community, and public forest managers to monitor uses and sanction offenders—or, as Aldo Leopold might suggest, on the development of a stronger conservation ethic among users.

With a US population of 330 million people (and growing) as of the date of this writing, it is hard to imagine that open-access regimes on forestlands themselves will be a prominent part of the future forest governance landscape. However, the notion that open access leads to tragic outcomes is not limited to land and the resources found on it; it also applies to issues such as carbon emissions into a shared atmosphere and the consequent effects on global climate stability. Forests are now seen as key to mitigating the tragedy of collective carbon emissions, and both policies and markets have been developed in recent years to advance forests as climate solutions. Carbon offset schemes, wood-based energy production, and even the creation of novel engineered wood products are all elements of an emerging "bioeconomy" believed by many to help advance both climate solutions and economic activity. Recent developments suggest that this is more than a passing fad. Carbon offset programs associated with both state and regional climate policies are driving potentially significant changes in forest ownership and management. A key challenge for these programs going forward will be demonstrating to both researchers and the broader public that they are in fact contributing to additional and permanent carbon sequestration in ways that avoid leakage and that continue to conserve other values of broad public importance.

## Common Property

Common property arrangements, in stark contrast to open access, are governed by a defined community, the members of which possess both rights to the land and obligations to steward it. When rules of exclusion and use are created and enforced by members of the community, and when those rules are recognized by higher levels of government, common property tenure holds great promise for achieving economic, social, and environmental benefits over the long term. Nevertheless, a prevailing paradigm in environmental management tends to conflate common property with open access, an unfortunate misunderstanding that can be traced back in part to the legacies of John Locke and Garrett Hardin. Although the common property tenure model would seem to be at odds with deeply rooted American concepts of individual private property, glimpses of it can be found in various settings. Historically, Spanish and Mexican land grants in the Southwest, town forests in New England, informal Appalachian forest "commons," and Indigenous territories nationwide were all examples of common property-type arrangements.

The extent of both Hispano land grants and tribal lands have been vastly diminished since the early nineteenth century and many tribal territories have been subjected to a parcelization and individualization process via the Dawes Act and similar laws, yet both systems continue to persist to the present day. Despite being subjected to decades of control by a bureaucracy committed to a narrow vision of forest management, tribal lands nevertheless served as sites of adaptive fire management during the twentieth century, and the varied processes of tribal land repatriation and co-management hold promise for extending the terrain on which scientific and traditional knowledges are intertwined. Following a series of seemingly capricious and self-contradictory alternations in federal tribal policy, federally recognized tribes have increased their capacity, self-determination, and legally recognized rights of sovereignty. These trends are reflected in the implementation of more culturally appropriate forest management practices on tribal lands as well as the increased participation of tribal populations in off-reservation cooperative management. The informal practices that had long been favored on these lands are increasingly garnering recognition in formal policies.

Community-owned forestlands have seen a resurgence nationwide since the late twentieth century, in many cases spurred by the changes associated with the restructuring of corporate timberland ownership. To some extent, the rise of collaborative and community-based governance processes on public lands also represents a commons-inspired approach. Although tenure rights to the lands in question are not formally possessed by the nonfederal participants, collaborative efforts nevertheless depend heavily on understandings of "community" and the lived connections to individual places. This has allowed for the participation of a more diverse array of interests and voices in the processes associated with forest decision-making than was typical in the past. To the extent that place-based collaborative efforts influence public land decision-making, they suggest at least a partial (and informal) replacement of the expert-driven model of governance with a model that integrates multiple knowledges and perspectives. The persistence of collaborative models into the future is uncertain, given that for nearly three decades now they have largely existed in a grey area between formality and informality, being widely institutionalized despite lacking a clear policy mandate or formal definition. By contrast, community forest tenure and tribal forest authority have grown out of clearer policy foundations. There is also a growing attention to urban forests as places of significant community-level engagements with forests and as sites that can potentially reproduce or alleviate social inequities.

## Public Forests

Government-owned lands are defined by their public ownership, by their accessibility to the general public (under rules set by government entities), and, in many cases, by their management according to scientific principles that privilege formal expertise over other forms of knowledge. These lands are ubiquitous in the United States, including urban parks, county and state forestlands, and varied federal agencies responsible for hundreds of millions of acres nationwide. Public forests span a wide range of management emphases, from remote landscapes managed as wilderness to timberlands intensively managed for commercial wood fiber production to parks and forests that attempt to accommodate varied and potentially conflicting recreational, economic, and conservation uses simultaneously. From a governance perspective, government-owned lands have the advantage of being highly equitable in the sense that access is not restricted to individual owners or members of a defined community, yet the enforcement of rules tends to prevent the tragedy of open-access scenarios from occurring.

Formal policies in the early years of forest conservation emphasized the capacity of the government to directly address identified needs through the deployment of government expertise and capacity. In just a few decades, the federal government amassed a permanent national forest estate exceeding 150 million acres and put in place an agency of foresters and researchers to administer it; states took over smaller but nevertheless substantial expanses of forestland in some of the nation's most abused and neglected landscapes. States and the federal government worked in tandem to suppress fires, an action they believed would help conserve valuable standing timber resources and encourage corporate owners to invest in long-term forest ownership and management. Informal practices such as the use of broadcast fire were suppressed and replaced by a formal policy of total suppression. These multiple-use lands began as a model in stark contrast to both private lands and strictly preserved parks; as time went on, such distinctions faded in some parts of the landscape.

Although the government ownership model was considered the gold standard for conservation for many years, scholars and practitioners have leveled criticism in recent decades against the legacy of scientific management with its simplifying tendencies and disregard for local values and traditional knowledges. The end of the twentieth and beginning of the twenty-first centuries saw the federal land management agencies forced to make an abrupt pivot from commodity-focused management to ecosystem management—a change that reflected the diverse social values pertaining to forests, a more complex policy landscape, and the expanding means by which diverse interests could participate in the planning of federal plans and projects. What followed was not only a policy shift but a paradigm shift, one in which social and ecological complexity could no longer be sidelined in the pursuit of commodity outputs. The federal forest agencies have since experimented with varied means of accommodating diverse values, voices, and interests on a limited land base and a dwindling non-fire budget. The development of collaborative and partnership-oriented governance models has now become institutionalized as federal managers engage with new social and ecological realities.

Still, questions abound regarding the ability of federal forest managers to undo the effects of a century of fire suppression and forest rationalization. Public forest managers, forest users, and residents confront a volatile mixture of altered forest conditions, climate-driven changes to disturbance regimes, and the proliferation of valuable human infrastructure in the forest interface. This challenge grows even more complicated now that government lands are no longer seen as self-contained islands but as integral parts of landscapes that contain myriad other ownerships, owners, uses, and cross-boundary processes. State and county forests, meanwhile, are showing signs that they, too, will be scrutinized more closely than they have been in the past; in some cases, these forests have also been devoted to new uses. The adoption of third-party forest certification has been widespread across state and county forestlands, implying that national and even global forest management principles will inform the management of forests that were long governed solely by local authorities under the direction of formal policies. States and counties are also increasing their influence on federal lands through the contribution of monetary and staff resources, encouraged by recent federal policy changes that invite states and other entities to partner to achieve restoration outcomes at greater scales. In spite of their occasional visibility, campaigns to divest public lands over large areas have not met with success to date. Rather, public lands appear poised to continue their slow but steady evolution as territories integrated into larger social and ecological landscapes through the collective efforts of diverse networks of decision-making and implementation.

## Private Forests

Private property in the US is often associated with Lockean ideals of independence from both social and ecological obligations. The combination of individual (family) forestlands and various private corporate models makes up the lion's share of forest ownership nationwide. As in the other models profiled here, there is tremendous diversity in private forestlands, and this diversity is expressed along several dimensions: the size and ecological condition of individual properties; the gender, racial, ethnic, and other categories of owners; the ownership structure (e.g., individual, family trust, industrial, TIMO, REIT); the values and objectives driving ownership; and the management approaches applied. Private forests have long been objects of concern for professional foresters and of fascination for forestry researchers. Until the widespread adoption of sustained-yield techniques in the second half of the twentieth century (and even later adoption of best management practices), private forestlands were often seen as prone to overexploitation, poor stewardship practices, and abandonment as well as serving as sources of sometimes uncontrollable and deadly wildfires. Gifford Pinchot and many other conservation leaders made frequent calls for national regulation of private forestry practices even as they attempted to utilize professional advice, fire control, and changes to tax policy to encourage long-term stewardship of private forestlands. Although conservation leaders never got the national forestry legislation they wanted, the formal policy regime for private forestlands came to rely more heavily on policy sticks by the second half of the twentieth century. Federal law threatened punishment for actions that harmed declining species and several states enacted mandatory rules for forest practices. The federal government attempted to drive improvements in the nation's water quality through a mix of policy sticks and carrots, allowing states to play the lead role in setting and enforcing standards for non-point pollution produced by forest roads and harvest operations.

Owners of private forestland enjoy many rights and privileges under constitutional, statutory, and common-law principles, yet their freedoms are not absolute; both state and federal governments can require them to take certain actions or refrain from others to protect broad public interests. In some states, landowners may be required to allow members of the public to continue longtime informal practices by accessing lakes or non-timber forest products on their properties; in other states, landowners may be required to maintain their lands undeveloped and undivided to comply with zoning laws. Many states regulate forest practices directly through either comprehensive policies or piecemeal legislation, and practices are governed indirectly via numerous wildlife, water, and other policies at various levels. An ongoing debate concerns the separation of authorities between state and federal governments regarding the limits of federal regulation, illustrated most clearly by the ever-changing scope of waters subject to the Clean Water Act. Although this point of legal tension is between state and federal governments, the implications are greatest for private landowners whose properties contain potentially "navigable" waters as defined by regulations and court interpretations.

Federal policy sticks have recently been complemented (but not replaced) by various carrots to incentivize forest cover conservation and address concerns related to wildlife habitat, carbon sequestration, and other issues that were never contemplated by Pinchot or his contemporaries. At state and local levels, the private land paradigm continues to enjoy widespread deference, even in states such as Oregon that have garnered reputations for government intervention in private land decision-making. Although broader public interests may be reflected in zoning, herbicide, or forest practice policies, the reach of such policies on private lands is relatively modest overall. The contemporary private forestland policy emphasis is now squarely on carrots that emphasize voluntary participation and payments for conducting activities (or providing outputs) considered important.

One paradigmatic change that has been felt on private forest estates is that of seeing private forest parcels as elements of larger landscapes—landscapes characterized by ecological processes that transcend property boundaries. Forest owners and managers nationwide are faced with growing wildlife, water, fire, insect, and disease concerns that are made even more complicated by patterns of global climate change, the introduction of novel species, and the continued fragmentation and development of forests. Recent federal policy is increasingly including private lands in large-scale collaborative restoration projects, and states are now thinking and planning at larger scales to prioritize investments for greatest impact.

## Beginning and Ending with Complexity

Those who participate in decision-making for forests in the early twenty-first century might feel as if they are crossing in and out of multiple worlds as they move through the various rooms and floors that make up the architecture of US policy and governance. One is the world of sustained-yield forestry that began in the late nineteenth century as the first forest conservation agencies and policies came into being. Another is the world of the 1970s-era faith in science and planning to solve complex problems. Still other worlds reflect romantic preservationist visions or misunderstandings of common property, all with a foundation in the world of European colonial concepts of land, property, and racial hierarchy. In the wider landscape outside the house, an ecological unraveling supercharged by climate disruption is taking place—signaling an emerging world whose contours have only clearly come into view since the end of the twentieth century.

In the world of sustained-yield forestry, the problems to be solved were those of waste and inefficiency. Forests were being exploited irrationally, and valuable resources were turning to ash and smoke in uncontrollable conflagrations. Poor logging practices were destroying watersheds and putting future timber supplies in doubt. The science of ecology was nascent at best, and understandings of forests as complex systems were nonexistent. Policies and paradigms in this world found success through simplification. The German forestry tradition had already devised mathematically precise tools for calculating forest growth and harvest of wood, and Gifford Pinchot and others in the scientific forestry tradition applied those tools to American forests. The power of government ownership, training, and capacity was put to the task of rationalizing forests and eliminating wildfires, insects, and diseases. In the end, sustained yield was achieved—but this was a fleeting victory, one only made possible through simplifying the forest and excluding the voices and values of all Americans but an elite few.

This is not to deny the achievements of Pinchot or of the Progressive movement that he and others of his time brought into being. Conservation leaders of this period took the actions that appeared to be necessary to them, and in doing so they successfully replaced wasteful and damaging patterns of land use with practices that could be sustained over longer periods. They prevented millions of acres of forests from being appropriated through corruption and created one of the world's preeminent forest management, research, and outreach agencies. The forest products harvested under their direction provided wood to a growing economy and supplied economic activity in rural and hinterland regions nationwide. And although the focus on fire exclusion may have been shortsighted, it also responded to the very real threat of deadly and uncontrollable blazes associated with the worst timber harvesting approaches of the time.

Gifford Pinchot and his contemporaries were also not the only leaders and thinkers whose visions failed to take in the complexity of American forests. John Muir and the tradition he inspired—a tradition that includes our national parks as well as designated wilderness areas—found success in part through telling inspiring yet incomplete stories. These stories said that nature had a purity that could only be defiled through human inhabitation and use. Pure nature could be preserved, in this telling, by denying human history and creating spaces set apart from human culture.[1] The nation's parks and wilderness areas protect some of our most awe-inspiring landscapes from industrial development and allow people worldwide the opportunity to access and experience them. Yet the designation of these protected areas often came at the price of the livelihoods and rights of those previously inhabiting those landscapes and at the price of disseminating incomplete understandings of forests and their histories.

Likewise, John Locke's story of the contrast between land held in common and land held in private offers a story that is compelling in its simplicity despite its deeply flawed factual and historical bases. Locke provided a justification for individuals—and not merely royalty—to own land and allowed the flourishing of a private property tradition that has produced both innovations and material benefits to society. Yet his story also justified the dispossession of Indigenous populations and failed to consider the inescapable social and ecological contexts within which land is always situated. Garrett Hardin shone a light on the tragic outcomes of open-access property systems, but his unfortunately misnamed story of "commons" management continues to be taught in college classrooms nationwide despite its evidentiary (and ethical) failings. Even the architects of the 1970s-era environmental laws that helped to democratize public land management may have been guilty of oversimplification in their beliefs that scientific analysis and rational planning could predict and resolve complex social-environmental dilemmas.

The contemporary inhabitants of this oddly constructed edifice are facing dilemmas of a scale and complexity that would have been difficult for earlier generations to imagine. Rather than confronting problems of inefficiency, the inhabitants of the contemporary world confront problems that cannot be addressed through a top-down process of rationalization. Whereas leaders in earlier worlds simplified forests to eliminate waste, those in the modern world confront forests developed in an era of fire exclusion that are now ill prepared to withstand the rapid changes brought about by climate disruption and novel species introductions. In earlier worlds, a stark divide between private and public lands promised to generate both economic activity and resource conservation. Now residential development is

*Figure C.1* This piece of guerilla art by Todd McGrain was installed in Portland, Oregon's Mt. Tabor Park in 2021. It celebrates the life of York, an African American man enslaved by William Clark who formed part of Lewis and Clark's "Corps of Discovery." Photograph by author.

consuming forestland nationwide, and there is a growing recognition that conservation challenges cannot be confronted by focusing only on public lands. Whereas earlier worlds were filled with stories that could be told without straying beyond White men and marketable commodities, the contemporary world includes multiple histories, multiple voices, and diverse relationships with forests. These multiple stories and experiences have a rightful place in our collective forest policymaking and governance processes (Figure C.1). Some of them may even hold important insights to help us collectively navigate current challenges.

The house we inhabit was crafted over varied historical periods, and its design reflects problems and solutions that made sense to the political and thought leaders of those times. We have rightly come to reject many of the assumptions and commitments of those leaders even as their conservation legacies live on through our policies, practices, paradigms, and forms of participation. In the past, we tended to add on to the old structure each time an expansion became necessary. As the magnitude of our current environmental crisis comes into view, we face questions of what renovations are needed to withstand the coming storm—and how to ensure that the house makes room for all.

## Note

1 The recreation loophole in this story has still never been fully explained or resolved.

# Author biography

**Jesse Abrams** is Assistant Professor of Natural Resource Policy and Sustainability at the Warnell School of Forestry and Natural Resources and Savannah River Ecology Laboratory, University of Georgia. He holds PhD and MS degrees in Forest Resources from Oregon State University and a BA in Environmental Studies from New College of Florida. His research and teaching focus on the varied policy and governance influences on forests across ownership categories. His academic and professional experiences include time spent studying, managing, and enjoying forests in places as varied as south Florida, the Missouri Ozarks, Northern Arizona, New Mexico, the Massachusetts Islands, the western Cascades, and Wallowa County, Oregon. He has also taught and conducted research in Mexico, Ecuador, Brazil, and Argentina. For these opportunities and others, he is eternally grateful.

# Index

For Product Safety Concerns and Information please contact our EU
representative GPSR@taylorandfrancis.com
Taylor & Francis Verlag GmbH, Kaufingerstraße 24, 80331 München, Germany

www.ingramcontent.com/pod-product-compliance
Ingram Content Group UK Ltd.
Pitfield, Milton Keynes, MK11 3LW, UK
UKHW021831101126
477531UK00016B/136
* 9 7 8 0 3 6 7 4 8 9 5 7 1 *